AF710265

Labelled Deduction

APPLIED LOGIC SERIES

VOLUME 17

Managing Editor

Dov M. Gabbay, *Department of Computer Science, King's College, Londen, U.K.*

Co-Editor

John Barwise, *Department of Philosophy, Indiana University, Bloomington, IN, U.S.A.*

Editorial Assistant

Jane Spurr, *Department of Computer Science, King's College, London, U.K.*

SCOPE OF THE SERIES

Logic is applied in an increasingly wide variety of disciplines, from the traditional subjects of philosophy and mathematics to the more recent disciplines of cognitive science, computer science, artificial intelligence, and linguistics, leading to new vigor in this ancient subject. Kluwer, through its Applied Logic Series, seeks to provide a home for outstanding books and research monographs in applied logic, and in doing so demonstrates the underlying unity and applicability of logic.

The titles published in this series are listed at the end of this volume.

Labelled Deduction

edited by

DAVID BASIN

Albert-Ludwigs-Universität Freiburg, Germany

MARCELLO D'AGOSTINO

Università di Ferrara, Italy

DOV M. GABBAY

King's College, London, United Kingdom

SEÁN MATTHEWS

Logica GmbH, Frankfurt am Main, Germany

and

LUCA VIGANÒ

Albert-Ludwigs-Universität Freiburg, Germany

KLUWER ACADEMIC PUBLISHERS

DORDRECHT / BOSTON / LONDON

A C.I.P. Catalogue record for this book is available from the Library of Congress.

ISBN 0-7923-6237-3

Published by Kluwer Academic Publishers,
P.O. Box 17, 3300 AA Dordrecht, The Netherlands.

Sold and distributed in North, Central and South America
by Kluwer Academic Publishers,
101 Philip Drive, Norwell, MA 02061, U.S.A.

In all other countries, sold and distributed
by Kluwer Academic Publishers,
P.O. Box 322, 3300 AH Dordrecht, The Netherlands.

Printed on acid-free paper

All Rights Reserved
© 2000 Kluwer Academic Publishers
No part of the material protected by this copyright notice may be reproduced or
utilized in any form or by any means, electronic or mechanical,
including photocopying, recording or by any information storage and
retrieval system, without written permission from the copyright owner

Printed in the Netherlands.

Contents

Preface

Labelled Deduction is an approach to providing frameworks for presenting and using different logics in a uniform and natural way by enriching the language of a logic with additional information of a semantic or proof-theoretic nature. Labelled Deduction systems often possess attractive properties such as modularity in the way that families of related logics are presented, parameterized proofs of metatheoretic properties, and ease of mechanizability. It is thus not surprizing that Labelled Deduction has been applied to problems in computer science, artificial intelligence, mathematical logic, cognitive science, philosophy, and computational linguistics, for example formalizing and reasoning about dynamic 'state oriented' properties, such as knowledge, belief, time, space, and resources.

This book is a collection of state-of-the-art research on Labelled Deduction. It is a followup of the First International Workshop on Labelled Deduction, LD'98, that was hosted by the University of Freiburg in September 1998, and where some of the contributions of this volume were first presented.

We would like to thank those who contributed to the success of LD'98, especially Ina Eckert for her organizatorial work. We also thank the referees for this volume: Jair Minoro Abe, Bráulio Coelho Ávila, Matteo Baldoni, Patrick Blackburn, Krysia Broda, Serenella Cerrito, Marta Cialdea Mayer, Detlef Fehrer, Marcelo Finger, Didier Galmiche, Chiara Ghidini, Guido Governatori, Reiner Hähnle, Felip Manyà, Maarten Marx, Christof Monz, Andreas Nonnengart, Nicola Olivetti, Alessandra Raffaetà, Antonino Rotolo, Prahladavaradan Sampath, Stefan Schlobach, Roberto Sebastiani, Luciano Serafini, Kazushige Terui, Leendert van der Torre, Alberto Zanardo.

D. BASIN, M. D'AGOSTINO, D. GABBAY, S. MATTHEWS, L. VIGANÒ

Contributing Authors

Vincent Balat is a PhD student in the computer science department of the École Normale Supérieure de Paris, France, and his research topics are linear logic and intuitionism.

Matteo Baldoni is a researcher at the Dipartimento di Informatica at the Università degli Studi di Torino, Italy. His main research interests include automatic reasoning, knowledge representation, modal and non-monotonic extensions of logic programming; other research interests include extraction of fractal features for image recognition and classification.

David Basin is a full professor at the Institute for Computer Science of the Albert-Ludwigs-University of Freiburg, Germany. His research focuses on methodologies for the specification, verification, and construction of software and hardware systems and computer support for these activities.

Bernhard Beckert is a researcher at the Department of Computer Science, University of Karlsruhe, Germany. His research focuses on theorem proving with tableau methods, software verification, and formal methods in software engineering.

Patrick Blackburn is a lecturer at the Department of Computational Linguistics, University of Saarland, Germany. His research focuses on logic and its applications in the computing and cognitive sciences.

Krysia Broda is a senior lecturer in the Department of Computing at Imperial College, London, UK. Her current research interests are in first order theorem proving for a variety of non-classical and classical logics, and in integrating neural and symbolic computation.

Serenella Cerrito is an assistant professor at the Department of Computer Science at the University of Paris-Sud, France. Her research focuses on proof theory and automated deduction for non-classical logics, with a particular attention to the application of logic to databases.

Marta Cialdea Mayer is an associate professor at the University "Roma Tre", Italy. Her current research interests are in logic for knowledge representation, automated reasoning, modal and temporal logics.

Marcello D'Agostino is an assistant professor at the Department of Human Sciences of the University of Ferrara, Italy. His research focuses on the semantics and proof-theory of classical and non-classical logics, on labelled deduction, and on the computational complexity of proof systems.

Thom Frühwirth is an assistant professor at the Ludwig-Maximilians-University of Munich, Germany. His research focuses on the theory and applications of constraint programming and computational logic in general.

Dov M. Gabbay is a full professor at the Department of Computer Science of King's College, London, UK. His research focuses on temporal logic and change, on the foundations of practical reasoning, and on labelled deductive systems.

Didier Galmiche is an assistant professor at the department of Computer Science of University Henri Poincaré, Nancy, France. His research, developed in the LORIA laboratory, focuses on proof theory, logic and type theory, proof-search in intuitionistic and linear logics.

Guido Governatori is a research fellow in the School of Computing and Information Technology at Griffith University, Brisbane, Australia. His research interests are: proof systems for non-classical logics, labelled deductive systems, defeasible and non-monotonic logics, logics for normative reasoning, abstract argumentation.

Reiner Hähnle is an assistant professor at the Department of Computer Science, University of Karlsruhe, Germany. His research focuses on theorem proving with tableau methods, many-valued logic, and formal methods in software engineering.

Felip Manyà is an assistant professor at the Department of Computer Science, University of Lleida, Spain. His research focuses on logic programming and theorem proving in many-valued logics.

Maarten Marx teaches at the Department of Artificial Intelligence of the Free University, Amsterdam, The Netherlands. His research interests are modal and algebraic logic with applications in knowledge representation.

Seán Matthews is currently employed by the software consultancy Logica GmbH, Frankfurt am Main, Germany.

Szabolcs Mikulás is a research fellow at the Department of Computer Science at King's College, London, UK. His research interests are in algebraic and modal logic.

Alessandra Raffaetà is a PhD student at the Computer Science Department of the University of Pisa, Italy. Her research focuses on programming language design and semantics, logic programming, and temporal and spatial reasoning.

Stefan Schlobach is a research associate at the Department of Computer Science at King's College, London, UK. His current research is centered around description logics, with particular interest in the integration of uncertain and fuzzy knowledge.

Leendert van der Torre is a researcher at the Department of Artificial Intelligence of the Vrije Universiteit in Amsterdam, the Netherlands. His research interests are in formal approaches to practical reasoning, in particular normative reasoning (deontic logic) and decision making (qualitative decision theory).

Luca Viganò is an assistant professor at the Institute for Computer Science of the Albert-Ludwigs-University of Freiburg, Germany. His research focuses on the theory and applications of non-classical logics, of proof development systems, and of logical frameworks.

LABELLED PROOF SYSTEMS FOR INTUITIONISTIC PROVABILITY

Vincent Balat and Didier Galmiche

Abstract In this paper, we propose new labelled proof systems to analyse the intuitionistic provability in classical and linear logics. An important point is to understand how search in a non-classical logic can be viewed as a perturbation of search in classical logic. Therefore, suitable characterizations of intuitionistic provability and related labelled sequent calculi are defined for linear logic. An alternative approach, based on the notion of proof-net and on the definition of suitable labelled classical proof-nets, allows to directly study the intuitionistic provability by constructing intuitionistic proof-nets for sequents of classical linear logic.

1 INTRODUCTION

Many proof-search methods (sequent calculus, tableaux, resolution, connections) have been naturally developed in classical logic (**CL**) with a view to avoiding the possible redundancies, unsuccessful explorations or loops, as illustrated by naive search in sequent calculi. Some of them have been adapted to intuitionistic logic (**IL**) that can be viewed, from a proof-theoretic point of view, as a specialization of classical logic. Let us recall that a sequent calculus for **IL** can be directly obtained from the classical one by restricting the sequent conclusions to a single formula. Another equivalent multi-conclusion sequent calculus for **IL**, that is a specialisation of **CL** multi-conclusion sequent calculus with so-called special rules, has been proposed to more efficiently deal with intuitionistic proof-search from classical proof-search [13]. Then classical logic plays a pivotal role in search calculi and an interesting point consists in understanding how search in a non-classical logic can be viewed as a perturbation on search in classical logic. It has been recently studied in [30] by considering a proof-theoretic approach, based on $\lambda\mu$-terms [29], where the terms permit an axiomatisation of the perturbation which yields the non-classical search and also some semantical explanations about embedding in classical logic. In this setting, some questions arise: can we conclude from a classical proof of a given sequent that there exists an intuitionistic proof of it ? can we extract such a

D. Basin et al. (eds.), Labelled Deduction, 1–32.
© 2000 *Kluwer Academic Publishers. Printed in the Netherlands.*

proof from the classical one ? is it possible to consider the intuitionistic search as a perturbation of classical search and to have a completeness result, i.e., to determine the intuitionistic provability of a sequent from any classical proof ? This paper deals with these questions mainly in the context of classical linear logic (**CLL**) that can be seen as a resource-sensitive refinement of **CL** [21, 22]. A central point is the choice of the proof systems to analyse the intuitionistic provability from classical search. We know that natural deduction is a suitable deduction system for **IL** but not for **CL** for which sequent calculus seems better suited. The Curry-Howard correspondence allows to annotate natural deductions with terms (for **IL** it yields to typed λ-calculus) but for sequent calculus, it is not clear what the appropriate notations are. For instance, Parigot has introduced a variant of multi-conclusion natural deduction and the related $\lambda\mu$-calculus, which seem suited for handling **CL** [29]. In the setting of linear logic, similar problems of representation arise, knowing that the specific notion of *proof-net* can moreover be seen as a counterpart of natural deduction. The relationships between classical linear logic (**CLL**) and intuitionistic linear logic (**ILL**) have to be studied from both proof-theoretic and proof-search points of view. Let us recall that **ILL** can be seen as a model of Petri nets [14, 16] and also as a foundation of functional and logic linear programming languages. Moreover recent specification logics are based on fragments of **ILL** (like Lolli [25]) or **CLL** (like Forum [28]), the choice being motivated by specific operational aims, for instance sequentiality in **ILL** and concurrency in **CLL**. Therefore, the study of intuitionistic provability from classical provability in linear logic would have an important impact on proof-search and semantics analysis.

In section 2, we study the intuitionistic provability from classical search by defining a new labelled proof system. This labelled sequent calculus for **CL** is an alternative to the based-on $\lambda\mu$-terms approach given in [30], being not type-theoretic but proof-search oriented. In both cases, there is no completeness result, i.e one cannot conclude the intuitionistic provability from any given classical proof. The main part of this paper deals with characterizations of the intuitionistic provability in the setting of linear logic. Therefore, in section 3, we summarize the main characteristics of classical and intuitionistic linear logics and focus on a particular intuitionistic system called (multiplicative) **FILL** (Full Intuitionistic Linear Logic) that simultaneously embodies features of concurrent logical computations, induced by the par connective and the sequential properties of intuitionistic linear implication [26]. In section 4, we then propose two labelled sequent calculi for (multiplicative) **CLL** with related characterizations of intuitionistic provability. The first one is a direct extension of the labelled **CL** sequent calculus to **CLL** but the provability characterization does not fit well from a logical point of view. The second one has a more suitable and useful definition of labels w.r.t. linear logic specificities. Moreover it can be independently considered as a new proof system for **FILL** without terms or pattern calculus [7, 26] that is more adapted to proof-search. As in the classical case, these systems do not give a positive answer to the above mentioned completeness problem. In section 5, we consider a new labelled proof system, based on the notion of *proof-net* and its possible use for proof-search. It allows to directly construct an intuitionistic proof, if there exits, from a classical search. For that, we define (multiplicative) **FILL** proof-nets that are in fact classical proof-nets with labels that represent some so-called dependency paths. Then we design an algorithm that, for a given **CLL** initial sequent, builds a **FILL** proof-net and also a sequent proof, if they exist. Therefore, the intuitionistic provability question can be successfully studied from labelled proof-nets and the intrinsic completeness problem arising with the labelled sequent calculi is solved. Finally, section 6 presents some concluding remarks.

2 INTUITIONISTIC PROVABILITY IN CLASSICAL LOGIC

The relationships between **CL** and **IL** are strong and then one can try to use the classical proof-search to characterize the intuitionistic provability. The intuitionistic logic is studied because of its underlying constructivism and its connections with the λ-calculus and programming (Curry-Howard isomorphism). A sequent calculus for **IL** can be obtained from classical sequent calculus of **CL** (see appendix 1) by restricting sequents to a single conclusion (see appendix 2). It is important to recall that there exists a multi-conclusion sequent calculus for **IL** (see appendix 3), that is well adapted to study efficient proof-search [13, 31, 32]. Such a calculus is in fact a restriction of classical sequent calculus including some particular rules called *special rules*. It allows to naturally consider the intuitionistic provability like a perturbation of classical provability. Thus, we can consider the following question: is it possible to conclude to the existence of an intuitionistic proof from the existence of a classical proof ? An analysis of this problem has been developed in [30] where classical proofs are encoded in the $\lambda\mu$-calculus [29] that is an extension of λ-calculus related to the classical logic. Therefore, in a type-theoretic approach, a classical proof is represented with a $\lambda\mu$-term and a characterization of terms representing classical proofs that include an intuitionistic proof is given. But such a characterization is not complete, i.e., a sequent can be provable in **IL** but one cannot observe it from any classical proof. Another approach is the one of [11] with a tableau system that distinguishes classical and intuitionistic logics by comparing semantic labels after a proof is constructed. Is the analysis easier using $\lambda\mu$-terms, semantics labels or initial classical proofs ? Because we are more interested on proof-search than on provability, we propose to directly analyse the classical proofs to detect the existence of an underlying intuitionistic proof. Then we define a sequent calculus with labels that allows to memorize the necessary information to analyse intuitionistic provability.

2.1 A LABELLED SEQUENT CALCULUS FOR CL

We now present a new labelled system for classical proofs that is used to detect their intuitionistic character, with a very simple criterion[1]. The principle consists in annotating each formula with a specific label (or word), to keep a trace of the construction of the classical sequent proof. From a given labelled classical proof, the labels of the sequent axioms allow to characterize the intuitionistic provability.
Considering its multi-conclusion (propositional) version **IL** differs from **CL** only because of the so-called *special rules* $\rightarrow_R$ and $\neg_R$ [13, 32]. Let us analyse the $\rightarrow_R$ rule, respectively in multi-conclusion **CL** and **IL** versions :

$$\frac{\Gamma, A \vdash B, \Delta}{\Gamma \vdash A \rightarrow B, \Delta} \qquad \frac{\Gamma, A \vdash B}{\Gamma \vdash A \rightarrow B, \Delta}$$

For each application of the $\rightarrow_R$ rule, we mark all formulae of Δ with the 0 symbol and the A and B formulae with the x symbol. The positions of the symbols in the labels memorize the different applications of such a special rule. In fact, we forbid a (sub)formula of A and a (sub)formula of one formula of Δ to meet in an axiom sequent. We apply the same principle to the $\neg_R$ rule. We thus consider the alphabet $\Sigma = \{1, x, 0\}$ and the sequents have the following form $A_1^{v_1}, ...A_k^{v_k} \vdash B_1^{w_1}, ...B_m^{w_m}$ where $v_1, ...v_k, w_1, ..., w_m$ are words of same length on Σ.

Notation 1 *$u[p]$ denotes the letter of the word u at position p, the letters being annotated from 1, from right-hand side to left-hand side. $|u|$ denotes the length of the word u and ε denotes the empty word. If $a \in \Sigma$ and $v = v[n]...v[1] \in \Sigma^*$, then av denotes the word $av[n]...v[1]$. If $\Gamma = A_1, ..., A_k$ is a list of formulae and $v = v_1, ..., v_k$ is a list of words, Γ^v denotes the set of labelled formulae $A_1^{v_1}, ...A_k^{v_k}$, and Γ^{av} denotes $A_1^{av_1}, ...A_k^{av_k}$.*

At the beginning, the formulae of the sequent to prove are labelled with the empty word. The labelled sequent calculus for **CL** is given in Figure 1.1. When labels are omitted it means that they are unchanged.

Definition 2 *Let $\Gamma \vdash \Delta$ be a labelled sequent, A^u (resp. A^v) a labelled formula of Γ (resp. Δ), and p an integer such that $0 < p \leq min(|u|, |v|)$, the pair $(u[p], v[p])$ is called a* correspondence at position p *on the sequent.*

Definition 3 *A set of axiom sequents is said to be* intuitionistic *if, for any position p, there is no correspondence of the form $(x, 0)$ or $(0, x)$.*

---

[1]The initial version of this system was discussed and developed with G. Delzanno, when he visited LORIA in Nancy during the year 1996/97.

$$\frac{}{\Gamma, A^v \vdash A^w, \Delta}\ ax \qquad \frac{\Gamma \vdash A, \Delta \quad \Gamma', A \vdash \Delta'}{\Gamma, \Gamma' \vdash \Delta, \Delta'}\ cut$$

$$\frac{\Gamma, A^v, A^v \vdash \Delta}{\Gamma, A^v \vdash \Delta}\ c_L \qquad \frac{\Gamma \vdash \Delta, A^v, A^v}{\Gamma \vdash \Delta, A^v}\ c_R$$

$$\frac{\Gamma \vdash A^v, \Delta \quad \Gamma, B^v \vdash \Delta}{\Gamma, (A \to B)^v \vdash \Delta}\ \to_L \qquad \frac{\Gamma^{1u}, A^{xv} \vdash B^{xv}, \Delta^{0w}}{\Gamma^u \vdash (A \to B)^v, \Delta^w}\ \to_R$$

$$\frac{\Gamma, A^v \vdash \Delta \quad \Gamma, B^v \vdash \Delta}{\Gamma, (A \vee B)^v \vdash \Delta}\ \vee_L \qquad \frac{\Gamma \vdash A^v, \Delta}{\Gamma \vdash (A \vee B)^v, \Delta}\ \vee_{R1} \qquad \frac{\Gamma \vdash B^v, \Delta}{\Gamma \vdash (A \vee B)^v, \Delta}\ \vee_{R2}$$

$$\frac{\Gamma, A^v, B^v \vdash \Delta}{\Gamma, (A \wedge B)^v \vdash \Delta}\ \wedge_L \qquad \frac{\Gamma \vdash A^v, \Delta \quad \Gamma \vdash B^v, \Delta}{\Gamma \vdash (A \wedge B)^v, \Delta}\ \wedge_R$$

$$\frac{\Gamma \vdash A^v, \Delta}{\Gamma, \neg A^v \vdash \Delta}\ \neg_L \qquad \frac{\Gamma^{1u}, A^{xv} \vdash \Delta^{0w}}{\Gamma^u \vdash \neg A^v, \Delta^w}\ \neg_R$$

Figure 1.1 Labelled sequent calculus for **CL**

Theorem 4 *If a sequent $\Gamma \vdash \Delta$ has a proof (in the system Figure 1.1) such that the set of axiom sequents is intuitionistic, then there exists an intuitionistic proof of the sequent.*

Proof It is a direct consequence of the relationships between what a correspondence of the form $(x, 0)$ or $(0, x)$ means and the application of so-called special rules in the multi-conclusion sequent calculus of **IL** (see appendix 3). ■

Then, this characterization of intuitionistic provability depends on syntactic labels (that are words) and mainly on the notion of position in such labels.

2.2 THE COMPLETENESS PROBLEM

Let us first consider a classical proof that includes an intuitionistic proof.

$$\dfrac{\dfrac{\dfrac{\dfrac{}{\Gamma^1, A^x \vdash A^1, B^x, B^0}\ ax \quad \dfrac{}{\Gamma^1, A^x, B^1 \vdash B^x, B^0}\ ax}{\Gamma^1, A^x, \underline{A \to B}^1 \vdash B^x, B^0}\ \to_L}{\Gamma, A \to B \vdash \underline{A \to B}, B}\ \to_R \quad \dfrac{}{\Gamma, A \to B, B \vdash B}\ ax}{\Gamma, A \to B, \underline{(A \to B) \to B} \vdash B}\ \to_L$$

Let us notice that if the inference rules are applied in another order, it is no more possible to claim, by application of the previous criterion, that there exists an intuitionistic proof of such a sequent (because of a $(x, 0)$ correspondence on the formula A of one axiom sequent).

$$\dfrac{\dfrac{\dfrac{\dfrac{}{\Gamma^1, A^x \vdash B^x, A^0, B^0}\,ax}{\Gamma \vdash \underline{A \to B}, A, B}\,\to_R \qquad \dfrac{}{\Gamma, B \vdash A, B}\,ax}{\Gamma, \underline{(A \to B) \to B} \vdash A, B}\,\to_L \qquad \dfrac{}{\Gamma, (A \to B) \to B, B \vdash B}\,ax}{\Gamma, \underline{A \to B}, (A \to B) \to B \vdash B}\,\to_L$$

We have no completeness result, i.e, for a given sequent, we cannot conclude from any classical proof if it is also provable in **IL**. It is well illustrated with both previous proofs. It is also the case in the type-theoretic approach proposed in [30] where classical logic can be viewed as a type theory based on an extension of the Parigot's $\lambda\mu$-calculus [29]. Conditions for intuitionistic provability are expressed on classical $\lambda\mu$-terms and are also equivalent to admissible conditions on prefix unifiers in [32]. The use of such terms for the axiomatisation of the perturbation which yields non-classical search, allows to have semantical explanations of syntactic tricks to embed non-classical search in classical systems. Our equivalent approach, based on syntactic labels in a sequent calculus, is nevertheless simple and readable from both algorithmic and proof-search points of view.

When a classical proof contains an intuitionistic proof, we can use the information given by the labels to extract an intuitionistic proof. The principle consists in detecting the needless formulae in the classical proof by an analysis of the labels and then in erasing them. A simple algorithm can be proposed for this operation. The main advantages and limits of using such labelled systems will be illustrated in the case of linear logic, for which the based-on $\lambda\mu$-calculus approach could be also used at first to devise a natural deduction formulation of classical linear logic [8].

3 LINEAR LOGIC

Linear logic is a resource-sensitive logic that can be seen as a refinement of classical logic sequent calculus where contraction and weakening rules are forbidden [22]. Both classical linear logic (**CLL**) and its intuitionistic fragment, called intuitionistic linear logic (**ILL**), are studied as models of computation [2] and of semantics of parallelism [20].

3.1 CLASSICAL LINEAR LOGIC

The usual presentation of **CLL** consists in starting from the **CL** sequent calculus with suppression of the contraction and weakening rules. Then the equivalence between the additive and multiplicative presentations of the $\vee$ and $\wedge$ connectives disappears and then leads to distinguish a multiplicative $\vee$ ("par", denoted $\parr$), an additive $\vee$ ("with", denoted &), a multiplicative $\wedge$ ("times", denoted $\otimes$), and an additive $\wedge$ ("plus", denoted $\oplus$). The atoms have the form A or $A^\perp$. The definition of the negation $()^\perp$, that is involutive, is extended to all formulae from the following rules: $(A \parr B)^\perp = (A^\perp \otimes B^\perp)$, $(A \otimes B)^\perp = (A^\perp \parr B^\perp)$, $(A \oplus B)^\perp = (A^\perp \& B^\perp)$ and $(A \& B)^\perp = (A^\perp \oplus B^\perp)$. The symbols $\mathbf{1}, \mathbf{0}, \top$ and $\perp$ respectively represent the neutrals of the $\otimes$, $\oplus$, $\&$ and $\parr$ connectives. The linear implication ($\multimap$) is defined by $A \multimap B = A^\perp \parr B$. The ! ("of course") and ? ("why not") connectives (called exponentials) are used to reintroduce in a controlled way the weakening and the contraction rules (but as logical rules). Let us mention that the intuitionistic implication $A \rightarrow B$ can be, with such operators, decomposed into two operations, i.e. $!A \multimap B$. See appendix 4 to have the rules of propositional **CLL**. The Multiplicative Linear Logic (only with the multiplicative connectives) without constants, is denoted **MLL**.

3.2 INTUITIONISTIC LINEAR LOGIC

In linear logic, compared to classical logic, the definition of an intuitionistic fragment is not so natural and direct from the initial sequent calculus. The Intuitionistic Linear Logic (**ILL**) has a mono-conclusion sequent calculus (see appendix 5). Compared to the non-linear case, **ILL** is defined with a formula grammar that is different from the **CLL** one.
In fact, the rules corresponding to $\parr$, its neutral element $\perp$, the negation $^\perp$ and to ? led to consider a multi-conclusion representation. Then the grammar of formulae in **ILL** is $\phi ::= p \mid \phi \otimes \phi \mid \phi \multimap \phi \mid \phi \& \phi \mid \phi \oplus \phi \mid !\phi$, where p is an atomic formula or a constant $\top$, $\mathbf{0}$ or $\mathbf{1}$. In [24], Girard and Lafont illustrate the interest of linear logic, compared to **IL** to represent types of functions and linear λ-terms have been defined in this context [27]. Therefore, the $\otimes$ connective is a strict $\wedge$ while the $\&$ connective is a lazy $\wedge$. In this setting, the real meaning of the $\parr$ connective is not clear with this interpretation of proofs as functions. The relationship between **CLL** and **ILL** can be expressed through the following result

Theorem 5 *Let be S a sequent built from the* **ILL** *(without constants) grammar, if S is provable in* **CLL** *then it is provable in* **ILL**.

Proof Let us consider a sequent provable in **CLL**, that only uses the **ILL** grammar. We first show that all sequents, in such a **CLL** proof, have a single formula as conclusion. For that, we observe that the axioms are mono-conclusion and then for each rule of **CLL**, if the premises are mono-conclusion then the conclusion is also mono-conclusion. Secondly, we show that a **CLL** rule applied to a mono-conclusion sequent gives the same result than the **ILL** corresponding rule. ∎

Let us note that if we consider a version of **CLL** including the rules for the $\multimap$ connective then the **ILL** proof will be identical to the **CLL** proof. Works on **ILL** have been developed from linear λ-calculus [6, 8, 24], but the lack of the $\parr$ connective seems more motivated by technical reasons and then we could ask for an intuitionistic fragment of linear logic including it. In this context, the semantics of $\parr$ has to be clarified even if a natural interpretation seems to be connected to the concurrent execution of processes [1, 9].

3.3 FULL INTUITIONISTIC LINEAR LOGIC

A multi-conclusion intuitionistic multiplicative linear logic, called *Full Intuitionistic Linear Logic* (**FILL**) has been proposed by Hyland and de Paiva in [26]. That is a variant of intuitionistic linear logic whose logical connectives are all independent, i.e., not inter-derivable as they are in multiplicative **CLL**. It is analogous to the situation concerning the relationship between **IL** and **CL**. The interest of **FILL** is that it simultaneously embodies features of concurrent logical computations, induced by the $\parr$ connective, and the sequential properties of intuitionistic linear implication.

In **FILL** the negation $A^{\perp}$ is defined as $A \multimap \perp$ and it is not an involution. This system, that includes the multiplicative disjunction $\parr$, is a subsystem of **CLL** and a proof in **FILL** is a proof in **CLL** with a certain intuitionistic property. For instance, one has $A^{\perp} \parr B \vdash A \multimap B$ in **FILL** but not the converse. Moreover, the excluded middle $\vdash A^{\perp} \parr A$ cannot be proved in **FILL**. It is the same for $A^{\perp\perp} \vdash A$ and for $\vdash (A^{\perp} \parr A)^{\perp\perp}$. This last remark illustrates the difference from the situation in **IL** where $\vdash \neg A \vee A$ is not provable but $\neg\neg(\neg A \vee A))$ is provable. It comes from the use of contraction in **IL** and the effective difference between the $\vee_R$ and $\parr_R$ rules. Moreover, $(A\otimes B)\multimap(C \parr D) \vdash (A\multimap C) \parr (B\multimap D)$ is an example of a sequent provable in **CLL** but not in **FILL** (see [10] for more details).

Let us mention also that the interaction between $\parr$ and the linear implication $\multimap$ is such that cut-elimination failed outright for some intuitive formulation of **FILL**. This problem has motivated, from the term assignment system in [26], some alternative presentations as the ones of Bierman [7] and Braüner and de Paiva [10]. The latter is based on a notion of *dependency* to capture, with a side condition of $\multimap_R$, the underlying notion of intuitionistic implication and to make the cut-elimination to go through in a straightforward manner. In fact, one defines, given a proof of $\Gamma, B \vdash A, \Delta$ in **CLL**, when the succedent formula occurrence A depends on the antecedent formula occurrence B. This notion of dependency between formulae occurrences allows to express the intuitionistic property that characterizes **FILL** [10]. The corresponding sequent calculus is given in Figure 1.2. Let τ be a **CLL** proof, for each formula C at the right-hand side of the sequents in τ, one defines the set $Dep_{\tau}(C)$ of the formulae of left-hand side on which it depends. In the rules, τ represents the proof-tree, the root of which is the conclusion of the rule, τ_1 (resp. τ_2) is the sub-proof, the

$$\frac{}{A \vdash A}\ ax \qquad Dep_{\mathcal{T}}(A) = A$$

$$\frac{\Gamma \vdash A, \Delta \quad \Gamma', A \vdash \Delta'}{\Gamma, \Gamma' \vdash \Delta, \Delta'}\ cut \qquad Dep_{\mathcal{T}}(C) = \begin{cases} Dep_{\mathcal{T}_1}(C) & \text{if } C \in \Delta \\ Dep_{\mathcal{T}_2}(C)[A \mapsto Dep_{\mathcal{T}_1}(A)] & \text{if } C \in \Delta' \end{cases}$$

$$\frac{\Gamma, A \vdash B, \Delta}{\Gamma \vdash A \multimap B, \Delta}\ \multimap_R \qquad Dep_{\mathcal{T}}(C) = \begin{cases} Dep_{\mathcal{T}_1}(C)[A \mapsto \emptyset] & \text{if } C \in \Delta \\ Dep_{\mathcal{T}_1}(B)[A \mapsto \emptyset] & \text{if } C = A \multimap B \end{cases}$$

$$\frac{\Gamma \vdash A, \Delta \quad \Gamma', B \vdash \Delta'}{\Gamma, \Gamma', A \multimap B \vdash \Delta, \Delta'}\ \multimap_L \qquad Dep_{\mathcal{T}}(C) = \begin{cases} Dep_{\mathcal{T}_1}(C) & \text{if } C \in \Delta \\ Dep_{\mathcal{T}_2}(C)[B \mapsto (Dep_{\mathcal{T}_1}(A) \cup (A \multimap B))] & \\ & \text{if } C \in \Delta' \end{cases}$$

$$\frac{\Gamma, A \vdash \Delta \quad \Gamma', B \vdash \Delta'}{\Gamma, \Gamma', A \parr B \vdash \Delta, \Delta'}\ \parr_L \qquad Dep_{\mathcal{T}}(C) = \begin{cases} Dep_{\mathcal{T}_1}(C)[A \mapsto (A \parr B)] & \text{if } C \in \Delta \\ Dep_{\mathcal{T}_2}(C)[B \mapsto (A \parr B)] & \text{if } C \in \Delta' \end{cases}$$

$$\frac{\Gamma \vdash A, B, \Delta}{\Gamma \vdash A \parr B, \Delta}\ \parr_R \qquad Dep_{\mathcal{T}}(C) = \begin{cases} Dep_{\mathcal{T}_1}(C) & \text{if } C \in \Delta \\ Dep_{\mathcal{T}_1}(A) \cup Dep_{\mathcal{T}_1}(B) & \text{if } C = A \parr B \end{cases}$$

$$\frac{\Gamma, A, B \vdash \Delta}{\Gamma, A \otimes B \vdash \Delta}\ \otimes_L \qquad Dep_{\mathcal{T}}(C) = Dep_{\mathcal{T}_1}(C)[A, B \mapsto A \otimes B]$$

$$\frac{\Gamma \vdash A, \Delta \quad \Gamma' \vdash B, \Delta'}{\Gamma, \Gamma' \vdash A \otimes B, \Delta, \Delta'}\ \otimes_R \qquad Dep_{\mathcal{T}}(C) = \begin{cases} Dep_{\mathcal{T}_1}(C) & \text{if } C \in \Delta \\ Dep_{\mathcal{T}_2}(C) & \text{if } C \in \Delta' \\ Dep_{\mathcal{T}_1}(A) \cup Dep_{\mathcal{T}_2}(B) & \text{if } C = A \otimes B \end{cases}$$

$$\frac{\Gamma \vdash \Delta}{\Gamma, \mathbf{1} \vdash \Delta}\ \mathbf{1}_L \qquad Dep_{\mathcal{T}}(C) = Dep_{\mathcal{T}_1}(C)$$

$$\frac{}{\vdash \mathbf{1}}\ \mathbf{1}_R \qquad Dep_{\mathcal{T}}(\mathbf{1}) = \emptyset$$

$$\frac{}{\bot \vdash}\ \bot_R \qquad \text{nothing to define}$$

$$\frac{\Gamma \vdash \Delta}{\Gamma \vdash \bot, \Delta}\ \bot_L \qquad Dep_{\mathcal{T}}(C) = \begin{cases} Dep_{\mathcal{T}_1}(C) & \text{if } C \in \Delta \\ \emptyset & \text{if } C = \bot \end{cases}$$

Figure 1.2 A **FILL** sequent calculus

root of which is the first (resp. second) premise, if it exists. Moreover the set of formulae occurrences $\Gamma[\Delta \mapsto \Delta']$ is defined as $(\Gamma \setminus \Delta) \cup \Delta'$ if $\Gamma \cap \Delta \neq \emptyset$ and as Γ otherwise. This notion of dependency allows a definition of **FILL** proof.

Definition 6 *A* **FILL** proof *is a proof of* **CLL** *such that, for any* $\multimap_R$ *rule and any formula* C *of* Δ*, we have* $A \notin Dep_{\mathcal{T}_1}(C)$ *[10].*

Let us now consider the following proof:

$$\dfrac{\dfrac{\dfrac{\dfrac{}{C \vdash C}\,ax \quad \dfrac{}{B \vdash B}\,ax}{\underline{C \parr B} \vdash B, C}\,\parr_L \quad \dfrac{}{A \vdash A}\,ax}{C \parr B, \underline{B \multimap A} \vdash C, A}\,\multimap_L}{B \multimap A \vdash \underline{(C \parr B) \multimap C}, A}\,\multimap_R$$

that is not a **FILL** proof. Here, the formula A in the right-hand side of the sequent $C \parr B, B \multimap A \vdash C, A$ depends on $C \parr B$. In fact we have $Dep(A) = \{C \parr B, B \multimap A\}$. The presentation of this system does not include a term assignment decorating the usual sequent calculus proofs like in [7, 26]. In the next section, we consider the relationships between **CLL** and **FILL** from both proof-theoretic and proof-search points of view. Then we propose new formulations of **FILL** with sequent calculi with labels that reflect the dependency conditions inside the inference rules of **CLL**.

4 INTUITIONISTIC PROVABILITY IN LINEAR LOGIC

Let us now consider the intuitionistic provability from classical search in the case of linear logic. With a proof in **ILL**, we can propose a simple criterion that corresponds to observe the presence of the $\parr$ and ? connectives and to test if the sequents have a single conclusion (see theorem 5). In **FILL**, the structure of the proofs is exactly the same than in **CLL** but we have to take into account the dependency relations. Compared to the previous study, we do not try to determine if a **CLL** proof contains a **FILL** proof but if it is a **FILL** proof. Among the different presentations of multiplicative **FILL** [7, 10, 26], mainly motivated by semantics analysis and by simple and readable cut-elimination proofs, the one of Braüner and de Paiva [10] appears to be the more related to our approach. In fact, our proposals can be seen as new proof systems for **FILL** that could be useful for further works based on objects or processes [9]. We now present labelled proof systems for multiplicative **CLL**. The first one is a direct extension of the criterion defined for **CL** and the second one is more simple and appropriate to the case of linear logic. From now, we only consider the multiplicative part of the logical fragments.

4.1 A LABELLED SEQUENT CALCULUS

In this first proposal, the formulae are labelled with words but we have not only to simulate the use of special rules where some formulae are erased but also to effectively test the dependencies between formulae. As we consider transitive dependency chains that can have arbitrary length, we then use an infinite alphabet Σ including the 1, x, and 0 symbols. The position in the word is also used to determine where special rules have been applied. A formula

$$\frac{\Gamma^{1u::l_1}, A^{x1^n::\emptyset} \vdash B^{1v::l_2}, \Delta^{0w::l_3}}{\Gamma^{u::l_1} \vdash (A \multimap B)^{v::l_2}, \Delta^{w::l_3}} \multimap_R$$

$$\frac{\Gamma \vdash A^{a^n::\emptyset}, \Delta \quad \Gamma', A^{a^n::\emptyset} \vdash \Delta'}{\Gamma, \Gamma' \vdash \Delta, \Delta'} \, cut \qquad \text{with } a \text{ fresh letter}$$

$$\frac{\Gamma \vdash A^{a^n::\emptyset}, \Delta \quad \Gamma', B^{v::a^n::l} \vdash \Delta'}{\Gamma, \Gamma', (A \multimap B)^{v::l} \vdash \Delta, \Delta'} \multimap_L \qquad \text{with } a \text{ fresh letter}$$

$$\frac{}{A^{l_1} \vdash A^{l_2}} \, ax$$

$$\frac{\Gamma \vdash \Delta}{\Gamma, \mathbf{1}^l \vdash \Delta} \mathbf{1}_L \quad \frac{}{\vdash \mathbf{1}^l} \mathbf{1}_R \quad \frac{}{\bot^l \vdash} \bot_L \quad \frac{\Gamma \vdash \Delta}{\Gamma \vdash \bot^l, \Delta} \bot_R$$

$$\frac{\Gamma, A^l \vdash \Delta \quad \Gamma', B^l \vdash \Delta'}{\Gamma, \Gamma', (A \parr B)^l \vdash \Delta, \Delta'} \parr_L \qquad \frac{\Gamma \vdash A^l, B^l, \Delta}{\Gamma \vdash (A \parr B)^l, \Delta} \parr_R$$

$$\frac{\Gamma, A^l, B^l \vdash \Delta}{\Gamma, (A \otimes B)^l \vdash \Delta} \otimes_L \qquad \frac{\Gamma \vdash A^l, \Delta \quad \Gamma' \vdash B^l, \Delta'}{\Gamma, \Gamma' \vdash (A \otimes B)^l, \Delta, \Delta'} \otimes_R$$

Figure 1.3 A labelled sequent calculus for **CLL**.

is labelled by a list of words that allows several correspondences at the same position between two formulae.

Notation 7 *We denote $u :: l$ a list with u as head and l as tail. Lists are presented in extension with ; and for instance the list with the words u, v and w as elements is denoted $u; v; w$. Moreover $\emptyset$ denotes the empty list.*

At the beginning, each formula of the sequent to prove is labelled with the $\varepsilon :: \emptyset$ list (the empty word is the only element). The labelled sequent calculus is given in Figure 1.3, where labels are omitted when they are unchanged and n is the length of the head word for any conclusion formula. Let us note that, for a given n, the words a^n and 1^n allow to guarantee that the head words have all the same length. Moreover, in $\multimap_R$ we pad out the label of A with a safe number of 1. The notion of correspondence of definition 2 is extended to this new sequent calculus and leads to a suitable definition of *dependency chain* on sequences[2] of sequents.

Definition 8 *Let E be a sequence of sequents, a* dependency chain on E at position p between the letters α and β *is a sequence of pairs of the form $(\alpha, \alpha_1), (\alpha_1, \alpha_2), \ldots, (\alpha_n, \beta)$ ($n \in \mathbb{N}^*$), the elements of which are correspondences at position p in one of the sequents of E.*

Definition 9 *A sequence of sequents is said to be* intuitionistic *iff for any sequent there exists no position p with a dependency chain between x and 0.*

[2]We prefer to deal with sequences and not with sets mainly because of the further management of axiom sequents extracted from proofs.

Theorem 10 *Let P be a proof (in the system Figure 1.3), P is a* **FILL** *proof if and only if the sequence of its axiom sequents is intuitionistic.*

Proof The proof of this theorem is a directed adaptation of the proof we will develop in the next section for the second and main characterization. ■

We now present a proof in **CLL** of the sequent $C \multimap B, D \multimap A \vdash (C \parr D) \multimap B, A$, that is not a **FILL** proof, because of a dependency chain $(x, y), (y, 0)$.

$$
\dfrac{\dfrac{\dfrac{\dfrac{\dfrac{}{C^x \vdash C^z}\,ax \quad \dfrac{}{D^x \vdash D^y}\,ax}{(C \parr D)^x \vdash C^z, D^y}\,\parr_L \quad \dfrac{}{B^{1;z} \vdash B^1}\,ax}{(C \multimap B)^1, (C \parr D)^x \vdash D^y, B^1}\,\multimap_L \quad \dfrac{}{A^{1;y} \vdash A^0}\,ax}{(C \multimap B)^1, (D \multimap A)^1, (C \parr D)^x \vdash B^1, A^0}\,\multimap_L}{C \multimap B, D \multimap A \vdash (C \parr D) \multimap B, A}\,\multimap_R
$$

In our first criterion, the position p in a word is used to know for which applications of the $\multimap_R$ the letters appeared. It avoids to mix the dependency chains. But this principle has the disadvantages to complicate the system presentation by keeping several words for each formula and by creating words of the form y^n because we do not know at which position the y will be used. This proposal is then based on the notion of position but in several words.

4.2 ANOTHER LABELLED SEQUENT CALCULUS

An alternative, not directly connected to our initial criterion, consists in using a different marking for each special rule. In this case, the position p in a label has no more importance, the words can have different lengths and we only need one word per formula. The alphabet is then $\Sigma = \Sigma_1 \cup \Sigma_2$, with Σ_1 and Σ_2 being disjoint. In fact, Σ_1 is used for the extremities of the dependency chains and Σ_2 for the intermediate symbols.
We begin with the empty word on each formula of the given sequent. The rules of this new labelled sequent calculus are given in Figure 1.4. As before, labels are omitted when they are unchanged. As the notion of position has no more significance here, then the appropriate notions of *correspondence* and of *dependency chains* are defined in the following way:

Definition 11 *A* correspondence *on a sequent $\Gamma \vdash \Delta$ is a pair (α, β), where α is a letter in a label of a formula in Γ and β a letter of a label of the same formula in Δ.*

Definition 12 *Let E be a sequence of sequents, and let $S_0, ..., S_n$ be a subsequence of E, a* dependency chain *on E between α and β is a sequence of pairs of the form (α_0, α_1), (α_1, α_2),...,(α_n, α_{n+1}) ($n \in \mathbb{N}^*$) such that $\alpha_0 = \alpha$, $\alpha_{n+1} = \beta$, and (α_i, α_{i+1}) is a correspondence on S_i.*

$$\frac{\Gamma, A^x \vdash B^v, \Delta^{xw}}{\Gamma \vdash (A \multimap B)^v, \Delta^w} \multimap_R \qquad \text{with } x \text{ fresh letter of } \Sigma_1$$

$$\frac{\Gamma \vdash A^y, \Delta \quad \Gamma', A^y \vdash \Delta'}{\Gamma, \Gamma' \vdash \Delta, \Delta'} \; cut \qquad \text{with } y \text{ fresh letter of } \Sigma_2$$

$$\frac{\Gamma \vdash A^y, \Delta \quad \Gamma', B^{yv} \vdash \Delta'}{\Gamma, \Gamma', (A \multimap B)^v \vdash \Delta, \Delta'} \multimap_L \qquad \text{with } y \text{ fresh letter of } \Sigma_2$$

$$\frac{}{A^v \vdash A^w} \; ax$$

$$\frac{\Gamma \vdash \Delta}{\Gamma, \mathbf{1}^v \vdash \Delta} \mathbf{1}_L \quad \frac{}{\vdash \mathbf{1}^v} \mathbf{1}_R \quad \frac{}{\bot^v \vdash} \bot_L \quad \frac{\Gamma \vdash \Delta}{\Gamma \vdash \bot^v, \Delta} \bot_R$$

$$\frac{\Gamma, A^v \vdash \Delta \quad \Gamma', B^v \vdash \Delta'}{\Gamma, \Gamma', (A \parr B)^v \vdash \Delta, \Delta'} \parr_L \qquad \frac{\Gamma \vdash A^v, B^v, \Delta}{\Gamma \vdash (A \parr B)^v, \Delta} \parr_R$$

$$\frac{\Gamma, A^v, B^v \vdash \Delta}{\Gamma, (A \otimes B)^v \vdash \Delta} \otimes_L \qquad \frac{\Gamma \vdash A^v, \Delta \quad \Gamma' \vdash B^v, \Delta'}{\Gamma, \Gamma' \vdash (A \otimes B)^v, \Delta, \Delta'} \otimes_R$$

Figure 1.4 Another labelled sequent calculus for **CLL**.

4.2.1 *Some properties.* Let us begin to prove some properties that are useful to understand the behaviour of the labels in this setting.

Property 13 *Let P be a proof (in the system of Figure 1.4) of $\Gamma \vdash \Delta$, if the conclusion of a subproof contains a letter $\alpha \in \Sigma_2$ then α is always on the same side in all sequents of P.*

Let us also mention that if (α, β) is a correspondence on a sequent S inside a proof then all symbols in a dependency chain between α and β are in fact introduced in the subproof, the conclusion of which is S.

Property 14 *Let P be a proof (in the system of Figure 1.4) of $\Gamma \vdash \Delta$ with a dependency chain between α and β on the axiom sequents; for any subproof that contains this chain, if its conclusion has a correspondence (α, β) then the formula labelled with β depends (in the sense of 1.2) on the one labelled with α.*

Proof By induction on the depth of the proof tree.
Base case: $n = 1$; the property is verified for the axiom rules.
Induction case: Let us assume that the property is verified for n and consider the last inference rule applied in the proof.
- For the rules with one premise, it is trivial.
- For the rules with two premises, we have the following cases:
i) for the $\parr_L$ and $\otimes_R$ rules, we verify that the dependency chain is completely included in one of the subtrees.
ii) for the *cut* and $\multimap_L$, it is the same proof argument if the new introduced letter y is not in the dependency chain. Else if it appears in the chain and if

there is a correspondence (α, β) at the root, then the subchain between α and y (resp. y and β) is completely included in the left (resp. right) subtree.
By the induction hypothesis we have dependencies between the formulae labelled by α and y in the left-hand side premise and by y and β in the right-hand side premise. The result is then deduced from an analysis of the rule of the dependencies calculus. ■

Property 15 *Let P be a proof (in the system of Figure 1.4) of $\Gamma, E \vdash F, \Delta$, such that $E \in Dep(F)$; if the formulae E and F are in a correspondence (α, β), then there exists a dependency chain between α and β (on the sequence of axiom sequents).*

Proof By induction on the depth of the proof tree, with verification of the property for each rule. ■

From these properties, we can deduce the following results:

Lemma 16 *Let P be a proof (in the system of Figure 1.4) of a sequent $\Gamma \vdash \Delta$,*
(i) If there exists a dependency chain on the axiom sequents between two occurrences of the same letter of Σ_1, then the dependency condition (of definition 6) is not verified where this letter appears.
(ii) Let SP be a subproof, the last rule of which is $\multimap_R$, and such that there exists $A \in \Delta$ verifying $B \in Dep(A)$, then SP has a dependency chain between two occurrences of the same letter of Σ_2.

Proof Both results are direct consequences of the previous properties. ■

4.2.2 The intuitionistic characterization. These preliminary results are necessary to prove the next theorem. Its proof is based on the equivalence between the criteria we define and the calculus of dependencies of the multiplicative **FILL** system.

Definition 17 *A sequence of sequents is said to be* intuitionistic *if there is no dependency chain between two occurrences of the same letter of Σ_1.*

Theorem 18 *A proof P (in the system of Figure 1.4) is a* **FILL** *proof if and only if the sequence of axiom sequents, ordered from left to right, is intuitionistic.*

Proof The proof is a direct consequence of the previous lemma. ■

This proof that corresponds to the criterion of section 4.2, can be easily adapted to the one of section 4.1. Let us note that the order of premises in each rule of Figure 1.4 has to be respected during the proof construction. Let us consider again the example of section 4.1 with this new characterization. The proof is **CLL** but not **FILL** because of a dependency chain $(x, y), (y, x)$. Here $x \in \Sigma_1$ and $(y, z) \in \Sigma_2^2$.

$$
\dfrac{\dfrac{\dfrac{\dfrac{}{C^x \vdash C^z}\,ax \quad \dfrac{}{D^x \vdash D^y}\,ax}{(C \parr D)^x \vdash C^z, D^y}\,\parr_L \quad \dfrac{}{B^z \vdash B}\,ax}{C \multimap B, (C \parr D)^x \vdash D^y, B}\,\multimap_L \quad \dfrac{}{A^y \vdash A^x}\,ax}{\dfrac{C \multimap B, D \multimap A, (C \parr D)^x \vdash B, A^x}{C \multimap B, D \multimap A \vdash (C \parr D) \multimap B, A}\,\multimap_R}\,\multimap_L
$$

Let us now present an example that illustrates that a sequent can have a **CLL** proof that is a **FILL** proof but also another one that is not a **FILL** proof. Here, we only use one atomic formula A, the different occurrences of which that are distinguished with indices. Therefore we present two labelled proofs of the following sequent: $A_4 \parr A_5, A_8 \vdash A_1 \multimap (A_6 \otimes A_7), A_2 \otimes A_3$.

The first proof is not a **FILL** proof because of a correspondence (x, x) in one of the axioms.

$$
\dfrac{\dfrac{\dfrac{\dfrac{\dfrac{}{A_1^x \vdash A_2^x}\,ax \quad \dfrac{}{A_4 \vdash A_3^x}\,ax}{A_1^x, A_4 \vdash (A_2 \otimes A_3)^x}\,\otimes_{R-2,3} \quad \dfrac{}{A_5 \vdash A_6}\,ax}{A_1^x, A_4 \parr A_5 \vdash A_6, (A_2 \otimes A_3)^x}\,\parr_L \quad \dfrac{}{A_8 \vdash A_7}\,ax}{A_1^x, A_4 \parr A_5, A_8 \vdash A_6 \otimes A_7, (A_2 \otimes A_3)^x}\,\otimes_{R-6,7}}{A_4 \parr A_5, A_8 \vdash A_1 \multimap (A_6 \otimes A_7), A_2 \otimes A_3}\,\multimap_R
$$

One can easily verify that $A_1 \in Dep(A_2 \otimes A_3)$ in the premise of the $\multimap_R$ rule. In the following proof

$$
\dfrac{\dfrac{\dfrac{\dfrac{\dfrac{}{A_1^x \vdash A_6}\,ax \quad \dfrac{}{A_5 \vdash A_7}\,ax}{A_1^x, A_5 \vdash A_6 \otimes A_7}\,\otimes_{R-6,7} \quad \dfrac{}{A_4 \vdash A_3^x}\,ax}{A_1^x, A_4 \parr A_5 \vdash A_3^x, A_6 \otimes A_7}\,\parr_L \quad \dfrac{}{A_8 \vdash A_2^x}\,ax}{A_1^x, A_4 \parr A_5, A_8 \vdash (A_2 \otimes A_3)^x, A_6 \otimes A_7}\,\otimes_{R-2,3}}{A_4 \parr A_5, A_8 \vdash A_1 \multimap (A_6 \otimes A_7), A_2 \otimes A_3}\,\multimap_R
$$

there is no correspondence in the axiom sequents and consequently no dependency chain. Thus it is a **FILL** proof.

4.3 AN ALGORITHM FOR PROOF CONSTRUCTION

From the theorem 18 and this intuitionistic characterization, we can easily derive a corresponding algorithm. In fact, if we have a **CLL** proof built with the labelled sequent calculus then it is sufficient to analyse the labels of the axiom sequents and to do the test of the dependency chains. Such an algorithm could be presented in the following way:

For each sequent S in which a $x \in \Sigma_1$ is a label of a left-hand side formula, one call the procedure `Search_chain(`x`,`S`,`x`)` that is defined as by

Procedure `Search_chain(origin,sequent,current_letter)`

For all correspondences (`current_letter`, α),

- if $\alpha =$ `origin` then FAILURE

- else, for all axiom sequents S' on the right-hand side of `sequent`, call `Search_chain(origin,` S'`,` α`)`

It returns a failure message if and only if the given proof is not a **FILL** proof. In general, with the current examples, the size of the word is not too big and then the number of cases the algorithm has to deal with is reasonable. A study of the complexity could be interesting but would need some choices about implementation. Such investigations are out of the purposes of this paper.

4.4 SOME REMARKS

In fact, such a method allows to verify if a given **CLL** proof is also a **FILL** proof. An interesting point could be to analyse, in case of failure, how the labels can be used to propose some heuristics or proof transformations, to obtain more directly and efficiently some **FILL** proofs, if they exist, by classical search. For instance, in the proofs of section 4.2.2, we observe that the application of the $\otimes_R$ rule on a marked formula allows to eliminate one possibility of correspondence.
Bierman has analysed with Parigot's techniques the relationships between **ILL** and **CLL** to propose a classical linear λ-calculus. This work appears fruitful from the resulting programming language point of view and of more use than such a similar language based on proof-nets [8]. Therefore we could consider an approach similar to [30] by using a linear λ-calculus to encode the proofs and then analyse if they are "intuitionistic". This approach seems difficult to implement and to effectively use. Anyway, an alternative could be based on the notion of *proof-net* that can be seen as a counterpart of natural deduction in linear logic [23]. It could be an appropriate semantical and syntactical representation of proof-objects (or terms) for the analysis of intuitionistic provability.

5 PROOF-NETS AND INTUITIONISTIC PROVABILITY

We now propose an alternative approach to consider the intuitionistic provability in **CLL**. It is based on a specific structure of linear logic, i.e. *proof-net*, that was naturally introduced by Girard as a counterpart of natural deduction in linear logic [21].

5.1 SOME DEFINITIONS

A proof-net is a particular graph, the nodes of which are formulae of linear logic [23]. It is naturally defined for the multiplicative linear logic **MLL** (without constants) as recalled in this subsection and can be seen as an efficient tool for automated deduction in such a fragment [15]. There are also proof-nets definitions for other **CLL** fragments, like **MALL** (Multiplicative and Additive Linear Logic) [23] but they are more complicated to handle.

Definition 19 *A* proof-structure *is a graph, the vertices of which are formulae, inductively built from the following substructures (called* links*) :*

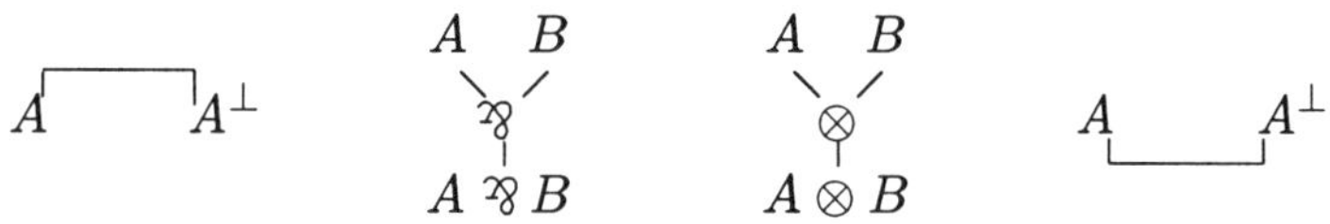

where A and B are formulae. In such a graph, any formula is the premise of at least one link and the conclusion of exactly one link.

A proof-structure can be also seen as the set of the subformulae of the formula or sequent to prove (represented as a decomposition tree) plus a set of axiom-links such that each atom of the tree is the conclusion of exactly one axiom-link. A proof-net, for a given sequent, is in fact a proof-structure that corresponds to legal proofs [21]. There are several equivalent definitions of what a proof-net is, that are generally based on criteria that characterize the proof-structures that are proof-nets [21], like for instance the Danos-Regnier characterization [12]. The construction of a proof-net can be seen as the search of the connections (axiom-links) that characterize the provability of a given sequent [15].
To simplify the different representations, we often omit the formulae labelling the nodes and only represents the connectives. Here we recall the inductive definition of proof-net, introduced by Bellin [5].

Definition 20 *A* **MLL** proof-net *and its* set of conclusions *are inductively defined in the following way:*

- *If A is any multiplicative formula,* $A \quad A^\perp$ *is a proof-net, with A and $A^\perp$ as conclusions* (axiom-link)
- *If Π_1 (resp. Π_2) is a proof-net with $\Gamma_1 \cup \{A\}$ (resp. $\Gamma_2 \cup \{A^\perp\}$) as conclusions, and if Π_1 and Π_2 are disjoint (not connected), then*
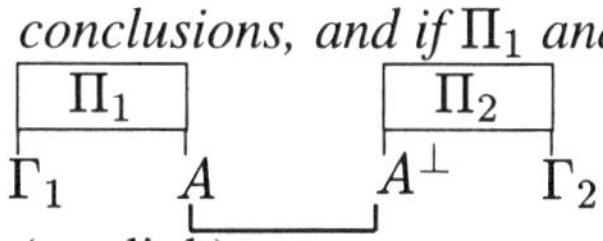

is a proof-net with $\Gamma_1 \cup \Gamma_2$ as conclusions (cut-link).

- *If* Π_1 *(resp.* Π_2*) is a proof-net with* $\Gamma_1 \cup \{A\}$ *(resp.* $\Gamma_2 \cup \{B\}$*) as conclusions, and if* Π_1 *and* Π_2 *are disjoint (not connected), then* 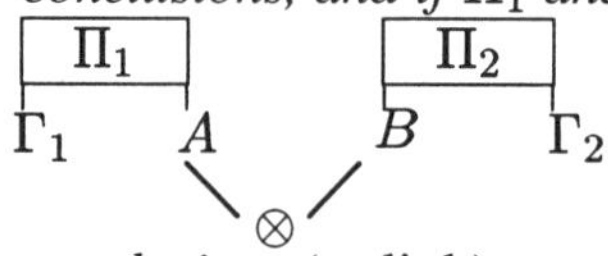 *is a proof-net with* $\Gamma_1 \cup \Gamma_2 \cup \{A \otimes B\}$ *as conclusions* ($\otimes$-link).

- *If* Π *is a proof-net with* $\Gamma \cup \{A, B\}$ *as conclusions, then* 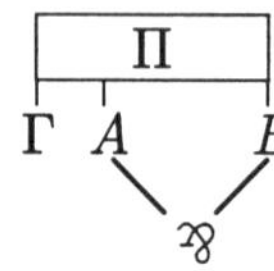 *is a proof-net with* $\Gamma \cup \{A \parr B\}$ *as conclusions* ($\parr$-link).

Some reductions operations can be defined on proof-nets that correspond to the notion of reduction in λ-calculus. There exists a theorem of cut-elimination for the proof-nets [23] and from now we only consider proof-nets without cut-links. Our goal is to define intuitionistic proof-nets from the classical proof-nets. Let us recall that the order of the premises in the proof-nets is only essential in the non-commutative logic [18].

5.2 ORIENTED PROOF-NETS

The construction of proof-nets from a given formula or sequent, is based on the use of negation and on the following equivalences: $(E \otimes F)^{\perp} = E^{\perp} \parr F^{\perp}$, $(E \parr F)^{\perp} = E^{\perp} \otimes F^{\perp}$, and $E \multimap F = E^{\perp} \parr F$. In **ILL** the linear implication cannot be defined from the $\parr$ connective because of the lack of negation. But in the spirit of using a classical proof to analyse the intuitionistic provability, it is interesting to keep the classical definition (and these equivalences) and to define the intuitionistic one from it. It is done by the introduction of a corresponding orientation to annotate each node of the structure with either the label I (for Input) or the label O (for Output) [5]. It allows to consider the classical structure and to add through some labels necessary information to detect, during the construction, if a proof-net is intuitionistic or not.

Definition 21 *Let* $\Gamma \vdash \Delta$ *be a* **ILL** *sequent, the orientation of the formula tree is defined in the following way:*
i) the formulae of Γ *(resp.* Δ*) are labelled with* I *(resp.* O*).*
ii) a formula that is premise of a $\parr$ *or* $\otimes$ *connective that does not come from an implication transformation, has the same label as the formula conclusion.*
iii) a formula on the right-hand (resp. left-hand) side of a $\parr$ *or* $\otimes$ *connective that comes from an implication transformation, has the same (resp. opposite) label as the formula conclusion.*

Then a proof-net with such labels or annotations is called an oriented proof-net. The example of Figure 1.5 presents an oriented proof-net corresponding to

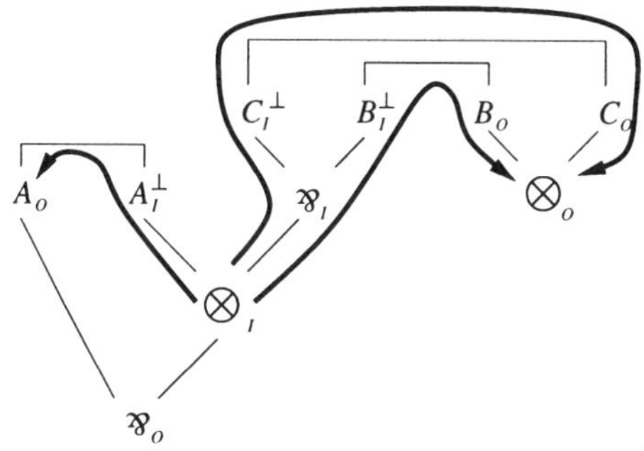

Figure 1.5 An oriented proof-net for $\vdash (A \parr (B \otimes C)) \multimap A,\ \ B \otimes C$.

the sequent $\vdash (A \parr (B \otimes C)) \multimap A,\ B \otimes C$. If we transform the $\multimap$ connective and annotate the formulae then we obtain the following sequent, that is classically equivalent to the initial one: $\vdash ((A_I^\perp \otimes (B_I^\perp \parr C_I^\perp)_I)_I \parr A_O)_O,\ (B_O \otimes C_O)_O$.

Property 22 *An oriented proof-net has at least one conclusion with label O.*

Proof By induction from the proof-nets definition. ■

Let us remark that there are proof-nets without conclusions labelled with I, for instance for provable sequents having no formula on the left-hand side.

Definition 23 *A* **ILL** proof-net *is an oriented proof-net without nodes of the following form:* $\begin{array}{c} O\ \ O \\ \parr \\ O \end{array}$ *or* $\begin{array}{c} I\ \ I \\ \otimes \\ I \end{array}$

Theorem 24 *A sequent is provable in* **ILL** *if and only if its associated oriented proof-net is a* **ILL** *proof-net.*

Proof It is a direct consequence of theorem 5. ■

5.3 FILL PROOF-NETS

The orientation defined in the previous section can be used to give a definition of a multiplicative **FILL** proof-net. The definition we propose is adapted from the one of Bellin for the multiplicative **FILL** + **MIX** [4].

Definition 25 *A* dependency path *is a path between a conclusion labelled with I and a conclusion labelled with O in a proof-net, that verifies that one always go upwards (resp. downwards) through I- (resp.O-) oriented nodes.*

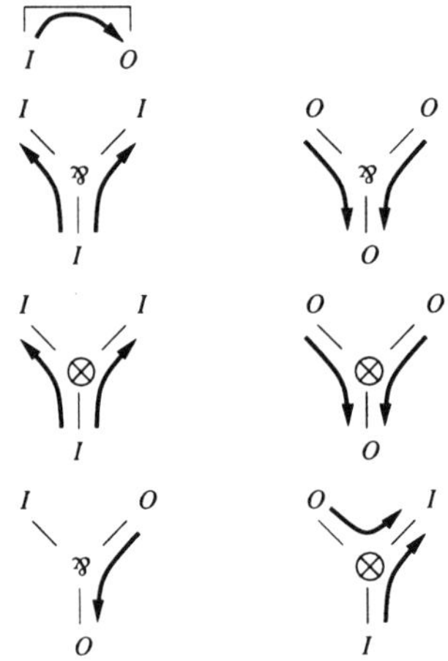

Figure 1.6 Dependency subpaths in **FILL** proof-nets

The possible dependency subpaths, for each type of node, are given in Figure 1.6. Let us recall that a substructure S' of a proof-structure S is a proof-structure that is a subgraph. Then a *subnet* of a proof-net is a substructure that is a proof-net [5]. Moreover, let P_n be a proof-net and A be one of its nodes, the *empire of* A is the largest subnet of P_n with A as conclusion.

Definition 26 *A* **FILL** proof-net *is an oriented proof-net such that for any node* $\begin{array}{cc} A_I & B_O \\ \multicolumn{2}{c}{\parr_O} \end{array}$, *there is no dependency path between* A *and a conclusion of the empire of* A *that is different from this occurrence of* B.

The Figure 1.7 presents two different proof-nets that correspond to the sequent proofs of section 4.2.2. The atomic formulae, that are A (resp. $A^{\perp}$) for the atoms labelled with O (resp. I), are not displayed. The first proof-net is not **FILL** because of the existence of a dependency path between the premise of the ⅋ labelled with I and a node labelled with O which is not a premise of the ⅋ connective.

5.4 FILL PROOF-NETS CONSTRUCTION

We have previously studied the problem of automated construction of proof-nets in different fragments of **CLL** like **MLL**, **MALL** or Non-Commutative **MLL** [17, 18, 19]. Some of the underlying principles could be applied for the case of multiplicative **FILL** fragment.

5.4.1 Principles for construction. We recall here the principles of our approach of proof-nets construction, taking the **MLL** fragment as illustration.

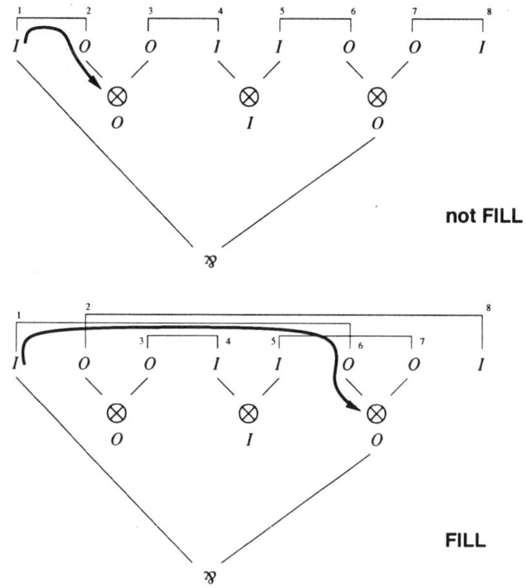

Figure 1.7 Proof-nets for $A_4 \parr A_5, A_8 \vdash A_1 \multimap (A_6 \otimes A_7), A_2 \otimes A_3$

More details can be found in [17, 19]. The construction is based on the inductive definition of proof-nets (see previous section) and instead of constructing a proof-structure and verifying with some criteria if it is a proof-net or not, subnets are built step by step to finally obtain a proof-net. Let us consider a **MLL** sequent of the form $\vdash \Delta$. At first, we decompose the formulae to obtain a decomposition tree of the sequent. The construction of a proof-net corresponds to the construction of axiom-links that added to the decomposition tree forms a proof-structure that is in fact a proof-net. But how to find the suitable axiom-links (or connections)? To find and justify the appropriate axiom-links, if they exist, we try to build, from some axiom-links (that are elementary proof-nets), some subnets that also are proof-nets, and to assemble them in the right way to finally obtain a proof-net, by using three procedures: one to select, in the decomposition tree, a leaf to treat; one to treat the leaves, of the decomposition tree, by constructing some axiom-links; one to extend the existing structures (by assembling subnets) with links. As we aim to reduce the number of disjoint subnets during this construction process, we choose the formula to treat in branches, some leaves of which have already been treated. Another strategy, based on permutability results and on the inductive definition of proof-nets consists in always dealing with the $\otimes$ connective before the $\parr$ connective, if possible, either to search the next leaf to treat or to assemble subnets. The search (-construction) of a proof-net follows these principles. To

start we have to choose a branch and the successive nodes to finally obtain a leaf and we consider the ⊗-nodes as a priority. Having selected such a leaf, we construct an initial subnet by construction of an axiom-link with a dual formula. Each time a new subnet is created, we want to assemble some of the subnets by extension with links in the following way:
if the node under the current formula is a

- ⊗-node, both premises of which belong to disjoint subnets, then mix them into one new net with an extension with this node.
- ⅋-node, both premises of which belong to the same subnet, then add the node to this subnet.
- ⅋-node, both premises of which belong to disjoint subnets, then keep this node in a waiting position.
- ⊗-node, only one of the premises being treated, then treat the other premise as a priority.

We go on by choosing a new formula to deal with, treating at first the ⊗ connectives and then the ⅋ connectives and trying to assemble subnets with the ⅋ connectives being in the waiting position. At the end, if the proof-net exists then all formulae are treated and the resulting structure (or net) is the final proof-net. More details that are not necessary here can be found in [17, 18, 19]. Such an algorithm can be used to define a connection-based proof-search method [15].

5.4.2 FILL vs CLL proof-nets. To take into account the criteria on paths in **FILL** proof-nets, we add a piece of information during the creation of each new node of the **CLL** net to memorize all the dependency chains between a conclusion with label I and a conclusion with label O of the current subnet. For that, we have to follow the construction rules given in Figure 1.6. It leads to the following algorithm, the essential points of which being summarized in Figure 1.8, where the dashed arrows are replaced by the plain arrows. Thus the intuitionistic provability is considered as a perturbation of classical proof-net construction.

During the creation of axiom-links: there exists a dependency path between the formula with label I and the one with label O.

During the creation of $\underset{I}{\overset{I\quad I}{⅋}}$ **or** $\underset{I}{\overset{I\quad I}{\otimes}}$ **links:** the dependency paths between one of the premises and a conclusion labelled with O are all replaced by dependency paths between the new conclusion labelled with I and the same conclusion labelled with O.

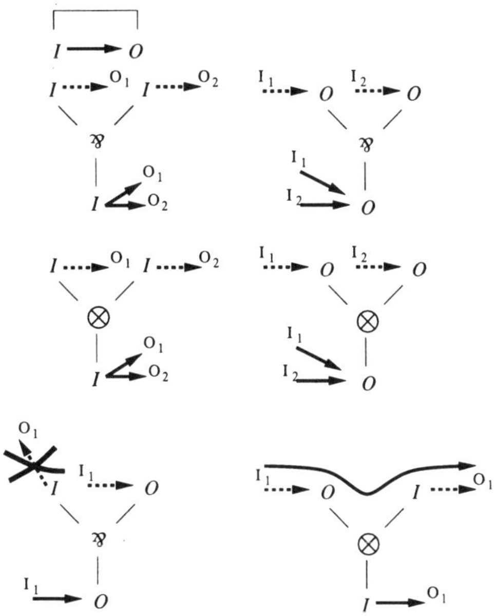

Figure 1.8 Update of dependency paths for **FILL** proof-nets.

During the creation of $\begin{array}{c} O \quad O \\ \parr \\ O \end{array}$ **or** $\begin{array}{c} O \quad O \\ \otimes \\ O \end{array}$ **links:** the dependency paths between a conclusion labelled with I and one of the premises are all replaced by dependency paths between the same conclusion labelled with I and the new conclusion labelled with O.

During the creation of $\begin{array}{c} I \quad O \\ \parr \\ O \end{array}$ **link:** if there exists a dependency path between the premise labelled with I and a conclusion O, different from the premise labelled with O, the proof-net is not **FILL**. Else, the dependency paths between a conclusion labelled with I and the premise labelled with O are all replaced by dependency paths between the same conclusion labelled with I and the new conclusion labelled with O; moreover the dependency paths, the origin of which being labelled with I, are suppressed.

During the creation of $\begin{array}{c} I \quad O \\ \otimes \\ I \end{array}$: the dependency paths between a conclusion labelled with I and a premise labelled with O are all replaced by

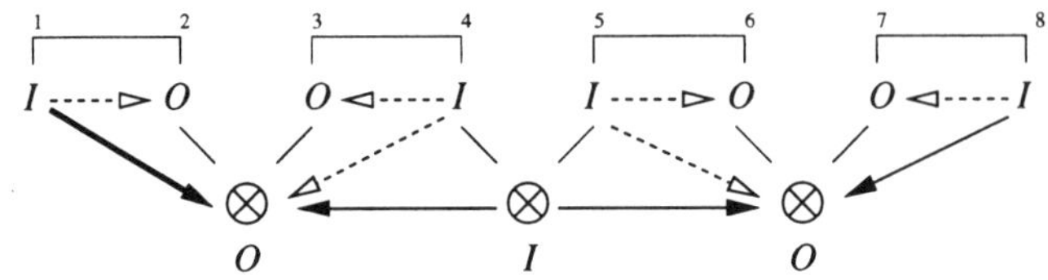

Figure 1.9 Dependency paths before the creation of the ⅋-link in Figure 1.7.

the paths of same origin but extended with the paths starting from the premise labelled with I; the dependency paths starting of the premise labelled with I are replaced by paths starting from the new conclusion labelled with I.

During the construction of classical **MLL** proof-nets, we memorize the information about dependency paths between labelled occurrences of formulae and we then dynamically verify the condition of definition 26 characterizing a multiplicative **FILL** proof-net. The algorithm always build such a proof-net if there exists.

5.4.3 *An example.* Let us come back to the examples of Figure 1.7. The algorithm builds the decomposition tree and then develops the proof-net from the leaves (or axiom-links), in a top-down approach. It tries to create an axiom-link between a node labelled with I (that corresponds to $A^{\perp}$) and a node labelled with O (that corresponds to A). Then it memorizes a path between these formulae respectively labelled with I and O. Let us assume that there is an axiom-link between the formulae 1 and 2 (see the first example). We go on with a $\otimes$ connective linked to one of these treated formulae and decide to consider the other premise. We then create the axiom-link between formulae 3 and 4 and a path from 4 to 3. Moreover, we can construct the $\otimes$-link and the two paths memorized are modified to point out on this $\otimes$-link. Therefore, the algorithm constructs the axiom-links and the $\otimes$-links and concludes with the construction of the ⅋-link. In this first example, the situation at the moment of the creation of ⅋-link is presented in Figure 1.9. In fact, it is impossible to create this link because of the existence of the bold arrow and this proof-net is not **FILL**. Then by backtracking there are other attempts of axiom-links construction leading to a proposal that allows to construct the ⅋-link and then to a **FILL** proof-net (see the second example). Let us mention that, in general, the number of possibilities for the axiom-links is limited.

5.4.4 *Main results.* To prove the correction of the algorithm, it is necessary to consider two main properties.

Property 27 *The algorithm memorizes all the dependency paths between a conclusion labelled with I and a conclusion labelled with O, and only these ones.*

Proof The property is verified, case by case, for each type of node. ■

Let us mention now a lemma about the existence of dependency path.

Lemma 28 *Let P_n be a proof-net and A a node of P_n labelled with I, there exists a dependency path between A and a conclusion of P_n, labelled with O.*

Proof The property 22 indicates that there always exists a conclusion labelled with O. This lemma is proved in a similar way, by induction on the proof-net definition. The axiom-link case is trivial. For the ⅋- and $\otimes$-links we can distinguish the case where the node I is the new one that has just been introduced. The significative case is the one of $\otimes$, when the induction hypothesis gives a path between a node labelled with I and the unique premise of this $\otimes$-link that is labelled with O. A second use of the induction hypothesis (on the other net) allows the resulting path to be connected to the previous one. ■

This result leads to the following property:

Property 29 *Let us consider a node A_I ⅋ B_O in a proof-net; then the existence of a dependency path between A_I and a conclusion labelled with O (different from B_O) is equivalent to the existence of a dependency path between A_I and a conclusion of the empire of A in the final proof-net, labelled with O.*

Proof The proof uses classical properties of proof-nets, that are not detailed here (see [21]). It follows a schema similar to the proof of property 27 and uses the lemma 28 to deal with the case of the $\otimes$-link, the premises of which do not have the same orientation. ■

Theorem 30 *Let $\Gamma \vdash \Delta$ be a* **CLL** *sequent, the algorithm builds a* **FILL** *proof-net for it iff it is provable in* **FILL**.

Proof It is a consequence of the algorithm properties and also of the previous properties and lemmas. ■

5.5 SOME REMARKS

The proof-nets construction algorithm for the multiplicative **FILL** has the property to simultaneously construct a sequent proof in a top-down (from axioms to final sequent) approach [17, 18]. Then we obtain an algorithm to directly construct a **FILL** proof if there exists and not only some criteria to verify if a **CLL** proof is also a **FILL** proof. As the previous algorithm finds a **FILL** proof if there exists, the use of proof-nets allows to solve the completeness problem w.r.t. intuitionistic provability we have mentioned in previous sections. This significant point illustrates, in the setting of linear logic, the interest of proof-net as an alternative tool for efficient automated deduction. We also observe the importance of the definition of labels for such a characterization.

It appears from the works of Asperti on logical analysis of distributed systems that **MLL** plus the **MIX** rule might be more suitable for the representation of concurrent logical computations [3]. This rule

$$\frac{\vdash \Gamma \quad \vdash \Delta}{\vdash \Gamma, \Delta}\ Mix$$

added to **MLL** has been also studied from the proof-net point of view by showing that the verification of correctness for proof-nets in this logic is equivalent to a successful termination of a concurrent game in the style of Petri nets [3]. Refinements and relationships with term assignments are proposed in [4] where a system of proof-nets for **MLL** (without constants) + **MIX** is given. From such a study of (multiplicative) classical linear logic, [4] has also proposed a proof-nets system for (multiplicative) **FILL** + **MIX**(without constants). Our results about intuitionistic provability can be extended and adapted to study the intuitionistic provability in **MLL** + **MIX**. It means that we can analyse it from the construction of a classical **MLL** + **MIX** proof-net with labels to detect the existence of a **FILL** + **MIX** proof-net. Such works could be significant for efficient proof-search in the context of analysis and proofs in concurrent and distributed systems.

6 CONCLUSION

In this paper, we have defined labelled sequent calculi to analyse and characterize the intuitionistic provability from proof-search in classical logic. Our first approach, in **CL** and **CLL**, is based on characterizations of the axiom sequents of a given labelled sequent proof. It does not use specific terms as classical or linear λ-terms to represent proofs (coming from natural deduction systems) as in [30]. We could argue that such sequent calculi seem more suitable from both proof-theoretic and proof-search points of view. In case of **CLL**, a recent study has defined a classical linear λ-calculus [8] and similar calculi could be potentially used to encode **CLL** proofs and to characterize the intuitionistic proofs, but it could be hard to handle mainly if provability is the main interest. Anyway, we cannot have a completeness result, i.e. the intuitionistic provability cannot be claimed from any given classical proof. Our labelled proof systems can be seen as new formulations, based-on labels, on the **FILL** system and could be independently interesting from the proof-theoretic point of view. We have proposed an alternative to this characterization in **CLL** that is based on the proof-net notion, that can be seen as a counterpart of natural deduction in linear logic. For that, we give a new labelled proof system and an algorithm that allow to build an intuitionistic proof-net (and a sequent proof) from a classical proof-net search, for which we have well-defined principles. With such a structure we obtain the completeness result missing with the other systems. This point illustrates the interest of the proof-net notion as an efficient tool for proof analysis and proof-search in linear logic. Because of the relationships between proof-nets and some term assignments systems [1] and also because of possible choices between intuitionistic or classical logical fragments to consider some programming paradigms, like proofs-as-programs or proofs-as-processes, our results and their possible extensions, for instance with the **MIX** rule or with the additive connectives, could be very fruitful. The result, in terms of proof-nets, is particular to linear logic but could encourage us to have similar structures or some graph-theoretic characterizations of provability in other substructural logics and therefore to derive new methods to analyse and search proofs. Significant points about the connected algorithmic and implementation problems will be studied in further work.

Vincent Balat
Ecole Normale Supérieure
45, rue d'Ulm
F-75005 Paris, France
Vincent.Balat@ens.fr

Didier Galmiche
LORIA UMR 7503
Université Henri Poincaré
Campus scientifique, BP 239
F-54506 Vandœuvre-lès-Nancy, France
Didier.Galmiche@loria.fr

References

[1] S. Abramsky. Computational interpretations of Linear Logic. *Theoretical Computer Science*, 111(1-2):3–58, 1993.

[2] V. Alexiev. Applications of Linear Logic to Computation: An Overview. *Bulletin of the IGPL*, 2(1):77–107, 1994.

[3] A. Asperti. Causal dependencies in multiplicative linear logic with MIX. *Math. Struct. in Comp. Science*, 5:351–380, 1995.

[4] G. Bellin. Subnets of proof-nets in multiplicative linear logic with MIX. *Math. Struct. in Comp. Science*, 7:663–699, 1997.

[5] G. Bellin and J. van de Wiele. Subnets of proof-nets in MLL^-. In J.Y. Girard, Y. Lafont, and L. Regnier, editors, *Advances in Linear Logic*, pages 249–260. Cambridge University Press, 1995.

[6] N. Benton, G. Bierman, V. de Paiva, and M. Hyland. A term calculus for intuitionistic linear logic. In *Int. Conference on Typed Lambda Calculi and Applications, LNCS 664*, pages 75–90, Utrecht, The Netherlands, March 1993.

[7] G.M. Bierman. A note on Full Intuitionistic Linear Logic. *Annals of Pure and Applied Logic*, 79:281–287, 1996.

[8] G.M. Bierman. Towards a classical linear λ-calculus (preliminary report). *Electronic Notes in Theoretical Computer Science*, 3, 1996.

[9] E. Boudinet and D. Galmiche. Proofs, concurrent objects and computations in a FILL framework. In *Workshop on Object-based Parallel and Distributed Computation, OBPDC'95, LNCS 1107*, pages 148–167, Tokyo, Japan, 1996.

[10] T. Braüner and V. de Paiva. Cut elimination for Full Intuitionistic Linear Logic. Unpublished draft, 1996.

[11] M. D'Agostino and D.M. Gabbay. A Generalization of Analytic Deduction via Labelled Deductive Systems. Part I: Basic substructural logics. *Journal of Automated Reasoning*, 13:243–281, 1994.

[12] V. Danos and L. Regnier. The structures of multiplicatives. *Arch. Math. Logic*, 28:181–203, 1989.

[13] R. Dyckhoff. Contraction-free sequent calculi for intuitionistic logic. *Journal of Symbolic Logic*, 57:795–807, 1992.

[14] U. Engberg and G. Winskel. Completeness results for linear logic on Petri nets. *Annals of Pure and Applied Logic*, 86:101–135, 1997.

[15] D. Galmiche. Connection methods in Linear Logic and Proof nets Construction. *Theoretical Computer Science*, 1999. Accepted for publication.

[16] D. Galmiche and D. Larchey-Wendling. Provability in intuitionistic linear logic from a new interpretation on Petri nets - extended abstract. *Electronic Notes in Theoretical Computer Science*, 17, 1998.

[17] D. Galmiche and B. Martin. Proof search and proof nets construction in linear logic. In *4th Workshop on Logic, Language, Information and Computation, Wollic'97*, Fortaleza, Brasil, August 1997. Logic Journal of IGPL, vol. 5-6, pp 883-885.

[18] D. Galmiche and B. Martin. Proof nets construction and automated deduction in non-commutative linear logic - extended abstract. *Electronic Notes in Theoretical Computer Science*, 17, 1998.

[19] D. Galmiche and G. Perrier. A procedure for automatic proof nets construction. In *International Conference on Logic Programming and Automated Reasoning, LNAI 624*, pages 42–53, St. Petersburg, Russia, July 1992.

[20] J.Y. Girard. Linear logic and parallelism. In *Mathematical Models for the Semantics of Parallelism, LNCS 280*, pages 166–182, Roma, Italia, October 1986.

[21] J.Y. Girard. Linear logic. *Theoretical Computer Science*, 50(1):1–102, 1987.

[22] J.Y. Girard. Linear logic: its syntax and semantics. In J.Y. Girard, Y. Lafont, and L. Regnier, editors, *Advances in Linear Logic*, pages 1–42. Cambridge University Press, 1995.

[23] J.Y. Girard. Proof nets: the parallel syntax for proof theory. In Ursini and Agliano, editors, *Logic and Algebra*, New York, 1995. M. Dekker.

[24] J.Y. Girard and Y. Lafont. Linear logic and lazy computation. In *TAPSOFT 87, LNCS 250*, pages 52–66, Pisa, Italia, March 1987.

[25] J. Hodas and D. Miller. Logic programming in a fragment of intuitionistic linear logic. *Journal of Information and Computation*, 110:327–365, 1994.

[26] M. Hyland and V. de Paiva. Full Intuitionistic Linear Logic (extended abstract). *Annals of Pure and Applied Logic*, 64:273–291, 1993.

[27] P. Lincoln and J. Mitchell. Operational aspects of linear lambda calculus. In *7th IEEE Symposium on Logic in Computer Science*, pages 235–246, Santa-Cruz, California, 1992.

[28] D. Miller. Forum: A multiple-conclusion specification logic. *Theoretical Computer Science*, 165(1):201–232, 1996.

[29] M. Parigot. $\lambda\mu$-calculus: an algorithmic interpretation of classical natural deduction. In *International Conference on Logic Programming and Automated Reasoning, LNAI 624*, pages 190–201, St. Petersburg, Russia, July 1992.

[30] E. Ritter, D. Pym, and L. Wallen. On the intuitionistic force of classical search. In *5th Int. Workshop TABLEAUX'96, LNAI 1071*, pages 294–313, Terrasini, Palermo, Italy, May 1996.

[31] H. Schellinx. Some syntactical observations on linear logic. *Journal of Logic and Computation*, 1(4):537–559, 1991.

[32] L.A. Wallen. *Automated Proof search in Non-Classical Logics.* MIT Press, 1990.

Appendix

1 A SEQUENT CALCULUS FOR CL

$$\frac{}{A \vdash A}\,ax \qquad \frac{\Gamma \vdash A, \Delta \quad \Gamma', A \vdash \Delta'}{\Gamma, \Gamma' \vdash \Delta, \Delta'}\,cut$$

$$\frac{\Gamma \vdash \Delta}{\Gamma, A \vdash \Delta}\,w_L \qquad \frac{\Gamma \vdash \Delta}{\Gamma \vdash \Delta, A}\,w_R \qquad \frac{\Gamma, A, A \vdash \Delta}{\Gamma, A \vdash \Delta}\,c_L \qquad \frac{\Gamma \vdash \Delta, A, A}{\Gamma \vdash \Delta, A}\,c_R$$

$$\frac{\Gamma \vdash A, \Delta \quad \Gamma, B \vdash \Delta}{\Gamma, A \to B \vdash \Delta}\,\to_L \qquad \frac{\Gamma, A \vdash B, \Delta}{\Gamma \vdash A \to B, \Delta}\,\to_R$$

$$\frac{\Gamma, A \vdash \Delta \quad \Gamma, B \vdash \Delta}{\Gamma, A \vee B \vdash \Delta}\,\vee_L \qquad \frac{\Gamma \vdash A, \Delta}{\Gamma \vdash A \vee B, \Delta}\,\vee_{R1} \qquad \frac{\Gamma \vdash B, \Delta}{\Gamma \vdash A \vee B, \Delta}\,\vee_{R2}$$

$$\frac{\Gamma, A, B \vdash \Delta}{\Gamma, A \wedge B \vdash \Delta}\,\wedge_L \qquad \frac{\Gamma \vdash A, \Delta \quad \Gamma \vdash B, \Delta}{\Gamma \vdash A \wedge B, \Delta}\,\wedge_R$$

$$\frac{\Gamma \vdash A, \Delta}{\Gamma, \neg A \vdash \Delta}\,\neg_L \qquad \frac{\Gamma, A \vdash \Delta}{\Gamma \vdash \neg A, \Delta}\,\neg_R$$

In the paper we use an equivalent presentation, without weakening rules (w_L and w_R) and where the axiom rule is replaced by $\frac{}{\Gamma, A \vdash A, \Delta}\,ax$.

2 A SEQUENT CALCULUS FOR IL

$$\frac{}{A \vdash A}\,ax \qquad \frac{\Gamma \vdash A \quad \Gamma', A \vdash B}{\Gamma, \Gamma' \vdash B}\,cut$$

$$\frac{\Gamma, A, A \vdash B}{\Gamma, A \vdash B}\,c \qquad \frac{\Gamma \vdash B}{\Gamma, A \vdash B}\,w \qquad \frac{}{\bot \vdash A}\,\bot_L$$

$$\frac{\Gamma \vdash A \quad \Gamma, B \vdash C}{\Gamma, A \to B \vdash C}\,\to_L \qquad \frac{\Gamma, A \vdash B}{\Gamma \vdash A \to B}\,\to_R$$

$$\frac{\Gamma, A \vdash C \quad \Gamma, B \vdash C}{\Gamma, A \vee B \vdash C}\,\vee_L \qquad \frac{\Gamma \vdash A}{\Gamma \vdash A \vee B}\,\vee_{R1} \qquad \frac{\Gamma \vdash B}{\Gamma \vdash A \vee B}\,\vee_{R2}$$

$$\frac{\Gamma, A, B \vdash C}{\Gamma, A \wedge B \vdash C}\,\wedge_L \qquad \frac{\Gamma \vdash A \quad \Gamma \vdash B}{\Gamma \vdash A \wedge B}\,\wedge_R$$

3 ANOTHER SEQUENT CALCULUS FOR IL

$$\frac{}{A \vdash A}\ ax \qquad \frac{\Gamma \vdash A, \Delta \quad \Gamma', A \vdash \Delta'}{\Gamma, \Gamma' \vdash \Delta, \Delta'}\ cut$$

$$\frac{\Gamma \vdash \Delta}{\Gamma, A \vdash \Delta}\ w_L \qquad \frac{\Gamma \vdash \Delta}{\Gamma \vdash \Delta, A}\ w_R \qquad \frac{\Gamma, A, A \vdash \Delta}{\Gamma, A \vdash \Delta}\ c_L \qquad \frac{\Gamma \vdash \Delta, A, A}{\Gamma \vdash \Delta, A}\ c_R$$

$$\frac{\Gamma \vdash A, \Delta \quad \Gamma, B \vdash \Delta}{\Gamma, A \to B \vdash \Delta}\ \to_L \qquad \frac{\Gamma, A \vdash B}{\Gamma \vdash A \to B, \Delta}\ \to_R$$

$$\frac{\Gamma, A \vdash \Delta \quad \Gamma, B \vdash \Delta}{\Gamma, A \vee B \vdash \Delta}\ \vee_L \qquad \frac{\Gamma \vdash A, \Delta}{\Gamma \vdash A \vee B, \Delta}\ \vee_R \qquad \frac{\Gamma \vdash B, \Delta}{\Gamma \vdash A \vee B, \Delta}\ \vee_R$$

$$\frac{\Gamma, A, B \vdash \Delta}{\Gamma, A \wedge B \vdash \Delta}\ \wedge_L \qquad \frac{\Gamma \vdash A, \Delta \quad \Gamma \vdash B, \Delta}{\Gamma \vdash A \wedge B, \Delta}\ \wedge_R$$

$$\frac{\Gamma \vdash A, \Delta}{\Gamma, \neg A \vdash \Delta}\ \neg_L \qquad \frac{\Gamma, A \vdash}{\Gamma \vdash \neg A, \Delta}\ \neg_R$$

4 A SEQUENT CALCULUS FOR CLL

$$\frac{}{A \vdash A}\ ax \qquad \frac{\Gamma \vdash A, \Delta \quad \Gamma', A \vdash \Delta'}{\Gamma, \Gamma' \vdash \Delta, \Delta'}\ cut$$

$$\frac{\Gamma \vdash \Delta}{\Gamma, \mathbf{1} \vdash \Delta}\ \mathbf{1}_L \qquad \frac{}{\vdash \mathbf{1}}\ \mathbf{1}_R \qquad \frac{}{\bot \vdash}\ \bot_L \qquad \frac{\Gamma \vdash \Delta}{\Gamma \vdash \bot, \Delta}\ \bot_R \qquad \frac{}{\Gamma, \mathbf{0} \vdash \Delta}\ \mathbf{0}_L \qquad \frac{}{\Gamma \vdash \top, \Delta}\ \top_R$$

$$\frac{\Gamma \vdash A, \Delta \quad \Gamma', B \vdash \Delta'}{\Gamma, \Gamma', A \multimap B, \vdash \Delta, \Delta'}\ \multimap_L \qquad \frac{\Gamma, A \vdash B, \Delta}{\Gamma \vdash A \multimap B, \Delta}\ \multimap_R$$

$$\frac{\Gamma, A \vdash \Delta \quad \Gamma', B \vdash \Delta'}{\Gamma, \Gamma', A \parr B \vdash \Delta, \Delta'}\ \parr_L \qquad \frac{\Gamma \vdash A, B, \Delta}{\Gamma \vdash A \parr B, \Delta}\ \parr_R$$

$$\frac{\Gamma, A, B \vdash \Delta}{\Gamma, A \otimes B \vdash \Delta}\ \otimes_L \qquad \frac{\Gamma \vdash A, \Delta \quad \Gamma' \vdash B, \Delta'}{\Gamma, \Gamma' \vdash A \otimes B, \Delta, \Delta'}\ \otimes_R$$

$$\frac{\Gamma, A \vdash \Delta}{\Gamma, A \& B \vdash \Delta}\ \&_{L1} \qquad \frac{\Gamma, B \vdash \Delta}{\Gamma, A \& B \vdash \Delta}\ \&_{L2} \qquad \frac{\Gamma \vdash A, \Delta \quad \Gamma \vdash B, \Delta}{\Gamma \vdash A \& B, \Delta}\ \&_R$$

$$\frac{\Gamma, A \vdash \Delta \quad \Gamma, B \vdash \Delta}{\Gamma, A \oplus B \vdash \Delta}\ \oplus_L \qquad \frac{\Gamma \vdash A, \Delta}{\Gamma \vdash A \oplus B, \Delta}\ \oplus_{R1} \qquad \frac{\Gamma \vdash B, \Delta}{\Gamma \vdash A \oplus B, \Delta}\ \oplus_{R2}$$

$$\frac{\Gamma \vdash A, \Delta}{\Gamma, A^{\bot} \vdash \Delta}\ neg_L \qquad \frac{\Gamma, A \vdash \Delta}{\Gamma \vdash A^{\bot}, \Delta}\ neg_R$$

$$\frac{\Gamma \vdash \Delta}{\Gamma, !A \vdash \Delta}\ \mathrm{w}_L \qquad \frac{\Gamma \vdash \Delta}{\Gamma \vdash ?A, \Delta}\ \mathrm{w}_R \qquad \frac{\Gamma, !A, !A \vdash \Delta}{\Gamma, !A \vdash \Delta}\ \mathrm{c}_L \qquad \frac{\Gamma \vdash ?A, ?A, \Delta}{\Gamma \vdash ?A, \Delta}\ \mathrm{c}_R$$

$$\frac{\Gamma, A \vdash \Delta}{\Gamma, !A \vdash \Delta}\ \mathrm{dir}_L \qquad \frac{\Gamma \vdash A, \Delta}{\Gamma \vdash ?A, \Delta}\ \mathrm{dir}_R \qquad \frac{!\Gamma, A \vdash ?\Delta}{!\Gamma, ?A \vdash ?\Delta}\ \mathrm{pro}_L \qquad \frac{!\Gamma \vdash A, ?\Delta}{!\Gamma \vdash !A, ?\Delta}\ \mathrm{pro}_R$$

5 A SEQUENT CALCULUS FOR ILL

$$\frac{}{A \vdash A}\ ax \qquad \frac{\Gamma \vdash A \quad \Gamma', A \vdash B}{\Gamma, \Gamma' \vdash B}\ cut$$

$$\frac{\Gamma \vdash A}{\Gamma, \mathbf{1} \vdash A}\ \mathbf{1}_L \qquad \frac{}{\vdash \mathbf{1}}\ \mathbf{1}_R \qquad \frac{}{\Gamma, \mathbf{0} \vdash A}\ \mathbf{0}_L \qquad \frac{}{\Gamma \vdash \top}\ \top_R$$

$$\frac{\Gamma \vdash A \quad \Gamma', B \vdash C}{\Gamma, \Gamma', A \multimap B, \vdash C}\ \multimap_L \qquad \frac{\Gamma, A \vdash B}{\Gamma \vdash A \multimap B}\ \multimap_R$$

$$\frac{\Gamma, A \vdash C}{\Gamma, A\&B \vdash C}\ \&_{L1} \qquad \frac{\Gamma, B \vdash C}{\Gamma, A\&B \vdash C}\ \&_{L2} \qquad \frac{\Gamma \vdash A \quad \Gamma \vdash B}{\Gamma \vdash A\&B}\ \&_R$$

$$\frac{\Gamma, A \vdash C \quad \Gamma, B \vdash C}{\Gamma, A \oplus B \vdash C}\ \oplus_L \qquad \frac{\Gamma \vdash A}{\Gamma \vdash A \oplus B}\ \oplus_{R1} \qquad \frac{\Gamma \vdash B}{\Gamma \vdash A \oplus B}\ \oplus_{R2}$$

$$\frac{\Gamma, A, B \vdash C}{\Gamma, A \otimes B \vdash C}\ \otimes_L \qquad \frac{\Gamma \vdash A \quad \Gamma' \vdash B}{\Gamma, \Gamma' \vdash A \otimes B}\ \otimes_R$$

$$\frac{\Gamma \vdash B}{\Gamma, !A \vdash B}\ \mathrm{w}_L \qquad \frac{\Gamma, !A, !A \vdash B}{\Gamma, !A \vdash B}\ \mathrm{c}_L \qquad \frac{\Gamma, A \vdash B}{\Gamma, !A \vdash B}\ \mathrm{dir} \qquad \frac{!\Gamma \vdash A}{!\Gamma \vdash !A}\ \mathrm{pro}$$

NORMAL MULTIMODAL LOGICS WITH INTERACTION AXIOMS

A Tableau Calculus and some (Un)Decidability Results

Matteo Baldoni

Abstract In this paper we present a *prefixed analytic tableau calculus* for a wide class of *normal multimodal logics*; the calculus can deal in a *uniform way* with any logic in this class. To achieve this goal, we use a *prefixed* tableau calculus *à la* Fitting, where we explicitly represent accessibility relations between worlds by means of a graph and we use the characterizing axioms as *rewriting rules*. Such rules create new paths among worlds in the counter-model construction. The prefixed tableau method is, then, used to prove (un)decidability results about certain classes of multimodal logics.

1 INTRODUCTION AND MOTIVATIONS

In recent years there has been a growing interest in the formalisms which allow knowledge to be represented in a *structured* way, supporting more sophisticate reasoning mechanisms. *Modal logics* are one of the most appreciated and studied of such tools [38].

On one hand, modal logics are useful to extend logic programming languages with new features [24, 28, 5, 3, 7]. On the other, they extend the classical framework by allowing the user to tackle in a pretty natural and task-oriented way problems that require some form of *distributed reasoning* [48]. One possible application of distributed reasoning is the construction of *teams of agents*, each of which has its own knowledge, reasoning style as well as its way of interacting with other agents. This kind of applications is of great interest to the Artificial Intelligence community, which is currently – deeply – investigating the use of modal logics and, in particular, *multimodal logics* [13] in agent modelling. Besides their adequacy in supporting reasoning in a multi-agent environment and to represent *knowledge, beliefs, dynamic changes, actions*, and *time* [35, 25, 52], the interest on the class of multimodal logics is due to their ability to express *complex modalities*, obtained by composing modal operators even of different types. This is also one of the main features of multimodal logis, see [13].

D. Basin et al. (eds.), Labelled Deduction, 33–57.
© 2000 *Kluwer Academic Publishers. Printed in the Netherlands.*

Let's now get into some more details, in order to explain how this paper places with respect to the work that has already been done on multimodal logics. One of the central points to be faced when one wants to define a body of (multi)modal knowledge is to define the meaning of the modal operators that can appear in the formulae. The easiest way for doing this is to use a *set of axioms*. Let's consider, for instance, the well-known *three wise men puzzle*:

Example 1 "A king wants to find the wisest out of his three wise men. In order to do this he arranges them in a circle and tells them that he will put a white or a black spot on their foreheads and that one of the three spots would certainly be white. The three wise men can see and hear each other but, of course, they cannot see their faces reflected anywhere. The king, then, asks to each of them to find out the color of his own spot: the first one who will be able to answer will be the wisest." *The formulation by multimodal logic is the following, where, as a shorthand, we use the metavariables X and Y, where $X, Y \in \{a, b, c\}$ and $X \neq Y$:*

(1) $[any](ws(a) \vee ws(b) \vee ws(c))$
(2) $[any](ws(X) \supset [Y]ws(X))$
(3) $[any](\neg ws(X) \supset [Y]\neg ws(X))$

$ws(X)$ *means* X *has a white spot on his forehead while* $[X]F$ *means that the wise man* X *believes* F*. The formulae are preceded by the modal operator* $[any]$ *which captures the information common to all the wise men and it is axiomatized by the axioms:*

$T(any)\ [any]\varphi \supset \varphi$
$4(any)\ [any]\varphi \supset [any][any]\varphi$

and by the interaction *axiom:*

$I(any, X)\ [any]\varphi \supset [X]\varphi$

where $X \in \{a, b, c\}$*. The formula (1) says that at least one of the wise men has a white spot, whereas formulae (2) and (3) mean that whenever one of them has (not) a white spot, the others know this fact. Moreover, whenever a wise man does (not) know something, the others know that he does (not) know this. That is, the following* interaction *axioms:*

$5(X, Y)\ \neg[X]\varphi \supset [Y]\neg[X]\varphi$
$4M(X, Y)\ [X]\varphi \supset [Y][X]\varphi$

(where $X \neq Y$ *and* $X, Y \in \{a, b, c\}$*) are assumed.*

This simple solution has, however, one serious limit: despite the fact that an axiom system is a calculus, it is *not appropriate* to automation, which, instead, is fundamental for practical applications. Other calculi, such as *resolution, sequent calculi,* and *tableau calculi* are better suited for this purpose, especially in the case that they satisfy the "subformula principle", which states that everything one needs for (dis)proving a formula is contained in the formula itself [20]. The advantage of axioms is that they allow to define characterizing properties of multimodal operators in an easy and intuitive way.

In this paper, we develop an *analytic tableau calculus* for a wide class of *normal multimodal logics* called *Incestual Multimodal Logics* (*IMLs* in the rest of the paper). The calculus is *modular* with respect to the axiom system of each logic in this class. The modularity is achieved by using *directly* the characterizing axioms of the logic as "*rewrite rules*" which create new paths among worlds in the counter-model construction (see Section 3).

The class of *IMLs* (first introduced in [13], where a general *determination* theorem was established) is characterized by a set of *interaction* axioms of the form:

$$G^{a,b,c,d} : \langle a\rangle[b]\varphi \supset [c]\langle d\rangle\varphi$$

a generalization of the *Geach* axiom $\Diamond\Box\varphi \supset \Box\Diamond\varphi$, where a, b, c, and d are *complex* labels built up from atomic ones using a composition operator ";" and a union operator "$\cup$", while φ is an arbitrary propositional modal formula (see Section 2). In [13] they are called a, b, c, d*-incestual axioms.* For instance, the axioms used in the three wise men puzzle are all incestual axioms. This class includes most of the well-known modal and multimodal systems and, among the others, those characterized by *serial*, *symmetric*, *transitive*, *dense*, *confluence*, and *Euclidean* accessibility relations. Differently from other multimodal systems, such as those studied in [35], these can be *non-homogeneous* (i.e., every modal operator is not restricted to belong to the same system) and can contain *interaction axioms* (i.e., every modal operator is not restricted to be independent from the others).

Although the development of automated deduction methods for modal and multimodal logics has received a lot of attention (see [25] for a survey) relatively few works address the problem of dealing with *interaction axioms*. For instance, in [23] a resolution for some modal and multimodal logics is presented, but, in the case of multimodal logics, the interaction axioms considered are simple and limited to axioms I and $4M$. In the case of tableaux [29, 15, 40, 33, 34, 18, 11, 12], most of the proposed calculi deal with modal but *not* with multimodal systems, and those developed for multimodal logics, see for instance [15, 34], only deal with specific cases (see Section 5 for a detailed discussion). Perhaps the most general proposals to deal with multimodal logics are the ones based on *translation methods* [47]. In particular, in [31] an optimized translation method for *inclusion multimodal logics*, a proper subclass of *IMLs*, is proposed. Again we refer to Section 5 for a comparison of our tableau method with translation methods.

The calculus proposed here has its origin in Nerode's tableau [42], which is closely related to the systems of *prefixed tableaux à la* Fitting [29]. In contrast to [29], worlds are not represented by prefixes (which describe paths in the model), but they are given an atomic name and the accessibility relationships among them are explicitly represented in a graph.

Prefixed tableau methods can be regarded as *Labelled Deductive Systems* (*LDSs*) [19, 30] that make implicitly use of labelled formulae. LDSs extend logics providing a "*labelling algebra*" that, in the case of modal logics, is defined as a binary first-order theory which, axiomatizing the semantics notion of accessibility relations, allows one to represent and reason syntactically about the underlying Kripke structure of possible worlds [49, 8, 10]. As pointed out by these authors, one of the most attracting issues of labelled systems is the possibility of dealing with classes of modal logics in a *uniform* and *modular* way. This is of practical advantages in the study of computational properties of classes of modal logics [9].

The work presented in this article was developed within this framework. It exploits a calculus that we developed so to be able to deal with a wide class of logics in order to prove some general properties that hold for the whole class. In particular, here we report the results that we have obtained about the *(un)decidability* of the class at issue. Prefixes play a fundamental role in this. In fact, for certain classes of multimodal logics, prefixes and rules for dealing with them have a simple interpretation in the *rewriting systems* and this allows one to import in an easy way some well-known results about formal languages.

The paper is organized as follows. Section 2 briefly summarizes both the syntax and the semantics of the class of *IMLs*. In Section 3 our prefixed tableau calculus is presented and an example of use is given. In Section 4 the problem of (un)decidability for the class of *IMLs* is addressed. Finally, Section 5 reports a comparison with related works.

2 A CLASS OF NORMAL MULTIMODAL LOGICS

We define a propositional multimodal language $\mathcal{L}$, containing a nonempty set VAR of *propositional variables*, the logical connectives "$\wedge$", "$\vee$", "$\supset$", and "$\neg$", and the *multimodal operator constructor* "[.]". Moreover, $\mathcal{L}$ contains a nonempty countable set MOD (the *alphabet of modalities*), disjoint from VAR, and two operators "$\cup$" (*union*) and ";" (*composition*), which allow to build up new labels for modalities, starting form the atomic labels in MOD. We also introduce a distinguished element "ε" to be the *neutral element* with respect to composition. The set LABELS of all modalities is defined as the least set that satisfies the following conditions:

- $\varepsilon \in$ LABELS;
- MOD $\subseteq$ LABELS;
- if $t, t' \in$ LABELS then $(t; t')$ and $(t \cup t')$ are in LABELS.[1]

A multimodal operator is built from the labels and the multimodal operator constructor: if $t \in$ LABELS, $[t]$ is a *universal* multimodal operator. We define $\langle t \rangle$ (*existential* multimodal operator) as $\neg[t]\neg$. The set of formulae of the language is constructed as usual by means of the propositional variables, the connectives, and the (universal and existential) multimodal operators. Examples of modalized formulae are $[(t; t') \cup t'' \cup \varepsilon]\varphi$ or $\langle t; t' \rangle\varphi$, where φ is an arbitrary formula.

We only consider *normal* multimodal logics, that is those whose axiomatization at least contains the axiom schemas for the classical propositional calculus, *modus ponens* and *necessitation* rules, and the axiom schema $K(t)$: $[t](\varphi \supset \psi) \supset ([t]\varphi \supset [t]\psi)$ for all multimodal operators. In particular, we focus on normal multimodal logics that are obtained by taking axiom systems that contain:

$$[\varepsilon]\varphi \quad \Leftrightarrow \quad \varphi \tag{2.1}$$
$$[t; t']\varphi \quad \Leftrightarrow \quad [t][t']\varphi \tag{2.2}$$
$$[t \cup t']\varphi \quad \Leftrightarrow \quad [t]\varphi \wedge [t']\varphi \tag{2.3}$$

where $t, t' \in$ LABELS, and a finite set of axiom schemas of the form:

$$G^{a,b,c,d} : \langle a \rangle [b]\varphi \supset [c]\langle d \rangle\varphi \tag{2.4}$$

where a, b, c, and d belong to LABELS. In [13] these logics are called *Incestual Multimodal Logics* (*IMLs*) and we will refer to (2.4) as an a, b, c, d-*incestual axiom schema*. Note that, by axioms (2.1), (2.2), and (2.3) the modalized formulae $[(t; t') \cup t'' \cup \varepsilon]\varphi$ and $\langle t; t' \rangle\varphi$ are equivalent to the formulae $[t][t']\varphi \wedge [t'']\varphi \wedge \varphi$ and $\langle t \rangle\langle t' \rangle\varphi$, respectively.

Let $\mathcal{A}$ be a set of a, b, c, d-incestual axioms. We denote by $\mathcal{I}^{\mathcal{A}}_{\mathcal{L}}$ the *IML* determined by the set $\mathcal{A}$ with $\mathcal{L}$ as underlying language, while we use $\mathcal{S}^{\mathcal{A}}_{\mathcal{L}}$ to denote its characterizing axiom system containing the axioms for normal modalities plus the set of axiom $\mathcal{A}$ together with the axioms (2.1), (2.2), and (2.3).

As it is remarked in [13], the fact that the parameters a, b, c, and d of an incestual axiom schema may be arbitrary expressions, built from atomic labels using the composition and union operators, makes axiom $G^{a,b,c,d}$ very general. In particular, the class of *IMLs* includes all the fifteen modal systems obtained

[1] For readability, we omit parentheses if they are unnecessary: we give "$\cup$" lower precedence than ";".

combining the axioms T, D, B, 4, and 5 [38] and their multimodal versions [35], as well as many extensions of K_n and $S4_n$ with *interaction axioms* such as those mentioned in [32, 23].

In order to define the meaning of a formula, we introduce the notion of *Kripke interpretation*. Formally, a Kripke interpretation M is a triple $\langle W, \mathcal{R}, V \rangle$, consisting of a non-empty set W of *possible worlds*, a mapping $\mathcal{R}$ from MOD to the powerset of $W \times W$ (it assigns to each atomic label of MOD a binary relation on W), and a *evaluation function* V, that is a mapping from $W \times$ VAR to the set $\{\mathbf{T}, \mathbf{F}\}$. In order to deal with any label $t \in$ LABELS, we extend the mapping $\mathcal{R}$ inductively as follows:

- $\mathcal{R}\varepsilon = I$, where $I = \{(w, w) \mid w \in W\}$ (the *identity* relation);
- $\mathcal{R}_{t;t'} = \mathcal{R}_t \circ \mathcal{R}_{t'}$, where "$\circ$" denotes the composition of binary relations;
- $\mathcal{R}_{t \cup t'} = \mathcal{R}_t \cup \mathcal{R}_{t'}$, where "$\cup$" denotes the union of binary relations.

We say that $\mathcal{R}_t$ is the *accessibility relation* of the modality $[t]$ and w' is *accessible* from w *by means of* $\mathcal{R}_t$ if $(w, w') \in \mathcal{R}_t$.

The meaning of a formula is given by means of a *satisfiability relation*, denoted by $\models$. Let $M = \langle W, \mathcal{R}, V \rangle$ be a Kripke interpretation, w a world in W and φ a formula, then, we say that φ is *satisfiable in the Kripke interpretation* M *at* w, denoted by $M, w \models \varphi$, if the following conditions hold:

- $M, w \models \varphi$ and $\varphi \in$ VAR iff $V(w, \varphi) = \mathbf{T}$;
- $M, w \models \neg\varphi$ iff $M, w \not\models \varphi$;
- $M, w \models \varphi \wedge \psi$ iff $M, w \models \varphi$ and $M, w \models \psi$;
- $M, w \models \varphi \vee \psi$ iff $M, w \models \varphi$ or $M, w \models \psi$;
- $M, w \models \varphi \supset \psi$ iff $M, w \not\models \varphi$ or $M, w \models \psi$;
- $M, w \models [t]\varphi$ iff for all $w' \in W$ such that $(w, w') \in \mathcal{R}_t$, $M, w' \models \varphi$;
- $M, w \models \langle t \rangle\varphi$ iff there exists a $w' \in W$ such that $(w, w') \in \mathcal{R}_t$ and $M, w' \models \varphi$.

Let $\mathcal{M}_{\mathcal{L}}$ be the set of all Kripke interpretations, as defined above. For each *IML* logic $\mathcal{I}_{\mathcal{L}}^{\mathcal{A}}$ we introduce a suitable notion of Kripke $\mathcal{A}$-interpretation, by adding some restrictions on the accessibility relations. More precisely, let $M = \langle W, \mathcal{R}, V \rangle$ be a Kripke interpretation and let $\mathcal{A}$ be a set of a, b, c, d-incestual axioms, we say M is a *Kripke* $\mathcal{A}$*-interpretation* if and only if for each axiom schema $\langle a \rangle [b]\varphi \supset [c]\langle d \rangle\varphi \in \mathcal{A}$, the following inclusion property on the accessibility relations, called a, b, c, d*-incestual property*, holds:

$$\mathcal{R}_b \circ \mathcal{R}_d^{-1} \supseteq \mathcal{R}_a^{-1} \circ \mathcal{R}_c \tag{2.5}$$

where $\mathcal{R}_t^{-1}$ is the *inverse relation* of $\mathcal{R}_t$. In other words, an a, b, c, d-incestual axiom corresponds to the *first-order* condition on accessibility relations [14]:[2]

$$(\forall w, w', w'')(\exists w^*) \quad ((w, w') \in \mathcal{R}_a \wedge (w, w'') \in \mathcal{R}_c \supset (w', w^*) \in \mathcal{R}_b \wedge (w'', w^*) \in \mathcal{R}_d) \tag{2.6}$$

The set of all Kripke $\mathcal{A}$-interpretations is denoted by $\mathcal{M}_{\mathcal{L}}^{\mathcal{A}}$ and it is a subset of $\mathcal{M}_{\mathcal{L}}$. Given a Kripke $\mathcal{A}$-interpretation $M = \langle W, \mathcal{R}, V \rangle$ of $\mathcal{M}_{\mathcal{L}}^{\mathcal{A}}$, we say that a formula φ of $\mathcal{I}_{\mathcal{L}}^{\mathcal{A}}$ is *satisfiable in* M if $M, w \models \varphi$ for some world $w \in W$. We say that φ is *valid in* M if $\neg\varphi$ is not satisfiable in M. Moreover, a formula φ is $\mathcal{A}$-*satisfiable* if φ is satisfiable in some Kripke $\mathcal{A}$-interpretation in $\mathcal{M}_{\mathcal{L}}^{\mathcal{A}}$ and $\mathcal{A}$-*valid* if it is valid in all Kripke $\mathcal{A}$-interpretations in $\mathcal{M}_{\mathcal{L}}^{\mathcal{A}}$ (written $\models_{\mathcal{A}} \varphi$). $\mathcal{S}_{\mathcal{L}}^{\mathcal{A}}$ is a *sound* and *complete* axiomatization with respect to $\mathcal{M}_{\mathcal{L}}^{\mathcal{A}}$ [13].

Remark 2 *Despite the fact that* IMLs *include a wide class of multimodal systems, it is worth noting that no set of inclusion properties of the form (2.5) can characterize the* induction axiom *that define both the* iteration *operator "$*$" of dynamic logic [36] and the* common knowledge *operator "$\mathcal{C}$" [32, 35]. In fact, let us consider the axioms*

$$[b]\varphi \supset \varphi \wedge [a][b]\varphi \tag{2.7}$$

$$\varphi \wedge [b](\varphi \supset [a]\varphi) \supset [b]\varphi \tag{2.8}$$

it is easy to see that the modal operator $[b]$ can represent both $[a^*]$ of dynamic logic and the common knowledge operator $\mathcal{C}$ (when a is the only agent). Axiom (2.7) is an incestual axiom (it is equivalent to $\langle \varepsilon \rangle [b]\varphi \supset [\varepsilon \cup a; b]\langle \varepsilon \rangle \varphi$) but axiom (2.8, the induction axiom, *is not (see also [13, 3]).*

Remark 3 *It is easy to see that each* IML *is* first-order definable *because it can be characterized by a class of Kripke structures that satisfy some properties expressed by means of* first-order formulae *of the form (2.6). The widest known class of first-order definable modal logics is the one characterized by sets of* Sahlqvist axioms *[51, 16]. In [14], the author generalizes the* Sahlqvist axiom *and related results to the multimodal case. The Generalized Sahlqvist axiom has the form $[a](\phi \supset \varphi)$, where a is an arbitrary parameter, φ is an affirmative formula, and ϕ is a formula that satisfies certain conditions. In [14] it is shown that it also corresponds to a first-order condition on accessibility relations that is effectively obtainable from the axiom itself. Moreover, the author supplies a* determination *theorem for multimodal logic systems defined by sets of this axioms. It is worth noting that the Generalized Sahlqvist axiom includes a, b, c, d-icestual axiom.*

---

[2]This property is named *incestual* because the "offsprings" (w' and w'') of a common "parent" (w) have themselves an "offspring" (w^*) in common [17].

3 A PREFIXED TABLEAU CALCULUS

Before introducing our tableau calculus, we need to introduce some preliminary notions. We define a *signed formula* Z as a formula prefixed by one of the two symbols $\mathbf{T}$ and $\mathbf{F}$ (*signs*). For instance, if φ is a formula then, $\mathbf{T}\varphi$ and $\mathbf{F}\varphi$ are signed formulae.

Definition 4 *Let $\mathcal{L}$ be a propositional multimodal language and let $\mathcal{W}_C$ be a countable non-empty set of* constant world symbols *(or* prefixes*). A* prefixed signed formula, *$w : Z$, is a prefix $w \in \mathcal{W}_C$ followed by a signed formula Z.*

Definition 5 *Let $\mathcal{L}$ be a propositional multimodal language and let $\mathcal{W}_C$ be a non-empty set of prefixes. An* accessibility relation formula, *$w\ \rho_t\ w'$, is a binary relation on $\mathcal{W}_C$, where $t \in$* LABELS.

Note that, we assume $\mathcal{W}_C$ contains always at least the prefix i, that is interpreted as the *initial world.* We say that an accessibility relation formula $w\ \rho_t\ w'$ is *true* in a tableau branch if it belongs to that branch.

A *tableau* is a *labelled tree* where each node consists of a *prefixed signed formulae* or of an *accessibility relation formula* and, intuitively, each tableau branch corresponds to the construction of a Kripke interpretation that satisfies the formulae belonging to it. The relation ρ_ε on a branch defines an equivalence relation among prefixes: when $w\ \rho_\varepsilon\ w'$ holds, w and w' can be regarded as representing the same worlds. By taking the reflexive, transitive and symmetric closure of the relation ρ_ε we define an equivalence relation among worlds. We denote by $\overline{w}$ the equivalence class of w with respect to this equivalence relation. For instance, in Example 7, Fig. 2.3, $\overline{w_2} = \overline{w_3}$ in both tableau branches because $w_2\ \rho_\varepsilon\ w_3$ is true in them (step 11.). A formula $w : \mathbf{T}\varphi$ ($w : \mathbf{F}\varphi$) on a branch of a tableau means that the formula φ is *true* (*false*) at the world $\overline{w}$ in the Kripke interpretation associated with that branch.

We say that a prefix w is *used* on a tableau branch if it occurs on the branch in some accessibility relation formula, otherwise we say that the prefix w is *new*. Moreover, given a label t, we say that an accessibility relation formula $w\ \rho_t\ w'$ is *available on* a branch S of a tableau if one of the following conditions hold:

1. $t = \varepsilon$ and $w = w'$;
2. $w_1 \in \overline{w}$, $w_2 \in \overline{w'}$ and $w_1\ \rho_t\ w_2$ is true in S;
3. $t = t'; t''$ and both $w\ \rho_{t'}\ w''$ and $w''\ \rho_{t''}\ w'$ are available on S, for some w'' used on the branch S;
4. $t = t' \cup t''$ and either $w\ \rho_{t'}\ w'$ is available on S or $w\ \rho_{t''}\ w'$ is available on S.

Note that if an accessibility relation formula is true in a tableau branch, it is also available on it (as a special case of the condition 2 above). Moreover, for any world w on a given branch, $w\ \rho_\varepsilon\ w$ is always available (condition 1). For instance, in Example 7, Fig. 2.3, $i\ \rho_c\ w_3$ is available on both tableau branches since $w_3 \in \overline{w_2}$ and $i\ \rho_c\ w_2$ is true in both tableau branches. Intuitively, $w\ \rho_t\ w'$ available on a branch of a tableau means that in the Kripke interpretation associated with that branch $(\overline{w}, \overline{w'}) \in \mathcal{R}_t$.

In order to simplify the presentation of the calculus we use the well-known *uniform notation* for signed formulae [29] (see Fig. 2.1). In the following, we will often use α, β, ν^t, and π^t as formulae of the corresponding type.

α	α_1	α_2
$\mathbf{T}(\varphi \wedge \psi)$	$\mathbf{T}\varphi$	$\mathbf{T}\psi$
$\mathbf{F}(\varphi \vee \psi)$	$\mathbf{F}\varphi$	$\mathbf{F}\psi$
$\mathbf{F}(\varphi \supset \psi)$	$\mathbf{T}\varphi$	$\mathbf{F}\psi$
$\mathbf{F}(\neg\varphi)$	$\mathbf{T}\varphi$	$\mathbf{T}\varphi$

β	β_1	β_2
$\mathbf{F}(\varphi \wedge \psi)$	$\mathbf{F}\varphi$	$\mathbf{F}\psi$
$\mathbf{T}(\varphi \vee \psi)$	$\mathbf{T}\varphi$	$\mathbf{T}\psi$
$\mathbf{T}(\varphi \supset \psi)$	$\mathbf{F}\varphi$	$\mathbf{T}\psi$
$\mathbf{T}(\neg\varphi)$	$\mathbf{F}\varphi$	$\mathbf{F}\varphi$

ν^t	ν_0^t
$\mathbf{T}([t]\varphi)$	$\mathbf{T}\varphi$
$\mathbf{F}(\langle t\rangle\varphi)$	$\mathbf{F}\varphi$

π^t	π_0^t
$\mathbf{F}([t]\varphi)$	$\mathbf{F}\varphi$
$\mathbf{T}(\langle t\rangle\varphi)$	$\mathbf{T}\varphi$

Figure 2.1 Uniform notation for propositional signed multimodal formulae.

A *tableau* is an attempt to build an interpretation in which a given formula is satisfiable. Starting from a formula φ, the interpretation is progressively constructed applying a set of *extension rules*, which reflect the semantics of the considered logic. At any stage, a branch of a tableau is a partial description of an interpretation. In our case, the tableau method tries to build Kripke interpretations, one for each branch: the worlds are formed by the prefixes that appear on the branch, the accessibility relations for the modalities are given by means of the accessibility relation formulae and the valuation function is given by means of the prefixed signed atomic formulae.

Definition 6 *Let $\mathcal{L}$ be a propositional multimodal language and let $\mathcal{A}$ be a set of incestual axioms. The extension rules for $\mathcal{I}_{\mathcal{L}}^{\mathcal{A}}$ are given in Fig. 2.2.*

The interpretation of the different kinds of extension rules is rather simple if we refer to the possible-worlds semantics. The rules for the formulae of type α and β are the usual ones of the classical calculus.

A formula of type ν^t is true at world w if ν_0^t is true in all worlds w' accessible from w by means of $\mathcal{R}_t$. Therefore, if $w : \nu^t$ occurs on an open branch, we can add $w' : \nu_0^t$ to the end of that branch for any w' such that $w\ \rho_t\ w'$ is available on that branch.

$$\frac{w : \alpha}{\begin{array}{c} w : \alpha_1 \\ w : \alpha_2 \end{array}}\ \alpha\text{-rule} \qquad \frac{w : \beta}{w : \beta_1 \mid w : \beta_2}\ \beta\text{-rule}$$

$$\frac{w : \nu^t \quad w\ \rho_t\ w'}{w' : \nu_0^t}\ \nu\text{-rule} \qquad \frac{w : \pi^t}{\begin{array}{c} w' : \pi_0^t \\ w\ \rho_t\ w' \end{array}}\ \pi\text{-rule}$$

($w\ \rho_t\ w'$ is *available* on the branch) (w' is *new* on the branch)

$$\frac{w\ \rho_{t;t'}\ w'}{\begin{array}{c} w\ \rho_t\ w'' \\ w''\ \rho_{t'}\ w' \end{array}}\ \rho_\alpha\text{-rule} \qquad \frac{w\ \rho_{t\cup t'}\ w'}{w\ \rho_t\ w' \mid w\ \rho_{t'}\ w'}\ \rho_\beta\text{-rule}$$

(w'' is *new* on the branch)

$$\frac{w\ \rho_a\ w' \quad w\ \rho_c\ w''}{\begin{array}{c} w'\ \rho_b\ w^* \\ w''\ \rho_d\ w^* \end{array}}\ \rho\text{-rule}$$

(where $\langle a\rangle[b]\varphi \supset [c]\langle d\rangle\varphi \in \mathcal{A}$, and
$w\ \rho_a\ w'$, $w\ \rho_c\ w''$ are *available* and
w^* is *new* on the branch)

Figure 2.2 Tableau rules for propositional incestual multimodal logics.

A formula of type π^t is true at the world w if there exists a world w' accessible from w by means of $\mathcal{R}_t$ at which π_0^t is true. Therefore, if $w : \pi^t$ occurs on an open branch, we can add $w' : \pi_0^t$ to the end of that branch, provided w' is new and $w\ \rho_t\ w'$ is true in it. Note that the label t of a formula ν^t or π^t can be an arbitrarily complex expression.

If an accessibility relation formula $w\ \rho_{t;t'}\ w'$ is true in a tableau branch, then $(\overline{w}, \overline{w'}) \in \mathcal{R}_{t;t'}$ holds in the Kripke interpretation represented by that branch. Therefore, $(\overline{w}, \overline{w'}) \in \mathcal{R}_t \circ \mathcal{R}_{t'}$ and, hence, there exists a world w'' such that $(\overline{w}, \overline{w''}) \in \mathcal{R}_t$ and $(\overline{w''}, \overline{w'}) \in \mathcal{R}_{t'}$. That is, $w\ \rho_t\ w''$ and $w''\ \rho_{t'}\ w'$ are true in that branch for some new world w''.

If an accessibility relation formula $w\ \rho_{t\cup t'}\ w'$ is true in a tableau branch, then $(\overline{w}, \overline{w'}) \in \mathcal{R}_{t\cup t'}$ holds in the Kripke interpretation represented by that branch. Therefore, $(\overline{w}, \overline{w'}) \in \mathcal{R}_t \cup \mathcal{R}_{t'}$ and, hence, $(\overline{w}, \overline{w'}) \in \mathcal{R}_t$ or $(\overline{w}, \overline{w'}) \in \mathcal{R}_{t'}$. That is, $w\ \rho_t\ w'$ or $w\ \rho_{t'}\ w'$ is true in that branch.

Finally, the case of ρ-rule. It is a general rule which allows us to deal with any incestual axiom. Let us suppose, for instance, that $\langle a\rangle[b]\varphi \supset [c]\langle d\rangle\varphi \in \mathcal{A}$ in our modal logic $\mathcal{I}_{\mathcal{L}}^{\mathcal{A}}$. If $w\ \rho_a\ w'$ and $w\ \rho_c\ w''$ are available on a tableau branch then, $(w, w') \in \mathcal{R}_a$ and $(w, w'') \in \mathcal{R}_c$ in the Kripke interpretation associated to that branch. Since the incestual axiom $\langle a\rangle[b]\varphi \supset [c]\langle d\rangle\varphi \in \mathcal{A}$ then, the corresponding a, b, c, d-incestual property (2.6) must hold, that is, there exists a world w^* such that $(w', w^*) \in \mathcal{R}_b$ and $(w'', w^*) \in \mathcal{R}_d$. Hence, $w' \; \rho_b \; w^*$ and $w'' \; \rho_d \; w^*$ are true in that Kripke interpretation for some new prefix w^*. The ρ-rule can be regarded as a *rewriting* rule which creates new paths among worlds according to the inclusion properties of the incestual modal logic.

We say that a tableau branch is *closed* if it contains $w : \mathbf{T}\varphi$ and $w' : \mathbf{F}\varphi$ for some formula φ such that $\overline{w} = \overline{w'}$. A tableau is *closed* if *every* branch in it is closed. Finally, let $\mathcal{L}$ be a propositional multimodal language, $\mathcal{A}$ a set of incestual axioms, and φ a formula. Then a closed tableau for $i : \mathbf{F}\varphi$ obtained by using the tableau rules of Fig. 2.2, is said to be a *proof* of φ.

1. $i : \mathbf{F}\langle a\rangle([b']p \wedge [b'']p) \supset [c]p$
2. $i : \mathbf{T}\langle a\rangle([b']p \wedge [b'']p)$
3. $i : \mathbf{F}[c]p$
4. $w_1 : \mathbf{T}([b']p \wedge [b'']p)$
5. $i \; \rho_a \; w_1$
6. $w_1 : \mathbf{T}[b']p$
7. $w_1 : \mathbf{T}[b'']p$
8. $w_2 : \mathbf{F}p$
9. $i \; \rho_c \; w_2$
10. $w_1 \; \rho_{b' \cup b''} \; w_3$
11. $w_2 \; \rho_\varepsilon \; w_3$

12a. $w_1 \; \rho_{b'} \; w_3$	12b. $w_1 \; \rho_{b''} \; w_3$
13a. $w_3 : \mathbf{T}p$	13b. $w_3 : \mathbf{T}p$
$\times$	$\times$

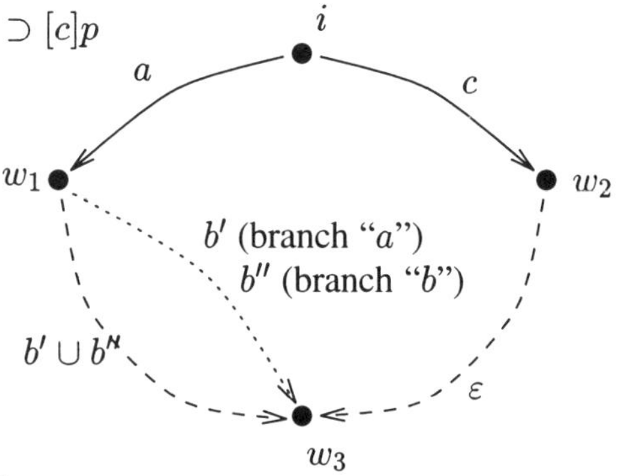

Figure 2.3 Tableau proof and counter-model construction for the Example 7.

Example 7 *Let us consider the* IML $\mathcal{I}^{\mathcal{A}}_{\mathcal{L}}$ *where* $\mathcal{A}$ *consists of the axiom schema* $\langle a\rangle[b' \cup b'']\varphi \supset [c]\langle\varepsilon\rangle\varphi$. *Then, the formula* $\langle a\rangle([b']p \wedge [b'']p) \supset [c]p$ *has a tableau proof which is reported in Fig. 2.3. We denote with "a" and "b" the two branches which are created by the application of* ρ_β*-rule to step 10. Explanation: 1.: the goal; 2. and 3.: from 1., by* α*-rule; 4. and 5.: from 2., by* π*-rule; 6. and 7.: from 4., by* α*-rule; 8. and 9.: from 3., by* π*-rule; 10. and 11.: from 5. and 9., by* ρ*-rule from axiom* $\langle a\rangle[b' \cup b'']\varphi \supset [c]\langle\varepsilon\rangle\varphi$ *since* $i \; \rho_a \; w_1$ *and* $i \; \rho_c \; w_2$ *are available; 12a. and 12b.: from 10., by* ρ_β*-rule; 13a.: from 6. and 12a., by* ν*-rule, 13b.: from 7. and 12b., by* ν*-rule. Since* $\overline{w_2} = \overline{w_3}$ *(*$w_2 \; \rho_\varepsilon w_3$ *belongs to the branch at step 11.) the branches "a" and "b" are close due to step 8. and steps 13a. and 13b., respectively.*

It is worth noting that, because of the use of prefixes, ρ-rule is able to factored out all the characteristics imposed by the incestual axioms of the considered

IML. In fact, α, β, π, and ν rules are just the rules for multimodal version of K.

Theorem 8 (Soundness and Completeness) *Let $\mathcal{I}^{\mathcal{A}}_{\mathcal{L}}$ be an* IML. *A formula φ of $\mathcal{L}$ has a tableau proof if and only if it is $\mathcal{A}$-valid.*

Proof The proof follows the well-known guideline of [29, 40, 33] and it can be found in [3]. ■

4 (UN)DECIDABILITY RESULTS

In the previous section we have defined a tableau method for the class of *IMLs*. The completeness result was obtained by means of a systematic tableau

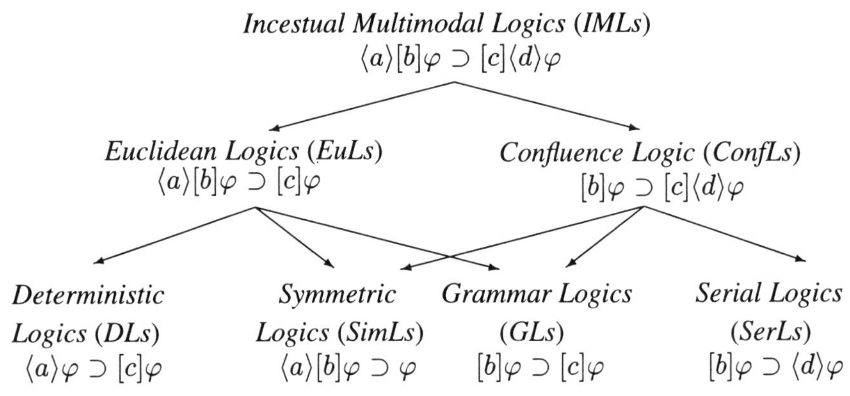

Figure 2.4 Subclasses of incestual multimodal logics: arrow means inclusion.

procedure that always finds a counter-model for a given formula if there exists one. As a result, the completeness establishes the *semi-decidability* of *IMLs*. On the other hand, we wonder if this class of logics is also *decidable*, that is if it is possible to define a *decision procedure* which works for the whole class of *IMLs*. This procedure should halt both if a counter-model exists and if a counter-model does not exist. Unfortunately, such an algorithm *does not* exist. In fact, the class of *IMLs* subsumes the class of *Thue Logics*, which have been shown *undecidable* in [26].

Nevertheless, if more restricted classes of *IMLs* are considered, some decidability results can be established. In particular, in this section, we focus our attention on the subclasses of the *IMLs* shown in Figure 2.4, which are characterized by finite sets of a, b, c, d-incestual axiom having a simpler form. These are obtained by setting, in any possible way, two parameters to the empty label in the axiom. We show that both the classes of *Deterministic Logics* (*DLs*) and of *Serial Logics* (*SerLs*) are *decidable*, whereas, the other two classes, *Symmetric Logics* (*SymLs*) and *Grammar Logics* (*GLs*) are *undecidable*. In order to prove these results, in the following, we introduce simplified ρ-rules, presented in Figure 2.5, for the different classes.

$$\frac{w\ \rho_a\ w' \quad w\ \rho_c\ w''}{w'\ \rho_\varepsilon\ w''}\ \rho\text{-rule}_{DL}$$
(where $\langle a\rangle\varphi \supset [c]\varphi \in \mathcal{A}$, and $w\ \rho_a\ w'$, $w\ \rho_c\ w''$ are *available* on the branch)

$$\frac{w\ \rho_a\ w'}{w'\ \rho_b\ w}\ \rho\text{-rule}_{SymL}$$
(where $\langle a\rangle[b]\varphi \supset \varphi \in \mathcal{A}$, and $w\ \rho_a\ w'$ is *available* on the branch)

$$\frac{w\ \rho_c\ w'}{w\ \rho_b\ w'}\ \rho\text{-rule}_{GL}$$
(where $[b]\varphi \supset [c]\varphi \in \mathcal{A}$, and $w\ \rho_c\ w'$ is *available* on the branch)

$$\frac{}{w\ \rho_b\ w^* \quad w\ \rho_d\ w^*}\ \rho\text{-rule}_{SerL}$$
(where $[b]\varphi \supset \langle d\rangle\varphi \in \mathcal{A}$, and w^* is *new* on the branch)

Figure 2.5 Simplified ρ-rules for *DLs*, *SymLs*, *GLs*, and *SerLs* (up to down, left to right).

The class of *serial logics* (*SerLs*) is characterized by finite sets of simplified incestual axioms of the form $[b]\varphi \supset \langle d\rangle\varphi$. Intuitively, *SerLs* can be characterized by Kripke interpretations in which for each world there exists a world reachable both by a path labelled by parameter b and one labelled by d. In other words, the axiom $[b]\varphi \supset \langle d\rangle\varphi$ implies that both the accessibility relations of b and d are serial.

Theorem 9 *The validity problem for the class of* SerLs *is decidable.*

Proof Let us assume that for any occurrence of a formula of the form $w : \nu^t$ we remember that, whenever a new prefix w', such that $w\ \rho_t\ w'$ is available, we also add $w' : \nu^t_0$. In this way, a branch cannot contain repetitions of formulae. Now, let us suppose that a systematic attempt at proving a formula φ goes forever. Then, there must be an infinite branch S containing infinitely many different prefixes because it does not contain repetitions and only a finite number of formulae can appear. There are two possibilities: either infinitely many prefixes are reachable from a same world and, hence, they have the same distance from the initial world i or there are only finitely many prefixes for any distance n from the initial world i. It is not the first case because we can assume to apply only once the π-rule for the same π^t formula and only once the ρ-rule for each world and axiom. Now, it is easy to observe that the maximum number of occurrences of modal operators for the formulae associated with a world decreases when the distance of the world increases. In fact, if a prefixed formula of the form $w : Z$ occurs on S it must have been added using the

ν or π-rule on some prefixed formula $w' : Z'$ where Z' has higher number of occurrences of modal operators. Hence, there is a number n such that all prefixes that have distance equal or more than n are associated with formulae that do not contain modal operators. From such worlds the only way to go on is to apply the ρ-rule. However, this cannot close a tableau and so we can stop our procedure. ■

The class of *deterministic logics* (*DLs*) is characterized by finite sets of simplified incestual axioms of the form $\langle a \rangle \varphi \supset [c]\varphi$. Intuitively, *DLs* can be characterized by Kripke interpretations in which the set of worlds reachable by a path labelled by parameter a is identical to the set of worlds reachable by a path labelled by parameter c.

Theorem 10 *The validity problem for the class of* DLs *is decidable.*

Proof For the sake of simplicity we give only a brief skecth of the proof that can be found in the extended version of this paper [4]. The proof shows that it is possible to define a systematic proof procedure that always terminates. If one observes the ρ-rule for the class at hand, Fig. 5, one can see that it never introduces new labels; then, it does not introduce new worlds. What it does is to "merge" classes of labels (worlds). The only rule that creates new worlds is the π-rule. Now, it is possible to show that the alternate application of the π-rule (let's informally call it "expansion because" it leads to the creation of new worlds), of the ρ-rule (let's informally call it "fusion" because it merges existing worlds) and of the others (which create new formulae within existing worlds) terminates. This proof is based on the observation that, since the initial formula is finite, the number of formulae associated to any given world is finite; furthermore, the process of expansion and creation will arrive to a point at which it will not be possible to apply any rule because all rules have been applied to all formulae already, because the fusion operation will create new classes that contain only formulae that have already been seen and because the fusion operation cannot be repeated forever, because at most everything will be merged in a same world. ■

The class of *grammar logics* (*GLs*) is characterized by finite sets of simplified incestual axioms of the form $[b]\varphi \supset [c]\varphi$. Intuitively, *GLs* can be characterized by Kripke interpretations in which the set of paths labelled by parameter c is included by the set of paths labelled by parameter b. *GLs* are particularly interesting from a computational point of view [26, 31] and they have been at the basis of different extensions of logic programming languages [3]. The class of *GLs* has been introduced by Fariñas del Cerro and Penttonen in [26]. The authors present a method to define multimodal logics from *formal grammars*, in order to simulate the behaviour of grammars [37]. Given a formal grammar $G = (V, T, P, S)$, we can define a *GL* (*based on* G) where a modal operator is associated with each nonterminal symbol in V and each terminal symbol in T (MOD $= V \cup T$). Such modal operators are characterized by a set of axioms $[t_1; \ldots ; t_n]\varphi \supset [s_1; \ldots ; s_m]\varphi$, one for each production rule $t_1 \cdots t_n \rightarrow s_1 \cdots s_m$ in P, where the t_i's and s_j's are either in V or in T.

Example 11 *Let* $G = (\{A\}, \{b\}, \{A \rightarrow \varepsilon, A \rightarrow A\ A, A \rightarrow b\ A\}, A)$ *be a formal grammar. Then, the inclusion modal logic* $\mathcal{I}_{\mathcal{L}}^{\mathcal{A}}$ *based on G contains the inclusion axioms* $[A]\varphi \supset \varphi$, $[A]\varphi \supset [A; A]\varphi$, *and* $[A]\varphi \supset [b; A]\varphi$.

The tableau method developed in the previous section allows us to give a precise correspondence between the membership problem for a given grammar and the validity problem in the corresponding *GL*. Since the problem of establishing if a word belongs to the language generated by an arbitrary (unrestricted) grammar is undecidable [37], we can conclude, from Theorem 12, that the class of *GLs* is *undecidable*.

Theorem 12 *Given a grammar* $G = (V, T, P, S)$, *let* $\mathcal{I}_{\mathcal{L}}^{\mathcal{A}}$ *be the* GL *based on* G. *Then, for any propositional variable* p *of* $\mathcal{L}$, $\models_{\mathcal{A}} [S]p \supset [s_1; \ldots ; s_m]p$ *if and only if* $S \Rightarrow_G^* s_1 \cdots s_m$, *where the* s_i*'s are in* $V \cup T$.

Proof See [3, 6] for a detailed proof. ■

The class of *symmetric logics* (*SymLs*) is characterized by finite sets of simplified incestual axioms of the form $\langle a\rangle[b]\varphi \supset \varphi$. Intuitively, *SymLs* can be characterized by Kripke interpretations in which the set of paths labelled by parameter b includes the set of inverse paths labelled by parameter a. The class of *SymLs* is strongly related with the class of *GLs*. In fact, as stated by the following proposition, it is possible to simulate any *GL* by means of a suitable *SymL*. As a consequence, the class of *SymLs* is also *undecidable*.

Proposition 13 *Let* $\mathcal{I}_{\mathcal{L}}$ *be a* GL *and let* ψ *be a formula in* $\mathcal{L}$. *Then, there exists a* SymL $\mathcal{I}_{\mathcal{L}'}^{\mathcal{A}'}$ *such that* ψ *is* $\mathcal{A}'$*-valid if* ψ *is* $\mathcal{A}$*-valid.*

Proof Let $\mathcal{A}$ be the collection of incestual axioms of the form $[b_i]\varphi \supset [c_i]\varphi$, for $i = 1, \ldots , n$, where n is the cardinality of $\mathcal{A}$. Let us define a *SymL* $\mathcal{I}_{\mathcal{L}'}^{\mathcal{A}'}$ as follows. $\mathcal{L}'$ is $\mathcal{L}$ extended by a set of n new atomic labels for modal operators t_i such that $t_i \neq t_j$ for $i \neq j$ and $\mathcal{A}'$ is the set of incestual axioms containing n pairs of axioms $\langle c_i\rangle[t_i]\varphi \supset \varphi$ and $\langle t_i\rangle[b_i]\varphi \supset \varphi$, one for each axiom $[b_i]\varphi \supset [c_i]\varphi$ in $\mathcal{A}$. Now, let us assume that ψ is $\mathcal{A}$-valid but not $\mathcal{A}'$-valid. Then, by definition, there must exist a Kripke $\mathcal{A}'$-interpretation M and a world w on it in which ψ is not satisfied. Now, since $\mathcal{R}_t \supseteq \mathcal{R}_c^{-1}$ and $\mathcal{R}_b \supseteq \mathcal{R}_t^{-1}$, it easy to see that the accessibility relation of M satisfies the constraint $\mathcal{R}_b \supseteq \mathcal{R}_c$.

Then, we can define a Kripke $\mathcal{A}$-interpretation M' based on M just by omitting the relations which involve the set of new atomic labels t_i's. Since the formula ψ does not use any new label ψ is again not satisfied in the world w of M'. But this is a contradiction with the fact that ψ is $\mathcal{A}$-valid. ∎

5 DISCUSSION AND OTHER RELATED WORK

In this paper we have presented a tableau calculus for the class of incestual multimodal logics. This class includes most of the well-known modal and multimodal systems discussed in the literature. In order to have a general framework able to cope with any kind of incestual multimodal logics, we have chosen the simplest way of representing models: prefixes are worlds, and relations between them are explicitly built up step by step by the rules of the calculus. Within this general framework we have proved some (un)decidability results about the class of logics at hand.

5.1 DECIDABILITY

The undecidability result for the class of *GLs* has already been shown in [26]. However, Fariñas del Cerro and Penttonen do not prove the correspondence between the membership problem for a given grammar and the validity problem in the corresponding *GL* (i.e. Theorem 12) for arbitrary grammars; they only prove this correspondence for a restricted class of the *GLs*, that are based on the *Thue systems* and that are called *Thue logics*. The Thue systems have also been used in [39] to define logics similar to those studied in [26], which, however, are not in the class on *GLs* since modalities enjoy some further properties like seriality and determinism. In [39] undecidability results are proved for this class of logics.

In [6, 3] a deeper study about the class of *GLs* is presented. In particular, a prefixed tableaux method is used to prove the undecidability of the class of *GLs* but also the undecidability of the subclasses of *GLs* based on *context sensitive* and *context free* grammars. Moreover, it is shown that the subclass of *GLs* based on *right-regular* grammars is decidable by means of the *filtration methods*, by defining an extension of the Fisher-Ladner closure [27]. A careful study of the filtration method for the multimodal logics case is presented in [14], where the author shows some limits of its use.

In this paper we do not mention about the number of multimodal operators to preserve and/or destroy decidability for the presented classes of multimodal logics. Some works on this line are [39] and [41]. In particular, in the second one, an example of undecidable *IML* with three multimodal operators, which with two is instead decidable, is made.

5.2 GRAPH VS PATH REPRESENTATION

Using prefixed formulae is very common in modal theorem proving (see [33] for an historical introduction on the topic). Our approach to prefixed tableaux is closely related to the approaches based on prefixes used in [29] and by other authors for modal (though non multimodal) systems [40, 33] and for dynamic logic [22]. In these works, prefixes are sequences of integers which represent each world as a *path* in the model, that connects the initial world to the one at hand. Thus, instead of representing *explicitly* worlds and accessibility relations of a Kripke interpretation in a *graph*, as we do in our approach by means of the accessibility relation formulae, a Kripke interpretation is represented as a set of paths which can be considered a *spanning tree* of the same graph. Although this representation might be more efficient, the disadvantage is that it requires a specific ν-rule for each logic. These rules code the properties of accessibility relations. Depending on the logic, the ν-rules may express complex relations between prefixes, which instead in our case are explicitly available from the representation. In particular, in [40] Massacci has proposed a "single step calculus" where ν-rules make use only of immediately accessible prefixes. His approach works for all the distinct basic normal logics obtainable from K by addition on any combination of the axioms T, D, 4, 5, and B in a modular way, but it still requires the definition of specific ν-rules. On the contrary, our calculus deals with all incestual modal logics (which include the logics considered in [29, 40, 33]), by means of a single ρ-rule. Moreover, the calculus is *modular* with respect to the characterizing axioms of the multimodal logic, i.e., it is enough to give the axioms to get the calculus.

Besides the disadvantage of requiring specific ν-rules and the fact that they do not work with multimodal systems, we think that it is difficult to extend the approach based on prefixes as sequences of integers to the whole class, even though it might be done for some subclasses of incestual axioms. In particular, it can be shown that a "generation lemma" ([40], page 732, and [33], Section 6.2) would not hold for such a calculus, i.e. it would not be true that, for any prefix occurring on a branch, all intermediate prefixes occur too. This property is at the basis of the completeness proof for the calculus in [40, 33].

Let us consider the following example. Assume the *IML* characterized by the axiom $[a;b]\varphi \supset [c]\varphi$. The formula $[a]p \wedge \langle c\rangle q \supset \langle a\rangle p$ is $\mathcal{A}$-valid (and, then, it has a tableau proof in our calculus). If σ is a prefix *à la* Fitting, that is a sequence of integer, we can image to denote by the prefix $\sigma.n_t$ the world reachable from σ by means of the accessibility relation $\mathcal{R}_t$. Then, we have:

1. $1. : \mathbf{F}([a]p \wedge \langle c\rangle q \supset \langle a\rangle p)$
2. $1. : \mathbf{F}\langle a\rangle p$
3. $1. : \mathbf{T}[a]p$
4. $1. : \mathbf{T}\langle c\rangle q$

5. $1.1_c. : \mathbf{T}q$

Now, by applying axiom $[a][b]\varphi \supset [c]\varphi$, the world "$1.1_c$" can also be represented by the sequence "$1.1_a.1_b$" and, thus

6. $1.1_a.1_b. : \mathbf{T}q$

whose subprefix "1.1_a" does not occur on the branch. On the other hand, this subprefix (world) is needed to apply the ν-rule to the formula at step 2. and 3. in order to close branch.

Moreover, adding explicitly subprefixes, as the one above, is not enough to solve the problem, since all prefixes representing the same world have to be identified. Assume that the *IML* characterized both by the axioms $[a]\varphi \supset [c]\varphi$ and $[b]\varphi \supset [c]\varphi$. Then, the formula $[a]p \wedge \langle c\rangle q \supset \langle b\rangle p$ is $\mathcal{A}$-valid. Using prefixes *à la* Fitting we have:

1. $1. : \mathbf{F}([a]p \wedge \langle c\rangle q \supset \langle b\rangle p)$
2. $1. : \mathbf{F}\langle b\rangle p$
3. $1. : \mathbf{T}[a]p$
4. $1. : \mathbf{T}\langle c\rangle q$
5. $1.1_c. : \mathbf{T}q$

Now, by applying axiom $[a]\varphi \supset [c]\varphi$ and axiom $[b]\varphi \supset [c]\varphi$ the world "1.1_c" will be denoted by both the prefixes "1.1_a" and "1.1_b":

6. $1.1_a : \mathbf{T}q$
7. $1.1_b : \mathbf{T}q$

and, then, applying twice the ν-rule to the formulae at steps 2. and 3. we have:

8. $1.1_b : \mathbf{F}p$
9. $1.1_a : \mathbf{T}p$

but the branch does not close because we cannot identify "1.1_b" and "1.1_a" which must represent the same world. To this purpose we would need to introduce equivalence classes among prefixes.

5.3 LABELLED SYSTEMS

Other tableau methods for propositional modal logics which make use of prefixed formulae are presented in [34, 18]. The system in [18] deals with all the fifteen propositional normal modal logics obtained by combining the axioms T, D, 4, 5, and B, while the system in [34] considers the propositional modal logics $K45$, $D45$, and $S5$ and the propositional modal logics $S5A$ and $S5P_{(n)}$. It was subsequently extended to deal with the above mentioned fifteen modal systems and the predicative case in [1]. These proof systems extend the calculus KE, a combination of tableau and natural deduction inference rules, which allows for a suitably restricted use of the cut rule [21]. In order to have a more efficient proof search, they generalize prefixes both by allowing the occurrence of variables and by using unification to show that two prefixes can name the same world. The main difference between the system in [34, 1] and the one in [18] is that the former uses only one type of path variable (single worlds) while the latter allows variables over single as well as sequences of worlds. Furthermore, in [34], only one ν-rule is used and unification is logic-dependent while, in [18], unification is independent of the logic but there is a different ν-rule for each logic.

One of the main features of these systems is the full permutability of the application of their rules. Unfortunately, our tableau method does not enjoy this property. In fact, similarly to the problem of applying the existential rules before the universal ones in the proof systems for classical logic, we need to apply the π-rules (or the ρ-rules) before the ν-rules. On the other hand, we deal with a wider class of logics. In particular, we think that it is hard to extend the unification method of prefixes so to deal with all the classes of logics that we considered for the same reasons given in Section 5.2 when discussing the proposals in [29, 40, 33].

The proposals in [34, 18, 11] address the problem of an efficient implementation of the tableau calculi for a wide class of modal logics. They generalize the prefixes by allowing occurrences of variables and they use unification to show that two prefixes are names for the same world. While a straightforward implementation of our calculus is unlikely to be efficient, the generality of the approach makes it suitable to study the properties of different classes of logics [6].

In [15] a general theorem prover for propositional modal logics is presented. This system, named TABLEAUX, uses a representation for the accessibility relations that is close to ours. In fact, in that work a tableau is a pair (Γ, R), where Γ is a set of prefixed formulae and R is a set of relations between worlds. Prefixes are constant symbols. TABLEAUX can deal with all the classical fifteen modal systems, and also with their multimodal versions. However, it does not deal with any interaction axiom as our does. This system uses three classes of tableau rules: the first is made of *simplification* rules, that are world independent and whose aim is to simplify the proof search; the second consists of the *transformation* rules and allows to introduce new operators in terms of the existing ones; finally, the third class of rules deals with formulae belonging to different worlds and can introduce modifications in the set R of relations.

In [50] a tableau system for the modal mu-calculus, an extension of the dynamic logic, is presented. Like in our proposal, the tableau calculus in [50] makes use of prefixed formulae and represents the underlying model construction during a proof as a graph (the transition system). However, as in dynamic logic, no axioms of the form we have considered are allowed.

More recently, in [12] Castilho et al. present a modal tableau calculus, whose underlying structure is not a tree but a root directed acyclic graph. Moreover, they define two kinds of rules. The first ones, the propagation rules, add formulas to nodes, the second ones, the structural rules, construct new edges and nodes. In particular, the structural rules can be regarded as rewriting rules which create new paths among the worlds (nodes) in a similar way as our ρ-rule. In this way, they can easily deal not only with the modal logics characterized by any combination of axioms T, D, 4, 5, and B but also with the ones whose accessibility relation is *dense* and *simple confluent*. However, they consider neither the multimodal case nor other axioms apart from De and G.

In the paper we introduced complex labels for the modal operators. In this case, we have only used a composition operator and an union operator. A further extension of our language in which also the *test* operator "?" and the *iteration* operator "$*$" of dynamic logic will be subject of future investigation. A basis for such a kind of extension could be the modal mu-calculus in [50].

5.4 TRANSLATION METHODS

Instead of developing specific theorem proving techniques and tools for modal logics, many authors have proposed the alternative approach of translating modal logics into classical first order logic, so that standard theorem provers can be used [47]. The translation methods are based on the idea of making explicit reference to the worlds by adding to all predicates an argument that represents the world where the predicate holds, so that the modal operators can be transformed into quantifiers of classical logic.

The *relational translation* is based on the direct simulation of the Kripke semantics by introducing a distinguished predicate symbol to represent the accessibility relation [45, 47]. This method has strong relationships with our approach. Indeed, we deal with inclusion properties of the accessibility relations, which are first-order axiomatizable, hence, the relational translation method can cope with them. On the other hand, as a drawback, the relational translation method destroys the structure of the formulae.

An alternative method is the *functional translation* [45, 2]. It is based on the idea of representing paths in the possible worlds structure by means of compositions of functions, which map worlds to accessible worlds. The most common properties, such as transitivity or reflexivity, are taken into account by an equational unification algorithm. An advantage of this approach is that it keeps the structure of the original formula. However the approach is suitable mainly for *serial logics*, for which optimization technique have been studied [46, 31], and it requires a different equational unification algorithm for each logic. A way to avoid the use of equational unification algorithms, retaining the advantages of the functional translation, has been developed in [43], where a mixed approach based on a relational and functional translation is defined. One of the aims of the author was to obtain Prolog programs starting from Horn clauses extended with modal operators [44]. This method requires that accessibility relation properties are first-order predicate logic definable. In particular, it provides a translation for the modal systems (all requiring seriality) KD, KT, $KD4$, $S4$, but it deals also with axioms like B and, then, with logics like KDB, $KD45$, $S5$ and the multimodal system $KD45_n$.

Acknowledgments

I would like to thank Prof. Alberto Martelli and Prof. Laura Giordano for the support they gave me in all these years. I would like also to thank the referees for the precious advice. I have received a financial support by *Consiglio Nazionale delle Ricerche* (*CNR*). I dedicate this work to the memory of my father.

Matteo Baldoni
Dipartimento di Informatica
Università degli Studi di Torino
Corso Svizzera, 185
I-10149 Torino, Italy
baldoni@di.unito.it

References

[1] A. Artosi, P. Benassi, G. Governatori, and A. Rotolo. Labelled Proofs for Quantified Modal Logic. In *Logics in Artificial Intelligence, JELIA '96*, volume 1126 of *LNAI*, pages 70–86. Springer-Verlag, 1996.

[2] Y. Auffray and P. Enjalbert. Modal Theorem Proving: An equational viewpoint. *Journal of Logic and Computation*, 2(3):247–297, 1992.

[3] M. Baldoni. *Normal Multimodal Logics: Automatic Deduction and Logic Programming Extension.* PhD thesis, Dipartimento di Informatica, Università degli Studi di Torino, Italy, 1998. Available at `http://www.di.unito.it/~baldoni`.

[4] M. Baldoni. Normal Multimodal Logics with Interaction Axioms: extended version. Technical report, Dipartimento di Informatica, Università degli Studi di Torino, 1999.

[5] M. Baldoni, L. Giordano, and A. Martelli. A Modal Extention of Logic Programming: Modularity, Beliefs and Hypothetical Reasoning. *Journal of Logic and Computation*, 6(5):597–635, 1998.

[6] M. Baldoni, L. Giordano, and A. Martelli. A Tableau Calculus for Multimodal Logics and Some (Un)Decidability Results. In H. de Swart, editor, *Proc. of the International Conference on Analytic Tableaux and Related Methods, TABLEAUX'98*, volume 1397 of *LNAI*, pages 44–59. Springer-Verlag, 1998.

[7] M. Baldoni, L. Giordano, A. Martelli, and V. Patti. A Modal Programming Language for Representing Complex Actions. In A. Bonner, B. Freitag, and L. Giordano, editors, *Proc. 1998 JICSLP'98 Post-Conference Workshop on Transactions and Change in Logic Databases, DYNAMICS'98*, pages 1–15, Manchester, UK, June 1998.

[8] D. Basin, S. Matthews, and L. Viganò. A Topography of Labelled Model Logics. In F. Baader and K. U. Schulz, editors, *Frontiers of Combining Systems — Proc. of the International Workshop*, volume 3 of *Applied Logic Series*, pages 75–92. Kluwer Academic Publisher, 1996.

[9] D. Basin, S. Matthews, and L. Viganò. Labelled Propositional Modal Logics: Theory and Practice. *Journal of Logic and Computation*, 7(6):685–717, 1997.

[10] D. Basin, S. Matthews, and L. Viganò. Natural Deduction for Non-Classical Logics. *Studia Logica*, 60(1):119–160, 1998. Special issue on *Natural Deduction* edited by F. Pfenning and W. Sieg.

[11] B. Beckert and R. Goré. Free Variable Tableaux for Propositional Modal Logics. In D. Galmiche, editor, *Proc. of the International Conference on Automatic Reasoning with Analytic Tableaux and Related Methods,*

TABLEAUX'97, volume 1227 of *LNAI*, pages 91–106. Springer-Verlag, 1997.

[12] M. A. Castilho, L. Faniñas del Cerro, O. Gasquet, and A. Herzig. Modal Tableaux with Propagation Rules and Structural Rules. *Fundamenta Informaticae*, 32(3/4):281–297, 1998.

[13] L. Catach. Normal Multimodal Logics. In *Proc. of the 7th National Conference on Artificial Intelligence, AAAI '88*, volume 2, pages 491–495. Morgan Kaufmann, 1988.

[14] L. Catach. *Les Logiques Multimodales.* PhD thesis, Universite Pierre et Marie Curie (Paris 6), Paris, 1989.

[15] L. Catach. TABLEAUX: A General Theorem Prover for Modal Logics. *Journal of Automated Reasoning*, 7(4):489–510, 1991.

[16] A. Chagrov and M. Zakharyaschev. *Modal Logic.* Clarendon Press, Oxford, 1997.

[17] B. F. Chellas. *Modal Logic: an Introduction.* Cambridge University Press, 1980.

[18] J. Cunningham and J. Pitt. Distributed Modal Theorem Proving with KE. In *Proc. of the 5th International Workshop on Theorem Proving with Analytic Tableaux and Related Methods, TABLEAUX '96*, volume 1071 of *LNAI*, pages 160–176. Springer-Verlag, 1996.

[19] M. D'Agostino and D. M. Gabbay. A generalization of analytic deduction via labelled systems. Part I: basic structural logics. *Journal of Automated Reasoning*, 13:243–281, 1994.

[20] M. D'Agostino, D. M. Gabbay, R. Hähnle, and J. Posegga, editors. *Handbook of Tableau Methods.* Kluwer Academic Publishers, 1999.

[21] M. D'Agostino and M. Mondadori. The Taming of the Cut. *Journal of Logic and Computation*, 4:285–319, 1994.

[22] G. De Giacomo and F. Massacci. Tableaux and Algorithms for Propositional Dynamic Logic with Converse. In W. McCune, editor, *Automated Deduction — CADE-15*, volume 1249 of *LNAI*, pages 613–627. Springer, 1996.

[23] P. Enjalbert and L. Fariñas del Cerro. Modal Resolution in Clausal Form. *Theoretical Computer Science*, 65(1):1–33, 1989.

[24] L. Fariñas del Cerro. MOLOG: A System that extends Prolog with Modal Logic. *New Generation Computing*, 4(1):35–50, 1986.

[25] L. Fariñas del Cerro and A. Herzig. Modal Deduction with Applications in Epistemic and Temporal Logics. In *Handbook of Logic in Artificial Intelligence and Logic Programming*, volume 4, pages 499–594. Oxford Science Publications, 1995.

[26] L. Fariñas del Cerro and M. Penttonen. Grammar Logics. *Logique et Analyse*, 121-122:123–134, 1988.

[27] M. J. Fischer and R. E. Ladner. Propositional Dynamic Logic of Regular Programs. *Journal of Computer and System Sciences*, 18(2):194–211, 1979.

[28] M. Fisher and R. Owens. An Introduction to Executable Modal and Temporal Logics. In *Proc. of the IJCAI'93 Workshop on Executable Modal and Temporal Logics*, volume 897 of *LNAI*, pages 1–20. Springer-Verlag, 1993.

[29] M. Fitting. *Proof Methods for Modal and Intuitionistic Logics*, volume 169 of *Synthese library*. D. Reidel, Dordrecht, Holland, 1983.

[30] D. M. Gabbay. *LDS - Labelled Deductive Systems (Volume 1 - Foundations)*. Clarendon Press, Oxford, 1996.

[31] O. Gasquet. Optimization of deduction for multi-modal logics. In *Applied Logic: How, What and Why?* Kluwer Academic Publishers, 1993.

[32] M. Genesereth and N. Nilsson. *Logical Foundations of Artificial Intelligence*. Morgan Kaufmann, 1987.

[33] R. A. Goré. Tableaux Methods for Modal and Temporal Logics. In D'Agostino et al. [20], pages 297–396.

[34] G. Governatori. Labelled Tableaux for Multi-Modal Logics. In *Proc. of the 4th Workshop on Theorem Proving with Analytic Tableaux and Related Methods, TABLEAUX '95*, volume 918 of *LNAI*, pages 79–94. Springer-Verlag, 1995.

[35] J. Y. Halpern and Y. Moses. A Guide to Completeness and Complexity for Modal Logics of Knowledge and Belief. *Artificial Intelligence*, 54:319–379, 1992.

[36] D. Harel. Dynamic Logic. In *Handbook of Philosophical Logic*, volume II, pages 497–604. D. Reidel Publishing Company, 1984.

[37] J. E. Hopcroft and J. D. Ullman. *Introduction to automata theory, languages, and computation*. Addison-Wesley Publishing Company, 1979.

[38] G. E. Hughes and M. J. Cresswell. *A New Introduction to Modal Logic*. Routledge, 1996.

[39] M. Kracht. Highway to the Danger Zone. *Journal of Logic and Computation*, 5(1):93–109, 1995.

[40] F. Massacci. Strongly Analytic Tableaux for Normal Modal Logics. In *Proc. of the CADE'94*, volume 814 of *LNAI*, pages 723–737. Springer-Verlag, 1994.

[41] R. Moddux. The equational theory of CA_3 is undecidable. *Journal of Symbolic Logic*, 45(2):311–316, 1980.

[42] A. Nerode. Some Lectures on Modal Logic. In F. L. Bauer, editor, *Logic, Algebra, and Computation*, volume 79 of *NATO ASI Series*. Springer-Verlag, 1989.

[43] A. Nonnengart. First-Order Modal Logic Theorem Proving and Functional Simulation. In *Proc. of International Joint Conference on Artificial Intelligence, IJCAI'93*, pages 80–85. Morgan Kaufmann, 1993.

[44] A. Nonnengart. How to use Modalities and Sorts in Prolog. In C. MacNish, D. Pearce, and L. M. Pereira, editors, *Proc. of the JELIA'94: Logics in Artificial Intelligence*, volume 838 of *LNAI*, pages 365–378, York, UK, 1994. Springer-Verlag.

[45] H. J. Ohlbach. Semantics-Based Translation Methods for Modal Logics. *Journal of Logic and Computation*, 1(5):691–746, 1991.

[46] H. J. Ohlbach. Optimized Translation of Multi Modal Logic into Predicate Logic. In A. Voronkov, editor, *Proc. of the Logic Programming and Automated Reasoning*, volume 822 of *LNAI*, pages 253–264. Springer-Verlag, 1993.

[47] H. J. Ohlbach. Translation methods for non-classical logics: An overview. *Bull. of the IGPL*, 1(1):69–89, 1993.

[48] M. A. Orgun and W. Ma. An overview of temporal and modal logic programming. In D. M. Gabbay and H. J. Ohlbach, editors, *Proc. of the First International Conference on Temporal Logic*, volume 827 of *LNAI*, pages 445–479. Springer-Verlag, 1994.

[49] A. Russo. Generalizing Propositional Modal Logics using Labelled Deductive Systems. In F. Baader and K. U. Schulz, editors, *Frontiers of Combining Systems — Proc. of the International Workshop*, volume 3 of *Applied Logic Series*, pages 57–74. Kluwer Academic Publisher, 1996.

[50] C. Stirling and D. Walker. Local model checking in the modal mu-calculus. *Theoretical Computer Science*, 89:161–177, 1991.

[51] J. Van Benthem. *Modal Logic and Classical Logic*. Bibliopolis, Naples, 1982.

[52] M. Wooldridge and N. R. Jennings. Agent Theories, Architectures, and Languages: A survey. In *Proc. of the ECAI-94 Workshop on Agent Theories*, volume 890 of *LNAI*, pages 1–39. Springer-Verlag, 1995.

THE SAT PROBLEM OF SIGNED CNF FORMULAS

Bernhard Beckert, Reiner Hähnle and Felip Manyà

Abstract *Signed conjunctive normal form* (signed CNF) is a classical conjunctive clause form using a generalised notion of literal, called *signed literal.* A signed literal is an expression of the form $S : p$, where p is a classical atom and S, its *sign*, is a subset of a domain N. The informal meaning is "p takes one of the values in S". Signed formulas are a logical language for knowledge representation that lies in the intersection of the areas *constraint programming* (CP), *many-valued logic* (MVL), and *annotated logic programming* (ALP). This central rôle of signed CNF justifies a detailed study of its subclasses including algorithms for and complexities of associated satisfiability problems (SAT problems). Although signed logic is used since the 1960s, there are only few systematic investigations of its properties. In contrast to work done in ALP and MVL, our present work is a more fine-grained study for the case of propositional CNF. We highlight the most interesting lines of current research: (i) signed versions of some main proponents of classical deduction systems including non-trivial refinements having no classical counterpart; (ii) incomplete local search methods for satisfiability checking of signed formulas; (iii) phase transition phenomena as known, for example, from classical SAT and the influence of the cardinality of N on the crossover point; (iv) the complexity of the SAT problem for signed CNF and its subclasses.

1 INTRODUCTION

Signed formulas are a logical language for knowledge representation that lies in the intersection of the areas *constraint programming* (CP), *many-valued logic* (MVL), and *annotated logic programming* (ALP).

Signed conjunctive normal form (signed CNF) is a classical propositional or first-order conjunctive clause form using a generalised notion of literal, called *signed literal.* A signed literal is an expression of the form $S : p$, where p is a classical atom and S, its *sign*, is a subset of a domain N. The informal meaning is "p takes one of the values in S".

When N is considered to be a truth value set, signed CNF formulas turn out to be a generic representation for finite-valued logics [17]: The problem of deciding the satisfiability of formulas (SAT problem) of any finite-valued

D. Basin et al. (eds.), Labelled Deduction, 59–80.
© 2000 *Kluwer Academic Publishers. Printed in the Netherlands.*

logic is in a natural way polynomially reducible to the problem of deciding satisfiability of formulas in signed CNF (signed SAT).

If N is equipped with an ordering, there is a natural notion of signed Horn formula (Definition 4). The particular case where N is lattice-ordered and S is an order filter is investigated in annotated logic programming [21] (there, S is called an *annotation*), therefore, annotated logic programs can be considered as particular signed logic formulas.

Third, $S : p$ can be interpreted as "p is constrained to the values in S" and, hence, as an instance of finite-domain constraint programming [20, 7].

Finally, it is also possible to embed signed formulas into classical monadic first-order logic by representing a signed literal $S : p$, where $S = \{i_1, \ldots, i_r\}$, as the classical formula

$$(\exists p)(s(p)) \wedge (\forall x)(s(x) \leftrightarrow (s(i_1) \vee \cdots \vee s(i_r)))$$

using a unary predicate symbol s.

Applications for deduction in signed logics derive from those of annotated logic programming (e.g., mediated deductive databases), constraint programming (e.g., scheduling), and many-valued logics (e.g., natural language processing). In addition, some problems usually denoted in classical clause logic can be formulated in a better or simply in a different way using signed logic: this comes from the disjunctive interpretation of signs that allows for a compact representation of certain finite-domain first-order properties; and there are additional dimensions along which one can calibrate, namely, the number and ordering of truth values as well as the form of the signs. This claim is supported by first experiments with combinatorial optimisation problems [6]. At the same time, computational complexity of signed logic is mostly comparable to classical logic (see Section 7). Altogether, signed logic constitutes an interesting trade-off between expressivity and complexity.

The central rôle of signed CNF justifies a detailed study of its subclasses, including algorithms for and complexities of associated SAT problems. In contrast to surveys of ALP [21] and MVL [16, 19], the present chapter constitutes a more fine-grained study into signed formulas within the framework of propositional logic and conjunctive normal form. Although some of the results described here are not yet formally published, the following has the character of a survey, because, given the limited space, we decided to trade in formal proofs for examples and explanations. The reader is invited to consult the technical references given throughout.

In the following section, syntax and semantics of signed CNF are defined formally. Of the remaining sections each captures a specific line of research. Sections 3 and 4 discuss signed versions of some main proponents of classical deduction systems including non-trivial refinements having no classical counterpart. Section 5 focuses on incomplete local search methods for satisfiability checking of signed formulas. In Section 6 we look into the phase transition phenomena well-known from classical satisfiability testing (and other NP-complete problems) and investigate the influence of the cardinality of N on the crossover point. Finally, in Section 7, results proven so far on the complexity of checking satisfiability of formulas in signed CNF (signed SAT) and its subclasses are collected.

2 PRELIMINARIES

2.1 SYNTAX

We assume that a signature, i.e., a denumerable set of propositional variables is given. To form signed literals, the propositional variables (atoms) are adorned with a sign that consists of a finite set of (truth) values.

Definition 1 *A* truth value set N *is a finite set* $\{i_1, i_2, \ldots, i_n\}$ *where* $n \in \mathbb{N}$. *The cardinality of* N *is denoted by* $|N|$. *A partial order* $\leq$ *is associated with* N, *which may be the empty order.*

Definition 2 *A* sign *is a set* $S \subseteq N$ *of truth values. A* signed literal *is of the form* $S : p$ *where* S *is a sign and* p *is a propositional variable. The* complement *of a signed literal* $S : p$, *denoted by* $\overline{S} : p$, *is* $(N \setminus S) : p$.

A signed clause *is a finite set of signed literals. A signed clause containing exactly one literal is called a* signed unit clause*; and a signed clause containing exactly two literals is called a* signed binary clause. *The empty signed clause is denoted by* $\Box$.

A signed CNF formula *is a finite set of signed clauses. A signed CNF formula whose clauses are binary is called a* signed 2-CNF formula.

The clauses of a signed CNF formula are implicitly conjunctively connected; and the literals in a signed clause are implicitly disjunctively connected. In the following we use $S_1 : p_1 \vee \cdots \vee S_k : p_k$ to represent a signed clause $\{S_1 : p_1, \ldots, S_k : p_k\}$.

Definition 3 *The* length *of a signed clause* C, *denoted by* $|C|$, *is its cardinality. The length of a signed formula* Γ, *denoted by* $|\Gamma|$, *is the sum of the lengths of its signed clauses.*

Definition 4 *For all* $i \in N$, *let* $\uparrow i$ *denote the sign* $\{j \in N \mid j \geq i\}$ *and let* $\downarrow i$ *denote the sign* $\{j \in N \mid j \leq i\}$ *where* $\leq$ *is the partial order associated with* N. *A sign* S *is* regular *if it is identical to* $\uparrow i$ *or to* $\downarrow i$ *for some* $i \in N$.

A signed literal $S : p$ *is a* regular literal *if (a) its sign* S *is regular or (b) its sign* $S = \overline{S'}$ *is the complement of a regular sign* S'.

A signed clause (a signed CNF formula) is a regular clause *(a* regular CNF formula*) if all its literals are regular.*

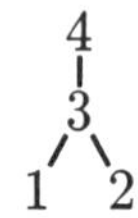

Figure 3.1 A partially ordered truth value set, see Example 5.

Example 5 *Let the truth value set $N = \{1, 2, 3, 4\}$ be ordered as shown in Figure 3.1, i.e., we use the standard order on natural numbers except that* 1 *and* 2 *are incomparable. Then the signs $\uparrow 1 = \{1, 3, 4\}$ and $\downarrow 1 = \{1\}$ are regular; and $\overline{\uparrow 1} = \{2\}$ and $\overline{\uparrow 3} = \{1, 2\}$ are complements of regular signs. The signs $\{3\}$ and $\{1, 4\}$ are neither regular nor complements of regular signs.*

The complement $\overline{\uparrow 3}$ of the regular sign $\uparrow 3$ is not *regular as it cannot be represented as $\uparrow i$ or $\downarrow i$ for any $i \in N$. Thus, a regular literal can have a sign that is not regular (but is the* complement *of a regular sign only).*

Whenever the (partial) order on the truth value set is not empty, polarities can be assigned to signed literals in a meaningful way, which gives rise to a generalised notion of Horn clauses.

Definition 6 *A regular sign S is of* positive *(resp.* negative*)* polarity *if it is of the form $\uparrow i$ (resp. $\downarrow i$) for some $i \in N$. A regular literal is of* positive (negative) polarity *if its sign is of positive (negative) polarity.*

A regular clause is a regular Horn clause *if it contains at most one literal of positive polarity and the signs of all its other literals are complements of signs with positive polarity. A regular CNF formula is a* regular Horn formula *if all its clauses are regular Horn clauses.*

Our notion of regular Horn formula coincides with that of a propositional *annotated logic program* [21].

Example 7 *Using the truth value set N and the associated ordering from the previous example, the clauses (1) $\uparrow 1 : p$, (2) $\uparrow 2 : p \vee \overline{\uparrow 3} : q$, and (3) $\uparrow 4 : q$ are Horn clauses. The regular clause $\uparrow 1 : p \vee \uparrow 2 : q$ is* not *a Horn clause as it contains more than one literal of positive polarity. Since $\downarrow 1 = \overline{\uparrow 2}$ but $\downarrow 4 \neq \overline{\uparrow i}$ for all $i \in N$, the clause $\downarrow 1 : p$ is Horn whereas $\downarrow 4 : p$ is* not *Horn (both clauses are regular).*

Definition 8 *A literal $S : p$ is* monosigned *if its sign $S = \{i\}$ is a singleton. A signed clause (a signed CNF formula) is* monosigned *if all its literals are monosigned.*

Classical two-valued CNF formulas are a special case of monosigned CNF formulas (using a truth value set N with two elements). Monosigned CNF formulas are (trivially) regular w.r.t. the empty ordering.

2.2 SEMANTICS

Definition 9 *An* interpretation *is a mapping that assigns to every propositional variable an element of the truth value set.*

An interpretation I satisfies *a signed literal $S : p$ iff $I(p) \in S$. It* satisfies *a signed clause C iff it satisfies at least one of the signed literals in C; and it* satisfies *a signed CNF formula Γ iff it satisfies all clauses in Γ.*

A signed CNF formula (a signed clause) is satisfiable *iff it is satisfied by at least one interpretation; otherwise it is* unsatisfiable.

Two signed CNF formulas (signed clauses) are equivalent *if they are satisfied by the same interpretations. They are* satisfiability equivalent *iff they are either both satisfiable or both unsatisfiable.*

By definition, the empty signed clause is unsatisfiable and the empty signed CNF formula is satisfiable.

As in classical logic, a Horn formula $C = \overline{\uparrow i_1} : p_1 \vee \cdots \vee \overline{\uparrow i_k} : p_k \vee \uparrow j : q$ is equivalent to the implication $\uparrow i_1 : p_1 \wedge \cdots \wedge \uparrow i_k : p_k \to \uparrow j : q$, i.e., an interpretation I satisfies C iff it does not satisfy one of $\uparrow i_1 : p_1, \ldots, \uparrow i_k : p_k$ or it satisfies $\uparrow j : q$.

Proposition 10 *For all propositional variables p and all signs $S_1, \ldots, S_k \subseteq N$ ($k \in \mathbb{N}$), the signed clauses*

$$S_1 : p \vee \cdots \vee S_k : p \vee D \quad \textit{and} \quad (S_1 \cup \cdots \cup S_k) : p \vee D$$

are equivalent.

The simplification expressed in Proposition 10 is often but not always useful, as its application changes the structure of signs and can, for example, destroy the regularity of a clause.

2.3 CLAUSE FORM TRANSLATION

One of the prominent features of signed CNF formulas is that any formula of any finite-valued logic can be translated in polynomial time into a satisfiability equivalent *signed* CNF formula (the transformation is structure preserving [17]); thus, the SAT problem of a finite-valued logic is polynomially reducible to the signed SAT problem.

In addition, every signed CNF formula can be translated in polynomial time into a satisfiability equivalent *regular* CNF formula with an arbitrary total order on N by the following simple trick: a signed clause containing literals of the form $S : p$ is first transformed into a monosigned clause by replacing $S : p$ with $\bigvee_{i \in S} \{i\} : p$ (using Proposition 10). Then all monosigned literal occurrences are eliminated by replacing a clause $C = \{i\} : p \vee D$ with three clauses $C_1 = \uparrow i : p \vee S : q$, $C_2 = \downarrow i : p \vee S : q$, and $C_3 = D \vee \overline{S} : q$, where

q is a new propositional variable not occurring anywhere else and S is an arbitrary regular sign (soundness of this transformation is a direct consequence of rule (3.1) below).

A *direct* polynomial time translation into satisfiability equivalent *regular* CNF formulas was given by Sofronie-Stokkermans [32] for the case that the set of truth values and its associated order form a distributive lattice; it exploits properties of distributive lattices and often produces much less clauses than the general method outlined above.

3 RESOLUTION

In this section we review in a uniform way resolution style calculi for signed CNF formulas and their subclasses that appeared in the literature [15, 17, 18, 28, 29, 25, 32, 33, 3]. The perhaps most straightforward, refutation complete version is formed by the rules below [28].

$$\frac{S_1{:}p \vee D_1 \qquad S_2{:}p \vee D_2}{(S_1 \cap S_2){:}p \vee D_1 \vee D_2} \qquad\qquad \frac{\emptyset{:}p \vee D}{D} \tag{3.1}$$

signed binary resolution   simplification

Note that, unlike classical resolution, the literal resolved upon does not necessarily vanish and a so-called *residue* remains. The following parallel resolution rule [15, 28] avoids building residues. Both versions (3.1) and (3.2) were originally thought to require the *merging rule* embodied in Proposition 10 for completeness; however, one can show that it is not necessary [18].

$$\frac{S_1{:}p \vee D_1 \quad \cdots \quad S_m{:}p \vee D_m}{D_1 \vee \cdots \vee D_m} \quad \text{if } S_1 \cap \cdots \cap S_m = \emptyset \tag{3.2}$$

signed parallel resolution

In the case of monosigned and regular CNF formulas over a totally ordered truth value set, completeness of signed binary resolution is preserved if rule applications generating a residue are not allowed; hence (3.1) can be simplified:

$$\frac{S_1{:}p \vee D_1 \qquad S_2{:}p \vee D_2}{D_1 \vee D_2} \quad \text{if } S_1 \cap S_2 = \emptyset \tag{3.3}$$

monosigned/regular binary resolution

Completeness of binary resolution, as well as of ordered resolution and hyperresolution, for monosigned CNF formulas is proved by Baaz and Fermüller [1]. If N is totally ordered, one obtains the hyperresolution-like refinements

(3.4) and (3.5) of regular binary resolution by combining several applications of rule (3.3) into one [17, 18].

$$\frac{\uparrow i_1 : p \vee D_1 \quad \cdots \quad \uparrow i_m : p \vee D_m \qquad \downarrow j : p \vee D}{D_1 \vee \cdots \vee D_m \vee D} \quad \text{if } (\max_{1 \leq k \leq m} i_k) > j$$

regular resolution

(3.4)

Using the *maximal* i_k in the rule above is not strictly necessary: admitting *any* $i_k > j$ yields a sound rule, but may lead to longer proofs. For regular formulas, (3.4) with $m = 1$ is the same as (3.3).

Example 11 *Let the truth value set be $N = \{1, 2, 3\}$ (with the natural order), and let Γ be the following regular CNF formula:*

$$\{\downarrow 1 : p_1 \vee \downarrow 2 : p_2,\ \uparrow 2 : p_1 \vee \downarrow 1 : p_2,\ \downarrow 1 : p_1 \vee \uparrow 3 : p_3,$$
$$\uparrow 3 : p_2 \vee \uparrow 2 : p_3,\ \uparrow 3 : p_2 \vee \downarrow 1 : p_3\}$$

The last three clauses resolve to $\downarrow 1 : p_1 \vee \uparrow 3 : p_2$ by rule (3.4), which in turn resolves to $\downarrow 1 : p_1$ with the first clause (by either rule (3.4) or (3.3)). From there, one obtains $\downarrow 1 : p_2$ with the second clause. In three more steps the empty clause can be derived.

$$\frac{\downarrow i_1 : p_1 \vee D_1 \quad \cdots \quad \downarrow i_m : p_m \vee D_m \qquad \uparrow j_1 : p_1 \vee \cdots \vee \uparrow j_m : p_m \vee E}{D_1 \vee \cdots \vee D_m \vee E}$$

provided $m \geq 1, i_l < j_l$ for all $1 \leq l \leq m$,
$D_1, \ldots, D_m, E$ contain only negative literals

regular negative hyperresolution

(3.5)

Sofronie-Stokkermans [32, 33] proved that, when clauses only contain positive regular literals or their complements and N is a distributive lattice, an analogue of rule (3.5) is complete where all negative literals of the form $\downarrow i : p$ are replaced with complements of positive literals, i.e., literals of the form $\overline{\uparrow i}:p$. When N is a lattice, the following calculus is complete [3]:

$$\frac{\begin{array}{c}\uparrow i : p \vee D_1 \\ \overline{\uparrow j} : p \vee D_2\end{array}}{D_1 \vee D_2} \qquad \frac{\begin{array}{c}\uparrow i : p \vee D_1 \\ \uparrow j : p \vee D_2\end{array}}{\uparrow (i \sqcup j) : p \vee D_1 \vee D_2} \qquad (3.6)$$

if $i \geq j$ — lattice-regular binary resolution

if neither $i \geq j$ nor $j \geq i$ — lattice-regular reduction

Note that, when N is totally ordered, the left rule of (3.6) is the same as (3.3) for regular formulas.

Refinements of regular binary resolution being complete for regular Horn formulas over a totally ordered truth value set are regular unit resolution [17] (this corresponds to the case $D_1 = \Box$ in rule (3.3)) and regular positive unit resolution [25] (where, in addition, the unit input clause must be a positive literal). Recently, we proved [3] that the rules below are complete for regular Horn formulas in case N forms an upper semi-lattice.

$$\frac{\begin{array}{c}\uparrow i : p \\ \uparrow j : p \vee C\end{array}}{C} \qquad \frac{\begin{array}{c}\uparrow i : p \\ \uparrow j : p\end{array}}{\uparrow (i \sqcup j) : p} \tag{3.7}$$

$$\text{if } i \geq j \qquad \text{if neither } i \geq j \text{ nor } j \geq i$$

lattice-regular positive unit resolution — lattice-regular reduction

Example 12 *Using the upper semi-lattice ordering and regular Horn clauses from Example 7, one may derive* $\uparrow 2 : p$ *from clauses (2) and (3) by lattice-regular positive unit resolution. The resolvent together with clause (1) gives* $\uparrow 3 : p$ *by lattice-regular reduction.*

Recall that lattice-based regular Horn formulas are propositional annotated logic programs. As a consequence, the various SLD-style resolution procedures developed for ALP [21, 23, 22] can be used as well. Note, however, that SLD resolution is optimised for first-order logic and is not very efficient on the propositional level.

We close this section with a brief remark on the techniques one can employ to prove completeness of the mentioned resolution calculi. It turns out that semantic tree arguments retain much of their clarity. The most straightforward approach is to use $|N|$-ary semantic trees [17]. Just as in classical resolution theory, more complex refinements are often better handled by inductive construction of a proof, where the number of atoms or atom occurrences in a formula supplies the induction parameter [18].

4 DAVIS-PUTNAM-LOVELAND PROCEDURES

In classical logic, among the most competitive propositional satisfiability solvers are variants of the Davis-Putnam-Loveland procedure (DPL) [9]. In this section we describe the extensions of DPL that have been proposed for signed and regular CNF formulas. They are complete proof procedures for testing the satisfiability of this kind of formulas and seem to be good candidates to implement signed satisfiability solvers.

4.1 THE SIGNED DPL PROCEDURE

The signed DPL procedure (Signed-DPL) is based on the following rules:

Signed one-literal rule: Given a signed CNF formula Γ that contains a signed unit clause $\{S : p\}$,

1. remove all clauses containing a literal $S' : p$ such that $S \subseteq S'$;
2. delete all occurrences of literals $S'' : p$ such that $S \cap S'' = \emptyset$;
3. replace all occurrences of literals $S''' : p$ with $(S''' \cap S) : p$.

Signed branching rule: Reduce the problem of determining whether a signed CNF formula Γ (that contains the propositional variable p) is satisfiable to the problem of determining whether there is an $i \in N$ such that $\Gamma \cup \{\{i\} : p\}$ is satisfiable.

Definition 13 *Given a signed CNF formula Γ that contains a unit clause $\{S{:}p\}$, let $simplify(\Gamma, S : p)$ denote the result of applying the signed one-literal rule to Γ using the unit clause $\{S : p\}$.*

The Signed-DPL procedure is shown in Figure 3.2. It first repeatedly applies the signed one-literal rule. Once the formula cannot be further simplified, it then applies the branching rule and recursively tries to solve each of $|N|$ sub-problems. As these sub-problems by construction contain a signed unit clause, the signed one-literal rule can be applied again. The procedure terminates when either a satisfiable sub-problem is found or all sub-problems have been shown to be unsatisfiable.

Intuitively, Signed-DPL constructs a proof tree using a depth-first strategy. The root node of that tree is labelled with the input formula; the other nodes are labelled with the formulas that result from a single application of the signed one-literal or the signed branching rule to the formula of their parent node. If all the leaves of the tree contain the signed empty clause, the input formula is unsatisfiable; otherwise, if at least one leaf is labelled with the empty signed CNF formula, the input formula is satisfiable.

Example 14 *Let the truth value set be $N = \{1, 2, 3\}$ with an arbitrary order; and let the signed CNF formula Γ consist of the following six clauses:*

$$\begin{array}{ll}
\{3\} : p_1 \vee \{1,2\} : p_2 & \{2,3\} : p_1 \vee \{1\} : p_2 \vee \{2,3\} : p_3 \\
\{1,3\} : p_1 \vee \{3\} : p_2 & \{2\} : p_1 \vee \{1,2\} : p_2 \\
\{3\} : p_2 \vee \{1\} : p_3 & \{3\} : p_2 \vee \{2,3\} : p_3
\end{array}$$

Figure 3.3 shows the proof tree created by Signed-DPL for input Γ. Edges corresponding to an application of the signed one-literal rule are labelled with the literal that is used for simplification.

```
procedure Signed-DPL
Input: a signed CNF formula Γ and a truth value set N = {i_1, ... , i_n}
Output: "satisfiable" or "unsatisfiable"

begin
 /* signed one-literal rule */
 while Γ contains a unit clause {S : p} do
   Γ := simplify(Γ, S : p)
 od;
 if Γ = ∅ then return "satisfiable" fi;
 if □ ∈ Γ then return "unsatisfiable" fi;

 /* signed branching rule */
 let p be a propositional variable occurring in Γ;
 for j = 1 to n do
   if Signed-DPL(Γ ∪ {i_j : p}) = "satisfiable" then
     return "satisfiable" fi
 od;
 return "unsatisfiable"
end
```

Figure 3.2 The Signed Davis-Putnam-Loveland procedure (Signed-DPL).

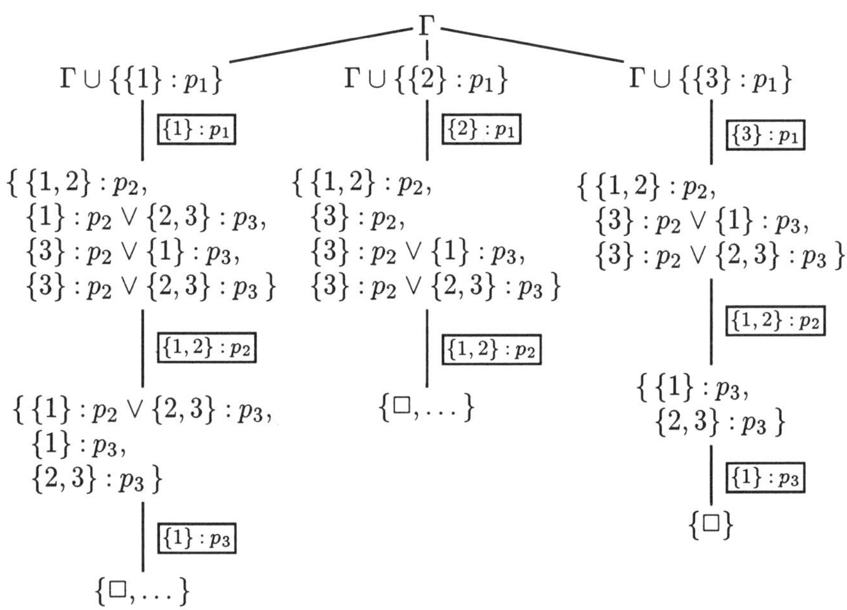

Figure 3.3 A proof tree created by Signed-DPL.

4.2 AN IMPROVED BRANCHING RULE FOR SIGNED-DPL

An application of the branching rule of Signed-DPL from the previous section always creates $|N|$ new sub-branches. In this section, we present an improved branching rule for Signed-DPL that in many cases creates less sub-branches [25].

Definition 15 *Let Γ be a signed CNF formula, and let p be a propositional variable occurring in Γ. Then, the set $N_\Gamma^p \subseteq N$ consists of those truth values that appear in Γ in literals of the form $S : p$.*

Truth values $i, j \in N_\Gamma^p$ are equivalent, denoted by $i \approx_p j$, if, for all literals of the form $S : p$ in Γ, $i \in S$ iff $j \in S$.

The partial order $\preceq_p$ on equivalence classes of N_Γ^p w.r.t. $\approx_p$ is defined by: $\bar{i} \preceq_p \bar{j}$ if, for all literals of the form $S : p$ in Γ, $\bar{i} \subseteq S$ implies $\bar{j} \subseteq S$.

The elements of maximal classes w.r.t. $\preceq_p$ are called maximal truth values *of p in Γ. A set $M = \{i_1, \ldots, i_m\} \subseteq N$ is called a* maximal truth value set *of p in Γ if it contains one element of each of the classes $\{\overline{i_1}, \ldots, \overline{i_m}\}$ that are maximal w.r.t. $\preceq_p$.*

It can happen that some truth values $i, j \in N$ occur in a formula Γ exactly in the same signs of literals of the form $S{:}p$, i.e., $i \approx_p j$ where $\approx_p$ is the equivalence relation from Definition 15. In that case, it suffices that the branching rule considers only one of the truth values i and j.

In addition, if the truth values of the equivalence class $\bar{j}$ occur (among other signs) in all signs in which the truth values of the equivalence class $\bar{i}$ occur, i.e., if $\bar{i} \preceq_p \bar{j}$, then the truth values in $\bar{i}$ can be ignored by the branching rule of Signed-DPL. This simplification is justified because, if an interpretation I satisfies Γ and $I(p) \in \bar{i}$, then Γ is as well satisfied by every interpretation I' that assigns a truth value from $\bar{j}$ to p and is identical to I for the other propositional variables. Consequently, it is sufficient if the branching rule of Signed-DPL considers only the elements of a maximal truth value set.

Example 16 *Let the set of truth values be $N = \{1, 2, \ldots, 7\}$, and let*

$$\begin{array}{lll} \{2,7\} : p \vee D_1 & \{1,3,6\} : p \vee D_2 & \{1,3,4\} : p \vee D_3 \\ \{1,2,3,7\} : p \vee D_4 & \{2,5,7\} : p \vee D_5 & \end{array}$$

be the clauses in the signed CNF formula Γ in which literals of the form $S : p$ occur. Then, the equivalence classes of N_Γ^p w.r.t. $\approx_p$ are $\{2,7\}$, $\{1,3\}$, $\{4\}$, $\{5\}$, and $\{6\}$. The maximal elements w.r.t. the order relation $\preceq_p$ are $\{2,7\}$ and $\{1,3\}$, because $\{4\} \preceq_p \{1,3\}$, $\{5\} \preceq_p \{2,7\}$, and $\{6\} \preceq_p \{1,3\}$. A maximal truth value set of p in Γ is $\{2,3\}$.

Proposition 17 *Let Γ be a signed CNF formula, let p be a propositional variable occurring in Γ, and let $\{i_1, \ldots, i_m\}$ be a maximal truth value set of p*

in Γ. *Then,* Γ *is satisfiable iff there is a* $k \in \{1, \ldots, m\}$ *such that* $\Gamma \cup \{i_k : p\}$ *is satisfiable.*

The branching rule of Proposition 17 can reduce the size of a Signed-DPL proof tree considerably. Consider, for example, the formula Γ from Example 14; $\{2, 3\}$ is a maximal truth value set of p_1 in Γ. Therefore, the leftmost branch of the proof tree for Γ shown in Figure 3.3 is actually redundant and is not constructed if the improved branching rule from Proposition 17 is used. When the branching rule is applied to the formula from Example 16, the number of new sub-branches is reduced from seven to two.

4.3 THE REGULAR DPL PROCEDURE

Signed-DPL is, of course, suitable for the (sub-)class of *regular* CNF formulas as well. However, there are some refinements and special techniques that can be applied. In this section we describe the regular Davis-Putnam-Loveland procedure (Regular-DPL) defined by Hähnle [18] for regular CNF formulas over a totally ordered truth value set. Regular-DPL was the first many-valued DPL-style procedure published and inspired some of the work reported in this survey.

The regular one-literal rule consists of only the first two parts of the signed one-literal rule from Section 4.1, which preserve regularity of the formula; the third part is not needed. The regular branching rule reduces the problem of checking whether a regular CNF formula Γ is satisfiable to the problem of checking whether one of the formulas $\Gamma \cup \{S : p\}$ and $\Gamma \cup \{\overline{S} : p\}$ is satisfiable where $S : p$ is a regular literal occurring in Γ. The branching factor is at most two when the *regular* branching rule is applied, but not all literals containing p are necessarily removed. In contrast to that, the *signed* branching rule removes all occurrences of p, but the branching factor can be as large as the cardinality of the truth value set.

Example 18 *Figure 3.4 shows the proof tree constructed by Regular-DPL for the regular CNF formula* Γ *from Example 11. As in Figure 3.3, edges corresponding to an application of the one-literal rule are labelled with the literal that is used for simplification.*

The performance of Regular-DPL depends (I) on the data structures used to represent formulas and (II) on the heuristic for selecting the next literal to which the branching rule is applied. Manyà et al. [26] describe an implementation of Regular-DPL that uses suitable data structures and incorporates the regular two-sided Jeroslow-Wang heuristic defined by Hähnle [18]. It is the only DPL-style procedure implemented so far in the framework of signed CNF formulas.

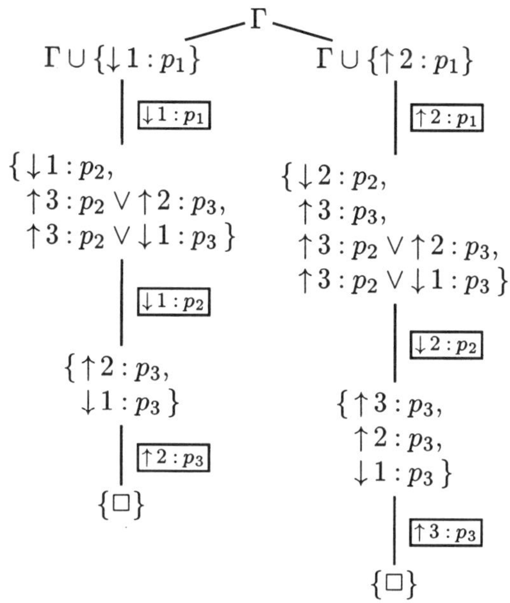

Figure 3.4 A proof tree created by Regular-DPL.

5 LOCAL SEARCH ALGORITHMS

Local search algorithms (LSAs) outperform deductive decision procedures for checking satisfiability of CNF formulas on some problem classes. In particular, this holds for satisfiable hard random 3-SAT instances, which the fastest implementations of DPL cannot solve within a reasonable time limit [30]. In this section, we describe the first LSA that deals with signed CNF formulas and we report some experimental results.

Regular-GSAT [4], an extension of GSAT [31] whose pseudo-code is shown in Figure 3.5, tries to find a satisfying interpretation for a regular CNF formula Γ (with a total order on truth values) performing a greedy local search through the space of interpretations. It starts with a randomly generated interpretation I. If I does not satisfy Γ, then it creates a set $\mathcal{S}$, formed by those variable-value pairs (p, k) that give rise to a maximal decrease (possibly zero or negative) in the total number of unsatisfied clauses of Γ when the truth value of I at p is changed to k. Next, a propositional variable p' appearing in $\mathcal{S}$ is randomly chosen. Then a truth value k' from $\{k \mid (p', k) \in \mathcal{S}\}$ is randomly chosen. Finally, I is updated to k' at p'. Such changes are repeated until either a satisfying interpretation is found or a pre-set maximum number of changes (MaxChanges) is reached. The whole process is repeated up to MaxTries times, if no satisfying interpretation is found before.

```
procedure Regular-GSAT
Input: a regular CNF formula Γ, MaxChanges, and MaxTries
Output: either a model of Γ, or "no satisfying interpretation found"

begin
  for i = 1 to MaxTries do
    I := a randomly generated interpretation for Γ;
    for j = 1 to MaxChanges do
      if I satisfies Γ then return I fi;
      S := {(p, k) | decrease in number of unsatisfied clauses of Γ
                     maximal, when I changed to k at p};
      select randomly p' ∈ {p | (p, k) ∈ S};
      select randomly k' ∈ {k | (p', k) ∈ S};
      I(p') := k'
    od
  od;
  return "no satisfying interpretation found"
end
```

Figure 3.5 The procedure Regular-GSAT.

Table 3.1 Comparison of running times for Regular-GSAT and Regular-DPL.

		Regular-GSAT				*Regular-DPL*	
V	*C*	*MaxTries*	*MaxChanges*	*time (secs.)*		*time (secs.)*	
				μ	σ	μ	σ
80	487	100	1000	0.68	0.63	1.10	0.85
120	720	200	2800	6.35	5.33	19.45	16.47
160	972	260	6200	25.98	21.48	290.16	325.40
200	1230	400	12000	99.97	88.49	3242.58	3000.16

Table 3.1 summarises an experiment performed in order to compare the performance of Regular-DPL and Regular-GSAT on satisfiable random regular (signed) 3-SAT instances of the hard region of the phase transition (see Section 6) with a different number of propositional variables and $|N| = 3$ [4]. Both procedures were applied to 100 satisfiable instances with 80, 120, 160 and 200 propositional variables. In order to obtain more accurate results, each instance was run 50 times with Regular-GSAT. The first column contains the number V of propositional variables and the second the number C of clauses of the instances tested. The remaining columns display the settings of MaxTries and MaxChanges employed, the average μ and the standard deviation σ of the time needed to solve the sets of instances considered. The run time of each instance solved with Regular-DPL corresponds to the time needed to solve that instance, whereas the run time of each instance solved with Regular-GSAT is the average run time over the 50 runs on that instance.

It is clear that local search algorithms for solving regular SAT problems scale better than Regular-DPL when the number of variables in the problem instances increases. This result suggests that local search algorithms, just as their classical counterparts, are good candidates for solving difficult satisfiable problems. First experiments with scheduling problems support this conjecture [6].

Local search algorithms are incomplete and cannot prove unsatisfiability. Recently, some impressive results were obtained by combining deterministic and complete satisfiability procedures (such as DPL) with randomisation to cope with the so-called "heavy-tailed" distribution phenomenon [14]. We expect this to generalise to signed logic as well.

6 PHASE TRANSITIONS

The phase transition phenomenon for the 3-SAT problem consists of two observations: (I) There is a sharp increase (phase transition) of the percentage of unsatisfiable random 3-SAT instances around a certain point when the ratio $\frac{C}{V}$ between the number C of clauses and the number V of variables is varied (at lower ratios, most instances are under-constrained and thus satisfiable, at higher ratios, most instances are over-constrained and thus unsatisfiable). (II) There is an easy-hard-easy pattern in the computational difficulty of solving problem instances as $\frac{C}{V}$ is varied; the hard instances tend to be found near the crossover point.

Phase transitions occur, among other NP-hard problems, in classical [27] and random regular 3-SAT problems [26]. In the present context, our interest in them is twofold: (i) The hard instances described below provide a first testbed to evaluate and compare satisfiability solvers for signed CNF formulas, and (ii) with an eye on knowledge representation with signed CNF formulas,

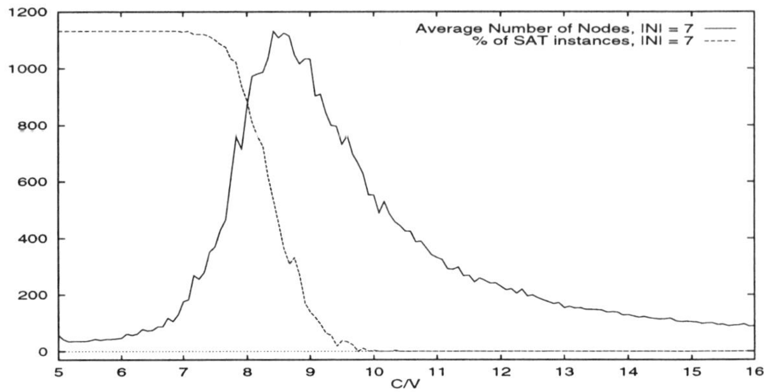

Figure 3.6 Phase transition in the random regular 3-SAT problem.

it would be valuable to know what impact the cardinality of N has on the crossover point.

Before we describe the phase transition phenomena in the signed case, we explain how random regular 3-SAT instances are generated. Given a fixed number C of clauses, a number V of propositional variables, and a totally ordered truth value set N, for one problem instance C non-tautological regular clauses are generated. Each regular clause is produced by uniformly choosing three literals with different propositional variables from the set of possible regular literals.

Manyà et al. [26] report on experiments performed on random regular 3-SAT instances with Regular-DPL (see Section 4.3). They observed both aspects (I) and (II) of phase transition. Figure 3.6 visualises this for the random regular 3-SAT problem, where $|N| = 7$ and $V = 60$. Along the vertical axis is the average number of nodes in the proof tree needed to solve a problem instance with Regular-DPL. Along the horizontal axis is the ratio $\frac{C}{V}$ in the test problems. One observes clearly the easy-hard-easy pattern as $\frac{C}{V}$ is varied. The dashed line indicates the percentage of instances found to be satisfiable (the 100 % mark is scaled to the maximum of the curve indicating hardness of problems).

Recent experiments indicate that the location of the crossover point increases logarithmically as a function of the cardinality of the truth value set [5]. Table 3.2 shows the location of the crossover point for different cardinalities of N. The following equation was derived from the experimental crossover points by using the Levenberg-Marquardt method for obtaining a non-linear regression model [5]:

$$L(|N|) = 6.30544 \ln^{0.391434} (|N|)$$

Table 3.2 Location of the crossover point for different cardinalities of N.

$\|N\|$	*crossover point*	$\|N\|$	*crossover point*	$\|N\|$	*crossover point*
2	4.25	10	9.08	30	10.16
3	6.08	15	9.50	40	10.33
4	7.08	20	9.75	50	10.41
5	7.75	25	10.00	60	10.50

7 COMPLEXITY OF THE SIGNED SAT PROBLEM

7.1 OVERVIEW

It is well-known that the classical SAT problem is NP-complete [8]. It is, however, polynomially solvable under certain restrictions. For example, there are linear-time algorithms for solving the classical SAT problem in case all clauses of the formula have at most one positive literal (Horn SAT) [10] and in case all clauses of the formula have at most two literals (2-SAT) [13].

Similar to the classical case, the *signed* SAT problem is NP-complete, but some of its sub-classes are polynomially solvable. In recent years, complexity results for the signed 2-SAT and signed Horn SAT problems have been established. These problems have the truth value set N (resp. $(N, \geq)$) as a second input parameter (besides the formula Γ to be tested for satisfiability). Thus, *signed SAT* is the problem of deciding for an arbitrary formula Γ over an arbitrary truth value set N, whether there is an interpretation over N satisfying Γ. One also considers decision problems where N is not an input parameter but fixed, which is denoted by attaching the fixed truth value set N as an index to the name of the decision problem. For example, given a fixed truth value set N, *signed* SAT_N is the problem of deciding for an arbitrary formula Γ over N whether there is an interpretation over N satisfying Γ.

NP-containment of the most general problem, signed SAT, is straightforward to show. The classical SAT problem is trivially reducible to signed $\text{SAT}_{\{0,1\}}$; therefore, the latter and signed SAT are both NP-complete. Further results are summarised in Table 3.3 and are discussed in Sections 7.2 and 7.3 below.

7.2 THE SIGNED 2-SAT PROBLEM

The signed 2-SAT_N problem for $|N| \geq 3$ and, therefore, the signed 2-SAT problem was proven to be NP-complete by Manyà [25] (as compared to the classical 2-SAT problem that can be solved in linear time); an alternative proof of NP-hardness of signed 2-SAT was later given by Beckert et al. [3]. Manyà [25] reduces the 3-colourability problem of graphs to signed 2-SAT_N to

Table 3.3 Known complexity results for signed SAT problems.

	SAT	*2-SAT*	*Horn SAT*
classical	NP-compl.	linear [13]	linear [10]
mono-signed	NP-compl.	linear [24]	—
regular, N totally ord.	NP-compl.	polynomial [25]	$\|\Gamma\| \log \|\Gamma\|$ [18, 25]
regular, N a distr. lattice, signs of form $\uparrow i$ *and* $\overline{\uparrow i}$	NP-compl.	NP-compl. [2]	$\|\Gamma\|\|N\|^2$ [33]
regular, N a lattice, signs of form $\uparrow i$ *and* $\overline{\uparrow i}$	NP-compl.	NP-compl.	polynomial [3]
regular, N a lattice, signs of form $\uparrow i$ *and* $\downarrow i$	NP-compl.	polynomial [2]	—
regular (arbitrary)	NP-compl.	NP-compl.	—
signed (arbitrary)	NP-compl.	NP-compl. [25, 3]	—

show its NP-hardness, whereas the NP-hard problem Beckert et al. [3] reduce to signed 2-SAT is classical SAT.

The *regular* 2-SAT problem is NP-complete as well; this can be shown by reducing the (general) signed 2-SAT problem to regular 2-SAT [2]. Under certain restrictions, however, satisfiability of regular 2-CNF formulas can be checked in polynomial time. This problem was first considered by Manyà [25] with the additional assumption that N is a totally ordered set. In that case, a refinement of Regular-DPL yields a quadratic-time procedure. A generalisation of this result was proved by Beckert et al. [2]: If N is a lattice and all occurring signs are of the form $\uparrow i$ or the form $\downarrow i$, then regular 2-SAT is polynomially solvable.

A further special case of the regular 2-SAT problem that can be solved in polynomial time is the monosigned 2-SAT problem. By examining the rules of monosigned binary resolution one can check that the number of possible resolvents for a given monosigned 2-CNF formula is polynomial in the number of distinct literals it contains. A quadratic-time procedure for solving monosigned 2-SAT was described by Manyà [25]. He later refined the result by showing that monosigned 2-SAT is solvable in time linear in the length of the formula using a reduction to classical 2-SAT [24].

7.3 THE REGULAR HORN SAT PROBLEM

A Horn fragment is naturally defined if (and only if) the truth value set N is totally ordered or at least a finite lattice.

If N is totally ordered, the problem of deciding whether a regular Horn formula Γ is satisfiable can be solved in time linear in $n = |\Gamma|$ in case $|N|$ is fixed, and in time linear in $n \log n$ otherwise [18]. Algorithms with the same complexity were described in [25]. An algorithm for a particular subclass of regular Horn formulas appeared before [11]; related results can be found in a paper by Escalada-Imaz and Manyà [12].

If N is a finite lattice, regular Horn SAT is solvable in time linear in the length of the formula and polynomial in the cardinality of N via a reduction to the classical Horn SAT problem [3]. For distributive lattices, the more precise bound $n \cdot |N|^2$ was found independently [33], which contains also some results on decidable first-order fragments of regular CNF formulas.

A closer inspection of the proofs in the cited papers yields immediately that all defined regular Horn SAT_N problems have linear complexity.

Acknowledgments

This work was partially supported by EC COST Action 15 *Many-valued Logics for Computer Science Applications.* The third author was partially supported by project TIC96-1038-C04-03 funded by the CICYT and his contribution was carried out during a visit at the University of Karlsruhe with a postdoctoral fellowship of the "Comissionat per a Universitats i Recerca" (1997BEAI400138).

Bernhard Beckert
University of Karlsruhe
Institute for Logic, Complexity and Deduction Systems
D-76128 Karlsruhe, Germany
beckert@ira.uka.de

Reiner Hähnle
University of Karlsruhe
Institute for Logic, Complexity and Deduction Systems
D-76128 Karlsruhe, Germany
reiner@ira.uka.de

Felip Manyà
Universitat de Lleida
Jaume II, 69
E-25001 Lleida, Spain
felip@eup.udl.es

References

[1] Matthias Baaz and Christian G. Fermüller. Resolution-based theorem proving for many-valued logics. *Journal of Symbolic Computation*, 19:353–391, 1995.

[2] Bernhard Beckert, Reiner Hähnle, and Felip Manyà. On the regular 2-SAT problem. University of Karlsruhe, Dept. of Computer Science. Available at `ftp://sonja.ira.uka.de/pub/beckert/Regular_2SAT.ps.gz`, 1999.

[3] Bernhard Beckert, Reiner Hähnle, and Felip Manyà. Transformations between signed and classical clause logic. In *Proceedings, 29th International Symposium on Multiple-Valued Logics (ISMVL), Freiburg, Germany*, pages 248–255. IEEE Press, Los Alamitos, 1999.

[4] Ramon Béjar and Felip Manyà. A comparison of systematic and local search algorithms for regular CNF formulas. In *Proceedings, 5th European Conference on Symbolic and Quantitative Approaches to Reasoning with Uncertainty (ECSQARU), London, England*, LNCS 1638, pages 22–31. Springer, 1999.

[5] Ramon Béjar and Felip Manyà. Phase transitions in the regular random 3-SAT problem. In *Proceedings, International Symposium on Methodologies for Intelligent Systems (ISMIS), Warsaw, Poland*, LNCS 1609, pages 292–300. Springer, 1999.

[6] Ramon Béjar and Felip Manyà. Solving combinatorial problems with regular local search algorithms. In *Proceedings, 6th International Conference on Logic for Programming and Automated Reasoning (LPAR), Tbilisi, Georgia*, LNCS 1705, pages 33–43. Springer, 1999.

[7] Thierry Castell and Hélène Fargier. Between SAT and CSP: Propositional satisfaction problems and clausal CSPs. In *Proceedings, European Conference on Artificial Intelligence (ECAI), Brighton, UK*, pages 214–218. John Wiley & Sons, 1998.

[8] Stephen Cook. The complexity of theorem-proving procedures. In *Proceedings, 3rd Annual ACM Symposium on Theory of Computing (STOC), Shaker Heights, USA*, pages 151–158. ACM Press, 1971.

[9] Martin Davis, George Logemann, and Donald Loveland. A machine program for theorem-proving. *Communications of the ACM*, 5(7):394–397, 1962.

[10] William Dowling and Jean Gallier. Linear-time algorithms for testing the satisfiability of propositional Horn formulæ. *Journal of Logic Programming*, 1(3):267–284, 1984.

[11] Gonzalo Escalada-Imaz and Felip Manyà. The satisfiability problem for multiple-valued Horn formulæ. In *Proceedings, International Symposium*

on Multiple-Valued Logics (ISMVL), Boston, USA, pages 250–256. IEEE Press, 1994.

[12] Gonzalo Escalada-Imaz and Felip Manyà. Efficient interpretation of propositional multiple-valued logic programs. In *Advances in Intelligent Computing*, LNCS 945, pages 428–439. Springer, 1995.

[13] S. Even, A. Itai, and A. Shamir. On the complexity of timetable and multicommodity flow problems. *SIAM Journal of Computing*, 5(4):691–703, 1976.

[14] Carla P. Gomes, Bart Selman, and Henry Kautz. Boosting combinatorial search through randomization. In *Proceedings, 15th National Conference on Artificial Intelligence (AAAI), Madison/WI, USA*, pages 431–437. AAAI Press, Menlo Park, 1998.

[15] Reiner Hähnle. Short CNF in finitely-valued logics. In *Proceedings, International Symposium on Methodologies for Intelligent Systems (ISMIS), Trondheim, Norway*, LNCS 689, pages 49–58. Springer, 1993.

[16] Reiner Hähnle. *Automated Deduction in Multiple-Valued Logics*, volume 10 of *International Series of Monographs in Computer Science*. Oxford University Press, 1994.

[17] Reiner Hähnle. Short conjunctive normal forms in finitely-valued logics. *Journal of Logic and Computation*, 4(6):905–927, 1994.

[18] Reiner Hähnle. Exploiting data dependencies in many-valued logics. *Journal of Applied Non-Classical Logics*, 6:49–69, 1996.

[19] Reiner Hähnle and Gonzalo Escalada-Imaz. Deduction in many-valued logics: A survey. *Mathware and Soft Computing*, 4(2):69–97, 1997.

[20] Reiner Hähnle, Ryuzo Hasegawa, and Yasuyuki Shirai. Model generation theorem proving with interval constraints. In F. Benhamou, W. Older, M. van Emden, and P. van Hentenryck, editors, *Proceedings, ILPS Post-Conference Workshop on Interval Constraints, Portland, USA*, 1995.

[21] Michael Kifer and V. S. Subrahmanian. Theory of generalized annotated logic programming and its applications. *Journal of Logic Programming*, 12:335–367, 1992.

[22] Sonia M. Leach and James J. Lu. Query processing in annotationed logic programming: Theory and implementation. *Journal of Intelligent Information Systems*, 6(1):33–58, 1996.

[23] James J. Lu. Logic programming with signs and annotations. *Journal of Logic and Computation*, 6(6):755–778, 1996.

[24] Felip Manyà. The 2-SAT problem in signed CNF formulas. *Multiple-Valued Logic. An International Journal*, 1999. To appear.

[25] Felip Manyà. *Proof Procedures for Multiple-Valued Propositional Logics.* Number 9 in Monografies de l'Institut d'Investigació en Intelligència Artificial. IIIA-CSIC, Bellaterra (Barcelona), 1999.

[26] Felip Manyà, Ramon Béjar, and Gonzalo Escalada-Imaz. The satisfiability problem in regular CNF-formulas. *Soft Computing: A Fusion of Foundations, Methodologies and Applications*, 2(3):116–123, 1998.

[27] David Mitchell, Bart Selman, and Hector Levesque. Hard and easy distributions of SAT problems. In *Proceedings, 10th National Conference on Artificial Intelligence (AAAI), San Jose, USA*, pages 459–465. MIT Press, 1992.

[28] Neil V. Murray and Erik Rosenthal. Signed formulas: A liftable meta logic for multiple-valued logics. In *Proceedings, International Symposium on Methodologies for Intelligent Systems (ISMIS), Trondheim, Norway*, LNCS 689, pages 275–284. Springer, 1993.

[29] Neil V. Murray and Erik Rosenthal. Adapting classical inference techniques to multiple-valued logics using signed formulas. *Fundamenta Informaticae*, 21(3):237–253, 1994.

[30] Bart Selman, Henry A. Kautz, and Bram Cohen. Noise strategies for local search. In *Proceedings, 12th National Conference on Artificial Intelligence (AAAI), Seattle, USA*, pages 337–343. AAAI Press, Menlo Park, 1994.

[31] Bart Selman, Hector Levesque, and David Mitchell. A new method for solving hard satisfiability problems. In *Proceedings, 10th National Conference on Artificial Intelligence (AAAI), San Jose, USA*, pages 440–446. MIT Press, 1992.

[32] Viorica Sofronie-Stokkermans. *Fibered Structures and Applications to Automated Theorem Proving in Certain Classes of Finitely-Valued Logics and to Modeling Interacting Systems.* PhD thesis, Johannes Kepler Universität Linz, Forschungsinstitut für symbolisches Rechnen, 1997.

[33] Viorica Sofronie-Stokkermans. On translation of finitely-valued logics to classical first-order logic. In *Proceedings, 13th European Conference on Artificial Intelligence (ECAI), Brighton, UK*, pages 410–411. John Wiley & Sons, 1998.

DISCIPLINE AS LOGIC: TREATING LABELS AS FIRST CLASS CITIZENS

Patrick Blackburn

Abstract This paper discusses modal labelled deduction from the perspective offered by *hybrid languages*. In essence, hybrid languages are modal languages in which the apparatus of labelled deduction is fully integrated into the object language. Hybrid languages enable us to define proof systems which in an obvious sense internalize labelled deduction, and when this is done, we shall find that *labelling discipline* emerges as *logic*. I show that this logical (or declarative) perspective on labelling discipline can be "lifted" to a full first-order discipline over labels, and conclude with some general remarks on hybridization.

1 INTRODUCTION

Labelled Deduction is a framework introduced by Dov Gabbay for handling the proof theories of a wide range of logics using unified mechanisms. One of its most striking successes has been its treatment of modal deduction, and it is not difficult to see why. For a start, the basic notation offered by labelled deduction brings modal logic to mind. Modern modal logic is based upon the Kripke satisfaction relation $\mathcal{M}, w \Vdash \varphi$, which says that the formula φ is satisfied at state w in model $\mathcal{M}$ (a model being a triple (W, R, V) where W is a set of states, R a transition relation between states, and V a valuation assigning subsets of states to atoms). Labelled deduction is built around the notation $l : \varphi$. Here the : constructor associates the label l with the formula φ, and this has a natural modal interpretation: regard labels as names for states and read : as Kripke satisfaction.

But labelled deduction offers more than notation: Gabbay has always emphasized the role played by *labelling discipline*. Labelled deduction is not simply about labelling formulas, rather it is about providing an explicit mechanism that controls how labels are manipulated (usually via what is called a *labelling algebra*) and this idea has important consequences for modal logic. The point is this. There is not one modal logic, but many (indeed non-denumerably many). These logics typically arise by varying properties of the transition relation between states; and while modal logicians have provided a reasonably detailed map of these logics, relatively little effort has been devoted to providing us-

D. Basin et al. (eds.), Labelled Deduction, 81–105.
© 2000 *Kluwer Academic Publishers. Printed in the Netherlands.*

able proof systems (traditionally, modal logicians have been content to supply Hilbert-style axiomatizations). And it is precisely here that labelled deduction has proved so modally natural: by varying the labelling algebra (often in quite simple ways) it is possible to provide perspicuous proof theories for a number of modal logics.

The purpose of the present paper is to discuss modal labelled deduction from a perspective which will probably be new to most readers, namely the perspective offered by *hybrid languages*. In essence, hybrid languages are modal languages in which the apparatus of labelled deduction is fully integrated into the object language. Now, in my view hybrid languages are one of the most interesting of the current generation of extended modal languages and are well worth getting to know quite independently of their link with labelled deduction (I'll mention several reasons why in the conclusion). But for most of this paper I want to focus on the aspect of hybrid languages most likely to be of interest to readers of this volume: hybrid proof theory. To this end, I am going to introduce two important types of hybrid language, sketch some of their key proof-theoretic properties, and indicate some of the territory they open up.

As I have already said, hybrid languages are essentially modal languages in which the apparatus of labelled deduction is fully integrated into the object language, and before getting into the technicalities it is worth being absolutely explicit about what this involves. Hybrid languages are built around a deceptively simple idea:

Treat labels as first-class citizens.

Sounds good — but what does it mean? Well, propositional modal logic is all about building complex formulas out of atomic formulas with the aid of $\neg$, $\wedge$, $\vee$, $\rightarrow$, $\Diamond$, and so on. That is, propositional modal languages don't contain anything except formulas. So if we want to *integrate* labels into the object language (as opposed to bolting them on artificially using ad-hoc mechanisms) we really have no choice: we must find a way of viewing labels as *formulas*.

This is surprisingly easy to do. Take an ordinary language of propositional modal logic (built over some collection of propositional variables p, q, r, and so on) and add a second type of atomic formula. These new atoms are called *nominals*, and are typically written i, j, k and l; nominals will be our labels. Both types of atom can be freely combined to form more complex formulas in the usual way; for example,

$$\Diamond(i \wedge p) \wedge \Diamond(i \wedge q) \rightarrow \Diamond(p \wedge q)$$

is a well formed formula. And now for the key semantic idea: insist that *each nominal be true at exactly one state in any model.* Thus a nominal labels a state by being true there and nowhere else. We have internalized the idea of labelling in the object language, and have done so without doing violence to the syntax or semantics of propositional modal logic.

This simple idea immediately gives rise to richer logics. Note, for example, that the previous formula is *valid*: if the antecedent is satisfied at a state w in a model $\mathcal{M}$, then the (unique) state labelled i must accessible from w, and hence there must be a state where both p and q are true. In effect we are seeing labelling discipline emerge as *logic*. Because we are treating labels as fully fledged *formulas* — that is, entities free to play a first-class role in the logical economy — crucial facts about labelling (such as the simple example given above) are reflected in the object language as validities.

Once we have nominals at our disposal, it becomes natural to form formulas of the form $i : \varphi$. Unsurprisingly, such a formula will be true at any state in a model if and only if φ is true at the (unique) state that the nominal i labels (we call these new i : prefixes *satisfaction operators*). But note well: unlike in Gabbay-style labelled deduction, expressions of the forms $i:\varphi$ are *not* metalinguistic constructions. They belong to the (hybrid) object language and they are perfectly ordinary formulas: it is completely legitimate to combine them with $\neg$, $\wedge$, $\vee$, $\rightarrow$, $\Diamond$, and so on. For example, an expression such as

$$i:\Diamond j \wedge j:\varphi \rightarrow i:\Diamond\varphi$$

is a well formed formula (and indeed, a validity). As I said above, labelling is *fully* integrated into the object language.

Now, the basic hybrid language we shall work with for the first half of the paper is simply the language I have just outlined: an ordinary language of propositional modal logic enriched with both nominals and satisfaction operators. But once we have hit upon the idea of treating labels as formulas, the way is open for further experiment. For example, why not allow ourselves to bind our labels using quantifiers? There is no reason not to: indeed, it turns out to be a natural way of obtaining a full first-order labelling discipline, as we shall discuss later.

Here's the plan of the paper. In Section 2, I shall define the syntax and semantics of the basic hybrid language, note some of its properties, and then define a simple tableau system which clearly mirrors the key ideas of Gabbay-style labelled deduction. Then, in Section 3, I shall consider how to handle stronger logics and the power of viewing labelling in logical terms will become apparent: complete tableau systems for a wide range of extensions can be obtained *automatically*. In Section 4, I shall introduce a more powerful hybrid language (one in which quantification over labels is possible) and finally, in Section 5, I shall discuss hybrid languages in more general terms.

Throughout the paper pointers to the (small but rapidly growing) literature on hybrid languages will be given, but several should be given right away. First, many of the proof theoretic ideas discussed below can be traced back to Seligman's work on deduction methods for Situation Theory (see [31, 32, 33]). Also relevant is Konikowska's work on modal logics of relative similarity (see, for example, [22]). More recently, work on practical deduction methods for hybrid languages has blossomed: see Blackburn [8] and Tzakova [35] for general approaches, and Demri [16] and Demri and Goré [17] for the case of nominal tense logic. The first half of the present paper is essentially an example-driven introduction to the approach taken in [8], and in order to keep it to a reasonable length, many important issues are simply not discussed. In particular, there is no discussion of sequent methods, nor of the link with Gabbay's work on "Rules for the Undefinable" (see [18]), nor any comparison with the methods of Seligman and Tzakova; but these topics are discussed in [8], and the reader who wants to know more should consult that. However for readers interested in learning how to *use* hybrid languages, the present paper is probably the better starting point; it is less abstract, and covers several topics (in particular, full first-order labelling discipline) not discussed in the earlier paper.

2 INTERNALIZING LABELLED DEDUCTION

In this section I define the *basic hybrid language* and show that it supports a tableau system in which the leading ideas of Gabbay-style labelled deduction are reflected in the object language.

Given a set of ordinary propositional symbols PROP $= \{p, q, r, \dots\}$, let NOM $= \{i, j, k, l, \dots\}$ be a nonempty set disjoint from PROP. The elements of NOM are called *nominals*; they are a second sort of atomic symbol which will be used to name states. The *basic hybrid language* (over PROP and NOM) is defined as follows:

$$\text{WFF} := i \mid p \mid \neg\varphi \mid \varphi \wedge \psi \mid \varphi \vee \psi \mid \varphi \rightarrow \psi \mid \Diamond\varphi \mid \Box\varphi \mid i{:}\varphi$$

Some remarks. First, as promised in the introduction, although they will be used to label states, viewed syntactically nominals are simply *formulas*. Second, note that if we ignore formulas of the form $i : \varphi$, we are simply working with an ordinary propositional modal language. Third, although it is possible to think of $:$ as a binary operator which combines a nominal with an arbitrary formula, it is more useful to think of the symbol sequence $i{:}$ (for any choice of nominal i) as a *unary* operator; we'll see why shortly. (Incidentally, in Blackburn and Seligman [10], Blackburn and Tzakova [14], and Areces, Blackburn and Marx [2], $i : \varphi$ is written $@_i\varphi$, a suggestive notation which emphasizes the unary perspective.) In what follows, formulas of the form $i{:}\varphi$ or $\neg i{:}\varphi$ will be called *satisfaction statements*. Note that $\neg i{:}\varphi$ is $\neg(i{:}\varphi)$; it *cannot* be $(\neg i){:}\varphi$, for this expression is not a formula.

The basic hybrid language is interpreted on *hybrid models*. A hybrid model $\mathcal{M}$ is a triple (W, R, V), where W is a non-empty set of states, R is a binary relation on W, and V is a *hybrid valuation* assigning subsets of W to atoms.

Following modal tradition, the pair (W, R) is called the *frame* underlying $\mathcal{M}$. But for present purposes the most important component is the hybrid valuation V, for it is here that the interpretation of nominals as labels is encoded. A hybrid valuation V on $\mathcal{F}$ is a function with domain PROP∪NOM and range *Pow*(W) such that for all $i \in$ NOM, $V(i)$ is a *singleton* subset of W. That is, just as in ordinary modal logic, we place no restrictions on how the elements of PROP are interpreted, but we insist that a valuation makes each nominal true at a *unique* state; the nominal labels this state by being true there and nowhere else. We call the unique state that belongs to $V(i)$ the *denotation* of i under V. We interpret the basic hybrid language as follows:

$\mathcal{M}, w \Vdash a$	iff	$w \in V(a)$, for all $a \in$ ATOM
$\mathcal{M}, w \Vdash \neg\varphi$	iff	$\mathcal{M}, w \nVdash \varphi$
$\mathcal{M}, w \Vdash \varphi \wedge \psi$	iff	$\mathcal{M}, w \Vdash \varphi$ and $\mathcal{M}, w \Vdash \psi$
$\mathcal{M}, w \Vdash \varphi \vee \psi$	iff	$\mathcal{M}, w \Vdash \varphi$ or $\mathcal{M}, w \Vdash \psi$
$\mathcal{M}, w \Vdash \varphi \rightarrow \psi$	iff	$\mathcal{M}, w \nVdash \varphi$ or $\mathcal{M}, w \Vdash \psi$
$\mathcal{M}, w \Vdash \Diamond\varphi$	iff	$\exists w'(wRw' \;\&\; \mathcal{M}, w' \Vdash \varphi)$
$\mathcal{M}, w \Vdash \Box\varphi$	iff	$\forall w'(wRw' \;\Rightarrow\; \mathcal{M}, w' \Vdash \varphi)$
$\mathcal{M}, w \Vdash i{:}\varphi$	iff	$\mathcal{M}, v \Vdash \varphi$, where v is the denotation of i under V.

Note that the first clause covers both propositional variables and nominals, but because valuations assign *singletons* to nominals, each nominal is satisfied at exactly one state in any model (namely, it's denotation). That is, nominals are a mechanism for referring to states in a propositional modal language: formulas are playing the role traditionally reserved for terms. Note that the satisfaction clause for formulas of the form $i : \varphi$ says: to evaluate $i : \varphi$, jump to the unique state named by i and evaluate φ there. It follows that a wff of the form $i{:}j$ expresses the identity of the states named by i and j. Similarly, a formula of the form $i : \Diamond j$ says that the state named by i R-precedes the state named by j.

If $\mathcal{M}, w \Vdash \varphi$ we say that φ is *satisfied* in $\mathcal{M}$ at w. We say that φ is satisfiable if and only if there a hybrid model $\mathcal{M}$ and a state w in $\mathcal{M}$ such that $\mathcal{M}, w \Vdash \varphi$. If φ is satisfied at all states in a hybrid model $\mathcal{M}$, then we say that φ is *globally satisfied* in $\mathcal{M}$ and write $\mathcal{M} \Vdash \varphi$. If $\mathcal{F}$ is a frame, and for all hybrid models $\mathcal{M}$ based on $\mathcal{F}$ we have $\mathcal{M} \Vdash \varphi$, then we say that φ is *valid* on $\mathcal{F}$ and write $\mathcal{F} \Vdash \varphi$. Finally, if φ is valid on all frames we call it a validity and write $\Vdash \varphi$. These concepts are extended to sets of formulas in the obvious way.

The basic hybrid language is very much a *modal* language. Although we have introduced labels into the object language, these are simply a second sort of atomic formula. Moreover, satisfaction operators are also modally natural. As has already been mentioned, $i : \varphi$ means "jump to the denotation of i and evaluate φ there", thus $i{:}$ seems to be some sort of unary modal operator. And this impression is correct: in fact, *each satisfaction operator is a normal modal operator*. To see this, note that for every nominal i,

$$\Vdash i:(\varphi \to \psi) \to (i:\varphi \to i:\psi).$$

That is, all satisfaction operators validate the modal normality (or distribution) schema. Moreover, we have that

$$\Vdash \varphi \text{ implies } \Vdash i:\varphi.$$

That is, the rule of necessitation holds for every satisfaction operator too, and so each satisfaction operator is a normal modal operator.

Moreover, the basic hybrid language is quite a *simple* modal language. In fact, its satisfiability problem (the problem of deciding whether a given formula is satisfiable) belongs to the same complexity class as the satisfiability problem for ordinary propositional modal logic:

Theorem 1 *The satisfiability problem for the basic hybrid language is* PSPACE-*complete.*

Proof PSPACE-hardness is clear, for the basic hybrid language contains the satisfiability problem for ordinary propositional modal logic as a special case, and this problem is known to be PSPACE-hard (see Ladner [23]). As for the upper bound, a game-theoretic argument showing the existence of a PSPACE algorithm for basic hybrid satisfiability is given in Areces, Blackburn and Marx [2]. ■

But for present purposes the most important fact about the basic hybrid language is that it supports proof theories which, in very obvious sense, internalize the leading ideas of Gabbay-style labelled deduction. Here's how to internalize labelled deduction in a Smullyan-style (see [34]) unsigned tableau system.[1]

Suppose we want to test whether a basic hybrid formula φ is valid. Now, if it is *not* valid then it must be possible to build a model for its negation at some state in some model. So, let i be a nominal that does not occur in φ (i is intended to act as a label for the falsifying state that is supposed to exist). We shall systematically search for a model for $\neg i:\varphi$; if we *can't* do this, then φ must be valid after all. We search for a model by building a tree with $\neg i:\varphi$ at its root. As in the Smullyan systems, a proof will simply be a tree whose nodes are decorated with object language formulas. That is: *no use will be made of metalinguistic labels, or indeed any additional metalinguistic apparatus;*

[1]This tableau system is the one introduced in Blackburn [8]; that paper also gives a sequent-style formulation of the system. It is also possible to internalize labelled deduction using natural deduction techniques; see Seligman [31] for details. For an approach based on Fitting-style prefix systems, see Tzakova [35].

our hybrid object language provides us with everything we need to define an orthodox tableau system.

The tree will be expanded in accordance with the following rules. These fall into four groups: rules for the booleans, rules for the satisfaction operators, rules for $\Diamond$ and $\Box$, and rewrite rules for coping with the fact that nominals really are *labels*. To state the rules, I shall use φ and ψ as metavariables over arbitrary wffs, s, t and u, as metavariables over nominals, and a as a metavariable over *new* nominals (that is, nominals that have not been used so far in the tableau construction). Here are the rules for the booleans:

$$\frac{\neg s{:}(\varphi \wedge \psi)}{\neg s{:}\varphi \mid \neg s{:}\psi} \qquad \frac{s{:}(\varphi \wedge \psi)}{\begin{array}{c} s{:}\varphi \\ s{:}\psi \end{array}}$$

$$\frac{\neg s{:}(\varphi \vee \psi)}{\begin{array}{c} \neg s{:}\varphi \\ \neg s{:}\psi \end{array}} \qquad \frac{s{:}(\varphi \vee \psi)}{s{:}\varphi \mid s{:}\psi}$$

$$\frac{\neg s{:}(\varphi \rightarrow \psi)}{\begin{array}{c} s{:}\varphi \\ \neg s{:}\psi \end{array}} \qquad \frac{s{:}(\varphi \rightarrow \psi)}{\neg s{:}\varphi \mid s{:}\psi}$$

$$\frac{\neg s{:}\neg\varphi}{s{:}\varphi} \qquad \frac{s{:}\neg\varphi}{\neg s{:}\varphi}$$

These rules have self-explanatory names: the $\neg\wedge$-rule and the $\wedge$-rule; the $\neg\vee$-rule and the $\vee$-rule; the $\neg\rightarrow$-rule and the $\rightarrow$-rule; and the $\neg\neg$-rule and the $\neg$-rule. Moreover, the content of these rules is also self-explanatory: they merely state what it means for formula above the horizontal line (the input to the rule) to be satisfied at the state labelled by a nominal s in a model, *and they do so using only the resources available in the hybrid object language.* Following traditional tableau terminology, the $\neg\wedge$-rule, the $\vee$-rule, and the $\rightarrow$-rule are called *branching rules*, for they yield two alternative outputs.

Now for the satisfaction operators:

$$\frac{\neg s{:}t{:}\varphi}{\neg t{:}\varphi} \qquad \frac{s{:}t{:}\varphi}{t{:}\varphi}$$

These are called the $\neg{:}$-rule and the $:$-rule respectively, and they could hardly be simpler: given a pair of nested satisfaction operators, the outermost can simply be discarded.

The interesting work starts with the modalities. Here is the $\neg\Diamond$-rule and the $\Diamond$-rule, followed by the $\neg\Box$-rule and the $\Box$-rule:

$$\frac{\neg s\!:\!\Diamond\varphi \quad s\!:\!\Diamond t}{\neg t\!:\!\varphi} \qquad\qquad \frac{s\!:\!\Diamond\varphi}{\begin{array}{c} s\!:\!\Diamond a \\ a\!:\!\varphi \end{array}}$$

$$\frac{\neg s\!:\!\Box\varphi}{\begin{array}{c} s\!:\!\Diamond a \\ \neg a\!:\!\varphi \end{array}} \qquad\qquad \frac{s\!:\!\Box\varphi \quad s\!:\!\Diamond t}{t\!:\!\varphi}$$

Note that the $\Diamond$-rule and $\neg\Box$-rules are stated using the metavariable a. This means that when we apply these rules to some formula in a tableau, we should choose a nominal that hasn't been used so far in the tableau construction process. Incidentally, these two rules are the only rules which introduce new nominals into a tableau, and for this reason they are called *existential rules*.

What do these rules say? Consider the $\Diamond$-rule. In effect, this decomposes the existential demand made by formulas of the form $s\!:\!\Diamond\varphi$ into two subdemands: (1) that there is a successor state to the state labelled s (we invent a brand new name a for this successor), and (2) that the state labelled a satisfies φ. The $\neg\Box$-rule works analogously. *In short, just as with all the other rules, we have expressed the satisfaction conditions using only the resources available in the object language.*

Now, these rules are discussed in more detail in [8], but one thing should be immediately clear to readers of the present volume: there is a striking analogy with the way Dov Gabbay handles the logic of $\Diamond$ in labelled deduction systems (see [19]). Here's Gabbay's $\Diamond$ rule:

$$\frac{s\!:\!\Diamond\varphi}{\textit{create } a,\ sRa,\ \textit{and } a\!:\!\varphi}$$

In fact, the only real difference is that Gabbay proceeds by manipulating labels *metalinguistically* (in effect, he makes use of a programming language containing expressions such as '*create*', '*and*', 'R', ':', and a supply of labels, to manipulate object language formulas) whereas in the present paper we work with an object language rich enough to state the required deduction step.

Let's turn to the $\Box$-rule and the $\neg\Diamond$-rule, the *universal rules*. Both rules are binary (they take *two* formulas as input) and one of the inputs to each rule (the minor premiss) is a near-atomic satisfaction statement of the form $s\!:\!\Diamond t$. Both rules should be read as follows: if a branch contains a pair of formulas of the form shown above the line, then we can extend that branch by adding the formula shown below the line. (The two input formulas are written side-by-side to indicate that there is no ordering intended on the input.) That is, if you can find a matching pair of formulas anywhere on the branch, no matter which occurs first, you can apply the rule. I like to think of these rules as demons scanning tableau branches, demons which are summoned by occurrences of $s:\Diamond t$ (such formulas state that t is a successor of s). Thus if both $s:\Box\varphi$ and $s:\Diamond t$ belong to a branch, then we are free to fire the $\Box$-rule and add $t:\varphi$ to the end of the branch.

We're almost there — but one task remains. So far nothing in the system reflects the fact that nominals are genuine *labels.* There are various ways of getting to grips with this fact (for example, Seligman [33] makes use of substitution rules in his hybrid sequent calculus) and I shall do so via the following rewrite rules:

Ref $\dfrac{[s \text{ occurs on the branch}]}{s:s}$

Sym $\dfrac{t:s}{s:t}$

Nom $\dfrac{s:t \quad t:\varphi}{t:\varphi}$

Bridge $\dfrac{s:\Diamond t \quad t:u}{s:\Diamond u}$

Ref and *Sym* should be clear (the *s occurs on the branch* in the statement of the *Ref* rule simply means that some formula on the branch in question contains an occurrence of s). The *Nom* rule reflects the fact that identical states carry identical information. Note that the obvious transitivity rule (that is, from $s:t$ and $t:u$ to deduce $s:u$) is merely a special case of *Nom* (in which φ is u). In short, the *Ref*, *Sym* and *Nom* rules directly reflect the fact that our labelled languages are strong enough to talk about state equality. Finally, *Bridge* regulates information about state successorship.

And that's the system. Let's consider some examples. First, let's prove a standard modal validity, the distribution axiom: $\Box(p \to q) \to (\Box p \to \Box q)$. We do so by choosing a nominal that does not occur in this formula (let's choose i), prefixing it with $\neg i:$ and then applying tableau rules:

1	$\neg i : (\Box(p \rightarrow q) \rightarrow (\Box p \rightarrow \Box q))$		
2	$i : \Box(p \rightarrow q)$		$1, \neg\rightarrow$
2′	$\neg i : (\Box p \rightarrow \Box q)$		Ditto
3	$i : \Box p$		$2', \neg\rightarrow$
3′	$\neg i : \Box q$		Ditto
4	$i : \Diamond j$		$3', \neg\Box, j$
4′	$\neg j : q$		Ditto
5	$j : p$		$3, 4, \Box$
6	$j : (p \rightarrow q)$		$2, 4, \Box$
7	$\neg j : p$	$j : q$	$6, \rightarrow$
	✠ 5, 7 ✠	✠ 4′, 7 ✠	

The annotations in the far right column should be self explanatory: they tell us which rule was applied and where. Note that at line 4 we also indicate which new nominal was introduced by the existential rule $\neg\Box$ (in this case, j). The annotations ✠ 5, 7 ✠ and ✠ 4′, 7 ✠ mark the fact that both branches of the tree have closed, with the numbers indicating where the conflicting items are to be found.

Now, the previous example showed how to use the tableau system to prove an ordinary modal formula; and in fact the system can prove *any* validity of the basic modal language. But it can do a lot more than that: the system captures not merely ordinary modal validity, but full hybrid validity. Let's consider some examples which show that we really have a calculus which gets to grip with the genuine logic of labelling.

Here's a simple example. We shall prove $i \wedge \varphi \rightarrow i : \varphi$, the *Introduction* axiom. This schema reflects an obvious fact about $:$, and plays an important role in the Hilbert-style axiomatization of hybrid logic given in Blackburn and Tzakova [14]. Any instance can be proved as follows (here j is a nominal that does not occur in φ):

1	$\neg j : (i \wedge \varphi \rightarrow i : \varphi)$	
2	$j : (i \wedge \varphi)$	$1, \neg\rightarrow$
2′	$\neg j : i : \varphi$	Ditto
3	$\neg i : \varphi$	$2', \neg :$
4	$j : i$	$2, \wedge$
4′	$j : \varphi$	Ditto
5	$i : j$	4, Sym
6	$i : \varphi$	4′, 5, Nom
	✠ 3, 6 ✠	

Here's another example. The schema

$$i : j \rightarrow (j : \varphi \rightarrow i : \varphi)$$

is valid. This schema is essentially a reflection in the object language of the idea captured by the *Nom* rule, and any instance of this schema can be proved as follows (here k is a nominal that does not occur in φ):

1	$\neg k:(i:j \to (j:\varphi \to i:\varphi))$	
2	$k:i:j$	$1, \neg\to$
2′	$\neg k:(j:\varphi \to i:\varphi)$	Ditto
3	$i:j$	$2, :$
4	$k:j:\varphi$	$2', \neg\to$
4′	$\neg k:i:\varphi$	Ditto
5	$j:\varphi$	$4, :$
6	$\neg i:\varphi$	$4', \neg:$
7	$i:\varphi$	$3, 5$ Nom
	✠ 6, 7 ✠	

We'll see some more examples of these rules in action shortly, but let's now state a basic (soundness and) completeness result. We shall say that a formula φ of the basic hybrid language is *tableau provable* if and only if there is a closed tableau with $\neg i:\varphi$ as its root node (where i is a nominal not occurring in φ). Then we have:

Theorem 2 *A formula φ is tableau provable iff it is valid.*

Proof Soundness is straightforward. Completeness is proved in Blackburn [8] using a Hintikka set argument. ■

3 STRONGER LOGICS

The tableau system just discussed generates the weakest hybrid logic: the set of formulas valid in *all* models. But when applying modal logic we are often interested only in models based on frames with certain properties. For example, when studying logics of knowledge we might insist on working with frames (W, R) such that R is transitive, reflexive, and symmetric; and when studying logics of time we might insist that R be transitive and irreflexive. In short, we will often be interested in stronger logics and stronger completeness theorems than the one just discussed.

It is precisely here that hybrid languages have some interesting advantages over orthodox modal languages. As we shall now discuss, nominals are powerful tools for defining properties of frames. Moreover, so long as we can define the properties we are interested in using formulas whose only atoms are nominals (that is, so long as we *don't* use any ordinary propositional variables) we *automatically* obtain the appropriate completeness theorem. These observations are not new. On the contrary, they have been one of the main motivations for studying hybrid languages, and the reader can find various statements of, and proofs of such results (usually for Hilbert-style proof systems) for a wide range of hybrid languages in (among other places) Bull [15], Passy and Tinchev [25], Gargov and Goranko [20], Blackburn and Tzakova [14], and Blackburn [8].

Let us say that a formula of the basic hybrid language is *pure* if and only if it contains no occurrences of ordinary propositional variables; that is, the only atoms that pure formulas contain are *nominals*. And, following traditional modal usage, let us say that a formula φ *defines* a class of frames F if and only if $\mathcal{F} \Vdash \varphi$ iff $\mathcal{F} \in \mathsf{F}$. That is, a formula defines the class of frames it is valid on. When we say that a formula defines a certain property (for example, transitivity) we mean it defines the class of all frames with that property.

What properties can we define using pure formulas? For a start, we can define many properties that are definable in the basic modal language. For example:

$i{:}\Diamond i$	*Reflexivity*
$i{:}\Box\Diamond i$	*Symmetry*
$\Diamond\Diamond i \rightarrow \Diamond i$	*Transitivity*
$\Diamond i \rightarrow \Diamond\Diamond i$	*Density*
$\Diamond i \rightarrow \Box i$	*Determinism*

(Determinism means that every state has at most one R-successor; that is, R is a partial function.) Moreover, we can define many properties that are *not* definable in orthodox propositional modal languages:

$i{:}\neg\Diamond i$	*Irreflexivity*
$i{:}\neg\Diamond\Diamond i$	*Asymmetry*
$i{:}\Box(\Diamond i \rightarrow i)$	*Antisymmetry*
$\Diamond\Diamond i \rightarrow \neg\Diamond i$	*Intransitivity*
$\Diamond i$	*Universality*
$j{:}\Diamond i \vee j{:}i \vee i{:}\Diamond j$	*Trichotomy*
$i{:}(\neg j \wedge \neg k) \rightarrow j{:}k$	*At Most 2 States*

Here *Universality* means $\forall x \forall y Rxy$ and *Trichotomy* means $\forall x \forall y (Rxy \vee x = y \vee Ryx)$. Note that the formula defining *At Most 2 States* is an encoding of the Pigeonhole Principle that can be generalized to any natural number n. For proofs that these properties are not definable in orthodox modal languages, see van Benthem [4].

It is easy to check from first principles that each of the formulas listed above defines the property claimed. However in Blackburn and Seligman [10] and Areces, Blackburn and Marx [2] the reader will find extensions of the well known *Standard Translation* of the ordinary propositional modal language into classical logic that cover the basic hybrid language. These extensions give us a simple (and purely mechanical) way of calculating the class of frames defined by any pure formula.

But the crucial observation is not about the frame defining powers of pure formulas; rather it is about the perfect match pure formulas exhibit between definability and completeness. In particular, *if we use pure formulas as additional axioms, they automatically yield a complete system with respect to the class of frames they define*. Before stating this result more precisely, let's consider a couple of examples.

As I mentioned above, the pure formula $\Diamond i \to \Box i$ defines determinism; that is, it is valid on precisely those frames (W, R) where R is a partial function. So in this case my claim amounts to the following: if we are free to introduce, at any stage of the tableau construction process, formulas of the form $l : \rho$, where l is a nominal and ρ is any instance of $\Diamond i \to \Box i$ (here "instance" means any formula obtained from this axiom by uniform substitution of nominals for nominals) then the enhanced system will be complete with respect to the class of deterministic frames. How plausible is this claim? Well, when working with deterministic frames, the formula $(\Diamond p \wedge \Diamond q) \to \Diamond(p \wedge q)$ is valid. Note that this formula is *not* pure; in fact it's an orthodox modal determinacy axiom. Hence, by the completeness result just claimed, we should be able to find a tableau proof of this formula — even though the axiom contains only nominals, and the formula to be proved contains none at all. And in fact we *can* do this:

1		$\neg i{:}((\Diamond p \wedge \Diamond q) \to \Diamond(p \wedge q))$		
2		$i{:}(\Diamond p \wedge \Diamond q)$		$1, \neg\!\to$
2′		$\neg i{:}\Diamond(p \wedge q)$		Ditto
3		$i{:}\Diamond p$		$2, \wedge$
3′		$i{:}\Diamond q$		Ditto
4		$i{:}\Diamond j$		$3, \Diamond, j$
4′		$j{:}p$		Ditto
5		$i{:}\Diamond k$		$3', \Diamond, k$
5′		$k{:}q$		Ditto
6		$i{:}(\Diamond j \to \Box j)$		*Axiom*
7	$\neg i{:}\Diamond j$	$i{:}\Box j$		$6, \to$
8	✠ 4, 7 ✠	$k{:}j$		$5, 7, \Box$
9		$j{:}k$		8, Sym
10		$j{:}q$		$5'$, 9, Nom
11		$\neg j{:}(p \wedge q)$		$2', 4, \neg\Diamond$
12		$\neg j{:}p$	$\neg j{:}q$	$11, \neg\wedge$
13		✠ 4′, 12 ✠	✠ 10, 12 ✠	

Here's a second example. As I mentioned above, the pure formulas

$$i{:}\neg\Diamond i \quad \text{and} \quad \Diamond\Diamond i \to \Diamond i$$

define irreflexivity and transitivity respectively. So in this case my claim amounts to the following: if we are free to introduce, at any stage of the tableau construction process, formulas of the form $l{:}\rho$, where ρ is any instance of either of these formulas (and once again, "instance" means any formula obtained from this axiom by uniform substitution of nominals for nominals) then the enhanced system will be complete with respect to the class of frames that are both transitive and irreflexive (that is, the class of strict pre-orders). Now, any strict pre-order is antisymmetric, and as we mentioned above, the formula $i : \Box(\Diamond i \to i)$ defines antisymmetry. Hence this formula should be tableau provable with the help of these two axioms. Here's how to do it:

1			$\neg k : i : \Box(\Diamond i \to i)$	
2			$\neg i : \Box(\Diamond i \to i)$	$1, \neg :$
3			$i : \Diamond j$	$2, \neg\Box$
3′			$\neg j : (\Diamond i \to i)$	Ditto
4			$j : \Diamond i$	$3', \neg\!\to$
4′			$\neg j : i$	Ditto
5			$i : (\Diamond\Diamond i \to \Diamond i)$	*Axiom*
6		$\neg i : \Diamond\Diamond i$	$i : \Diamond i$	$5, \to$
7	$3, 6, \neg\Diamond$	$\neg j : \Diamond i$	$i : i : \neg\Diamond i$	*Axiom*
8		✠ 4, 7 ✠	$i : \neg\Diamond i$	$7, :$
9			$\neg i : \Diamond i$	$8, \neg$
			✠ 6, 9 ✠	

A practical point. When doing proofs by hand, it is sometimes convenient to use two additional rules: *Labelled Modus Ponens* and *Big Bridge*. Labelled Modus Ponens is defined as follows: if $s : \varphi$ and $s : (\varphi \to \psi)$ are on a branch, then we are free to add $s : \psi$ to that branch. Big Bridge is the Bridge rule generalized as follows: if $s : \Diamond t$ and $t : \varphi$ are on a branch, then we are free to add $s : \Diamond\varphi$ to that branch. These rules often make it possible to cut down on the degree of branching. For example, here's how to prove the previous example with their help:

1	$\neg k : i : \Box(\Diamond i \to i)$	
2	$\neg i : \Box(\Diamond i \to i)$	$1, \neg :$
3	$i : \Diamond j$	$2, \neg\Box$
3′	$\neg j : (\Diamond i \to i)$	Ditto
4	$j : \Diamond i$	$3', \neg\!\to$
4′	$\neg j : i$	Ditto
5	$i : \Diamond\Diamond i$	3, 4, *Big Bridge*
6	$i : (\Diamond\Diamond i \to \Diamond i)$	*Axiom*
7	$i : \Diamond i$	5, 6, *Labelled Modus Ponens*
8	$i : i : \neg\Diamond i$	*Axiom*
9	$i : \neg\Diamond i$	$8, :$
10	$\neg i : \Diamond i$	$9, \neg$
	✠ 7, 10 ✠	

And in our earlier determinism example, the branching at line 7 can be avoided by an application of labelled modus (using lines 4 and 6). So Labelled Modus Ponens and Big Bridge can lead to more streamlined proofs. But they don't increase the power of the proof system (they are not needed to establish completeness).

Having seen some examples, let's state the desired completeness result more precisely. Let $\mathtt{Axiom}$ be a finite or countably infinite set of pure formulas, let $\mathtt{H} + \mathbf{Axiom}$ be the tableau system that results by using the formulas in $\mathtt{Axiom}$ as axioms. That is, for any formula ρ in $\mathtt{Axiom}$, and any nominals j, $j_1, \ldots, j_n$ that occur on a branch of a tableau, we are free to add either $j : \rho$ or $j : \rho[j_1/i_1, \ldots, j_n/i_n]$ to the end of that branch (here $i_1 \ldots, i_n$ are nominals in ρ).

Theorem 3 *Let* $\mathtt{Axiom}$ *be a finite or countably infinite set of pure formulas and let* Axiom *be the class of frames that* $\mathtt{Axiom}$ *defines. Then* $\mathtt{H} + \mathbf{Axiom}$ *is complete with respect to the frames in* Axiom.

Proof See Blackburn [8]. The reader may also find it useful to examine Theorem 4.10 of Blackburn and Tzakova [14]; this proves an analogous result for a Hilbert-style system using a Henkin argument. ■

Now this is a nice result — but what does it mean when viewed from the perspective of labelled deduction systems? Simply this:

Labelling discipline has become logic.

A *pure* formula is essentially a formula that is constructed from *labels*. And whenever we use a pure formula as an axiom, we are essentially writing down a genuinely declarative specification of a property of the transition relation; after all, such specifications are constructed from labels (nominals) using $\neg$, $\vee$, $\wedge$, $\Diamond$, and so on, where all these connectives receive their standard interpretation. Thus our specifications are *logical* in the strictest sense of the word. However the above theorem assures us that such specifications also have a clear connection to proof-theoretical reality: when we "run" them by using them as axioms, they have exactly the intended effect.

This theorem is far simpler than any comparable result for orthodox modal languages. For a start, when used as axioms, not all orthodox modal formulas are complete with respect to the frame classes they define. (Van Benthem [4] contains a striking counterexample: an orthodox modal formula which defines a first-order definable class of frames but which, when used as an axiom, is not complete with respect to any class of frames at all!) Moreover, while there are general results linking definability and completeness in orthodox modal languages, these results tend to be more complex than the result just stated. The best known examples are probably the Sahlqvist Theorems (see Sahlqvist [29]). These show that orthodox modal formulas satisfying certain syntactic properties (so-called Sahlqvist formulas) define first-order properties of frames, and that when they are used as axioms they are complete with respect to the class of frames they define. This is an important result — nonetheless the syntactic definition of Sahlqvist formulas is far more complex than the simple restriction to pure formulas needed for the previous theorem.

One work of warning: obviously matters can get more complex when we start working with stronger logics. For example, it is easy to give examples of stronger logics specified by pure formulas which lack the finite model property, and it is not much harder to give examples which are undecidable. Thus the completeness result just stated gives us no guarantees about the decidability of proof search in the resulting logic; rather, it simply makes clear that there is a large space of logics which are complete with respect to frames with specifiable properties. We are now going to enlarge this space even further.

4 FULL FIRST-ORDER LABELLING DISCIPLINE

Now that we are used to viewing labels as formulas, the way lies open for further enrichments. The most obvious is to regard nominals not as names but as *variables over states*, and to add quantifiers. That is, we are going to allow ourselves to form expressions like

$$\exists x \Diamond (x \wedge \exists y \Diamond (y \wedge \Diamond y \wedge p)).$$

This sentence is satisfied at a state w if and only if there is a state u that is accessible from w, and from u we can reach at least one reflexive state v where p is true. No formula with this property exists in ordinary modal languages, or even in the basic hybrid language.

Here's another example. The *Until* operator is *not* definable in ordinary propositional modal logic, or even in the basic hybrid language, but it *is* definable if we allow ourselves to quantify labels:

$$\textit{Until}\,(\varphi, \psi) \;:=\; \exists x (\Diamond (x \wedge \varphi) \wedge \Box (\Diamond x \rightarrow \psi)).$$

So adding quantifiers looks like an interesting idea; let's make their syntax and semantics precise and explore them further. Choose a denumerably infinite set SVAR $= \{x, y, z. \dots\}$, the set of *state variables*, disjoint from both PROP and NOM. We then stipulate that:

$$\text{WFF} \;:=\; x \mid i \mid p \mid \neg\varphi \mid \varphi \wedge \psi \mid \Diamond\varphi \mid \Box\varphi \mid i{:}\varphi \mid x{:}\varphi \mid \exists x\varphi \mid \forall x\varphi$$

That is, we have four new possibilities: all state variables are formulas; state variables can be used to build satisfaction statements; and (crucially) we can *bind* occurrences of state variables by building wffs of the form $\exists x\varphi$ and $\forall x\varphi$; in such a formula, all free occurrences of x in φ become bound. If a formula contains no free occurrences of any state variable it is called a *sentence*.[2]

[2]Precise definitions of such concepts as *free*, *bound*, *substitution*, and so on can be found in Blackburn and Tzakova [12]. But formal definitions are not needed here; experience with classical logic is a reliable guide.

Note that state variables are formulas, just as nominals are. And in the semantics presented below, state variables are going to be true at precisely one state, just as nominals are. In fact the only difference between these two types of label is that while state variables can be bound, nominals cannot. In effect, nominals will a play a role for us analogous to that of *parameters* in first-order proof theory.

Now for the semantics. We have to cope with the interpretation of free state variables, and the obvious way to do so is via Tarski-style assignment functions. So, given a model $\mathcal{M} = (W, R, V)$, an *assignment* on $\mathcal{M}$ is a function g : SVAR $\longrightarrow W$. (Clearly an assignment makes a state variable true at precisely one state.) We then relativize the clauses of the satisfaction definition for the labelled modal language to assignments, and add the new clauses we require. Here's a selection:

$\mathcal{M}, g, w \Vdash x$	iff	$w = g(x)$ where $x \in$ SVAR
$\mathcal{M}, g, w \Vdash x : \varphi$	iff	$\mathcal{M}, g, g(x) \Vdash \varphi$
$\mathcal{M}, g, w \Vdash \varphi \wedge \psi$	iff	$\mathcal{M}, g, w \Vdash \varphi$ and $\mathcal{M}, g, w \Vdash \psi$
$\mathcal{M}, g, w \Vdash \Diamond\varphi$	iff	$\exists(wRw' \;\&\; \mathcal{M}, g, w' \Vdash \varphi)$
$\mathcal{M}, g, w \Vdash \exists x\varphi$	iff	there is a $g' \overset{x}{\sim} g$ such that$\mathcal{M}, g', w \Vdash \varphi$
$\mathcal{M}, g, w \Vdash \forall x\varphi$	iff	for all g' such that $g' \overset{x}{\sim} g, \mathcal{M}, g', w \Vdash \varphi$.

The first four clauses are obvious generalizations of those given for the basic hybrid language. Now, the notation $g' \overset{x}{\sim} g$ means that g' is the assignment that differs from g, if at all, only in what it assigns to x; so in the fifth and sixth clauses we have essentially defined $\exists$ and $\forall$ to be classical quantifiers over states.

It is not difficult to see that this language offers full first-order expressive power. That is, if we regard a model (W, R, V) as a model for the first-order language containing a binary relation R, a unary relation symbol P for every propositional symbol $p \in$ PROP, and a constant c_i for every nominal $i \in$ NOM, then it is straightforward to translate every first-order formula of this language to an equivalent hybrid formula. This translation was known to Arthur Prior at least as early as 1967, and a generalization to arbitrary signatures can be found in Blackburn and Seligman [9]. Thus the ability to quantify labels enormously increases the expressive power at our disposal.

But can we cope with this power proof theoretically? Yes, and in a very simple way. First, we can re-use our original tableau rules for the booleans, satisfaction operators, and modalities, and also our rewrite rules, by making one small change: we should now interpret the metavariables s, t, and u in the statements of these rules as ranging over both nominals *and* state variables. Note, however, that in the statement of the existential rules, a will continue to be interpreted as a new *nominal*. Doing so means that we don't have to worry about a getting accidentally bound; that is, we really are using nominals as parameters.

Second, we add the following rules to cope with the quantifiers:

$$\frac{\neg s:\exists x\varphi}{\neg s:\varphi[t/x]} \qquad \frac{s:\exists x\varphi}{s:\varphi[a/x]}$$

$$\frac{\neg s:\forall x\varphi}{\neg s:\varphi[a/x]} \qquad \frac{s:\forall x\varphi}{s:\varphi[t/x]}$$

These rules should look familiar: they're essentially the classical tableaux rules for existential and universal quantifiers. Let's see how to use them.

Our first example shows that in certain respects $\forall$ behaves very classically; just as in first-order logic the hybrid formula

$$\forall x(\varphi \to \psi) \to (\varphi \to \forall x\psi)$$

is valid so long as as x does not occur free in φ. Here's a tableau proof:

1		$\neg i:(\forall x(\varphi \to \psi) \to (\varphi \to \forall x\psi))$	
2		$i:\forall x(\varphi \to \psi)$	$1, \neg\!\to$
$2'$		$\neg i:(\varphi \to \forall x\psi)$	Ditto
3		$i:\varphi$	$2', \neg\!\to$
$3'$		$\neg i:\forall x\psi$	Ditto
4		$\neg i:\psi[j/x]$	$3', \neg\forall, j$
5		$i:(\varphi \to \psi)[j/x]$	$2, \forall, j$
6	$\neg i:\varphi[j/x]$	$i:\psi[j/x]$	$6, \to$
	✠ 3, 6 ✠	✠ 4, 6 ✠	

(Note that the branch closure ✠ 3, 6 ✠ makes use of our assumption that x does not occur free in φ.)

So far so classical — but our new quantifiers have some surprises in store for us. For example, note that the (somewhat surprising looking) sentence $\exists xx$ is *valid*. Think about it: this claims that it is always possible to bind a state variable to the current state, which is clearly true. And we can prove it as follows:

1	$\neg i:\exists xx$	
2	$\neg i:i$	$1, \neg\exists$
3	$i:i$	*Ref*
	✠ 2, 3 ✠	

Second, it has long been known that the hybrid analog of the Barcan schema

$$\Diamond\exists x\varphi \to \exists x\Diamond\varphi$$

is valid; in fact, Bull [15] uses Barcan as part of his Hilbert-style axiomatization of (a variant of) the hybrid language we are working with here. Here's how to prove it:

1	$\neg i\!:\!(\Diamond\exists x\varphi \rightarrow \exists x\Diamond\varphi)$	
2	$i\!:\!\Diamond\exists x\varphi$	$1, \neg\rightarrow$
2′	$\neg i\!:\!\exists x\Diamond\varphi$	Ditto
3	$i\!:\!\Diamond j$	$2, \Diamond, j$
3′	$j\!:\!\exists x\varphi$	Ditto
4	$j\!:\!\varphi[k/x]$	$3', \exists, k$
5	$\neg i\!:\!\Diamond\varphi[k/x]$	$2', \neg\exists, k$
6	$\neg j\!:\!\varphi[k/x]$	$3, 5\neg\Diamond$
	✠ 4, 6 ✠	

But for our last example, let's prove something a little more exotic:

$$\forall x(\Diamond^n(x \wedge \varphi) \rightarrow \Box^m(x \rightarrow \varphi)).$$

(Here $\Diamond^n$ and $\Box^m$ mean n-length, and m-length prefixes of $\Diamond$ and $\Box$ operators respectively.) Now, what does this schema mean? Essentially this: if by following some sequence of transitions (the one encoded in the modality sequence $\Diamond^n$ in the antecedent) we can reach a state labelled x and bearing the information φ, then, no matter what other path we might follow, if we ever reach a state labelled x, we are *guaranteed* to find the information φ there — after all, the state labelled x is *unique*. This schema is used in a Hilbert-style system presented in Blackburn and Tzakova [12], and the question that now faces us is: how can we prove it using tableau methods?

Let's do so in two steps. First we'll strip off the n length prefix of diamonds by repeated application of the $\Diamond$ rule:

1	$\neg i\!:\!\forall x(\Diamond^n(x \wedge \varphi) \rightarrow \Box^m(x \rightarrow \varphi))$	
2	$\neg i\!:\!\Diamond^n(l \wedge \varphi[l/x]) \rightarrow \Box^m(l \rightarrow \varphi[l/x])$	$1, \forall, l$
3	$i\!:\!\Diamond^n(l \wedge \varphi[l/x])$	$2, \neg\rightarrow$
3′	$\neg i\!:\!\Box^m(l \rightarrow \varphi[l/x])$	Ditto
4	$i\!:\!\Diamond j_1$	$3, \Diamond, j_1$
4′	$j_1\!:\!\Diamond^{n-1}(l \wedge \varphi[l/x])$	Ditto
5	$j_1\!:\!\Diamond j_2$	$4', \Diamond, j_2$
5′	$j_2\!:\!\Diamond^{n-2}(l \wedge \varphi[l/x])$	Ditto
	$\vdots$	
$n+3$	$j_{n-1}\!:\!\Diamond j_n$	$n+2', \Diamond, j_n$
$n+3'$	$j_n\!:\!(l \wedge \varphi[l/x])$	Ditto
$n+4$	$j_n\!:\!l$	$n+3', \wedge$
$n+4'$	$j_n\!:\!\varphi[l/x]$	Ditto
$n+5$	$l\!:\!j_n$	$n+4$, Sym
$n+6$	$l\!:\!\varphi[l/x]$	$n+4', n+5$, Nom

Next, we'll strip off the m length prefix of boxes, by repeated application of the $\neg\Box$ rule:

$n+6+1$	$i:\Diamond k_1$	$3', \neg\Box, k_1$
$n+6+1'$	$\neg k_1:\Box^{m-1}(l \to \varphi[l/x])$	Ditto
$n+6+2$	$k_1:\Diamond k_2$	$3', \neg\Box, k_2$
$n+6+2'$	$\neg k_2:\Box^{m-2}(l \to \varphi[l/x])$	Ditto
	$\vdots$	
$n+6+m$	$k_{m-1}:\Diamond k_m$	$3', \neg\Box, k_m$
$n+6+m'$	$\neg k_m:(l \to \varphi[l/x])$	Ditto
$n+7+m$	$k_m:l$	$n+6+m', \neg\to$
$n+7+m'$	$\neg k_m:\varphi[l/x]$	Ditto
$n+8+m$	$k_m:\varphi[l/x]$	$n+6, n+7+m$, Nom

The last two lines close the branch and the tableau.

To conclude this section, some historical remarks. The strong hybrid languages we have been discussing in this section were actually the first type of hybrid languages to be studied. By 1967, Arthur Prior was making use of them, they were studied in depth in Bull [15] and Passy and Tinchev [24, 25], and since then practically every writer on hybrid languages has had something to say about them; see for example Seligman [31, 32, 33], Blackburn and Seligman [9, 10], Blackburn and Tzakova [12, 13] and Tzakova [35]. They are now well understood — and there is a clear sense in which they are fundamental: it's important to know that labelling discipline can be lifted to a full first-order level so easily.

Nonetheless, in recent years there has been a feeling that allowing unrestricted quantification over labels is too strong for many purposes, and there has been increasing interest in subsystems which allow labels to be bound, but only in more restricted ways. In particular, much attention has focussed on a binder called $\downarrow$ which only allows labels to be bound to the *current* state:

$$\mathcal{M}, g, w \Vdash \downarrow x\varphi \quad \text{iff} \quad \mathcal{M}, g', w \Vdash \varphi, \text{ where } g' \overset{x}{\sim} g \text{ and } g'(x) = w$$

That is, $\downarrow$ lets us create a label for the here-and-now; it is a genuinely *modal* way of binding labels. Key papers on $\downarrow$ include Goranko [21], Blackburn and Seligman [10], Blackburn and Tzakova [14], and Areces, Blackburn and Marx [2]. This last paper explores the expressivity of $\downarrow$-enriched hybrid languages in detail, and confirms that $\downarrow$ is a binder of fundamental importance.

Defining tableau rules for $\downarrow$ is straightforward. For example, here are the rules given in Blackburn [8]:

$$\frac{\neg s:\downarrow x\varphi}{\neg s:\varphi[s/x]} \qquad\qquad \frac{s:\downarrow x\varphi}{s:\varphi[s/x]}$$

For a Fitting-style prefix approach to $\downarrow$, see Tzakova [35].

5 CONCLUDING REMARKS

In this paper I have showed that by fully incorporating the mechanisms of labelled deduction into the modal object language, we can define proof systems which in an obvious sense internalize labelled deduction. I have also tried to make clear that when this is done, we find a perspective from which *labelling discipline* emerges as *logic*. Moreover, I have shown that this logical (or declarative) perspective on labelling discipline can be "lifted" to a full first-order discipline in a very simple way indeed: we need simply quantify over labels.

Underlying this work was the concept of a *hybrid language*, a modal language in which labels are incorporated as first-class citizens — that is, as *formulas*. Now, in my view there are a number of excellent reasons for being interested in hybrid languages, reasons entirely independent of their connection with labelled deduction. For start, hybrid languages arise naturally in applications. Modal logicians like to point out that notational variants of modal logics are often reinvented by workers in artificial intelligence, computational linguistics, and other fields — but it would often be more accurate to say that it is actually *hybrid languages* which are reinvented in this way. For example, while basic description languages (or terminological logics) such as $\mathcal{ALC}$ (see Schmidt-Schauß and Smolka [30]) correspond to multimodal languages, adding A-box elements to form the description language $\mathcal{ALCO}$ corresponds to the addition of nominals (see Blackburn and Tzakova [13]). Furthermore, while the basic Attribute Value Matrices (AVMs) used in computational linguistics are an obvious notational variant of deterministic multimodal logic, full AVM notation corresponds to deterministic multimodal logic enriched with nominals (see Blackburn [6], Blackburn and Spaan [11], Reape [28]). Finally, the interval based logic of Allen [1] can be viewed as a hybrid languages; see Blackburn [5]. In my view, the presence of hybrid languages in applied modal logic is unsurprising. Modal languages offer a simple notation for modeling many problems — but the ability to label a state is often of such fundamental importance that the move to hybrid languages (be it conscious or unconscious) is often unavoidable.

Other defences of hybrid languages are possible. For example, Areces, Blackburn and Marx [2] show that it is possible to bind labels in a way that characterizes the notion of "reachability" through a frame, Passy and Tinchev [25] show that they make it possible to handle the intersection constructor of PDL, Bull [15] uses them to get to grips with path based temporal logics, Blackburn [7] uses them to combine Priorean and Reichenbachian ideas on tense, while Prior [26], Prior and Fine [27], and Seligman [33] contain philosophical arguments supporting the use of hybridization. But the present paper began with labelled deduction, and with labelled deduction I want to end. So: what are the consequences of the present paper for modal labelled deduction?

In my view, primarily this. Labelled deduction is now established as one of the primary tools of modal proof theory. Nonetheless, its use is often presented simply as a (somewhat regrettable) fact of life: efficient and general approaches to modal deduction seem to require labelling in some form, and that's all there is to be said on the matter.

As is probably clear, I don't view labelling in this way at all. Far from being an unpleasant necessity forced on us for technical reasons, labelling is something to rejoice in. When viewed in logical terms, labelling emerges as a natural method of boosting the expressive power of modal languages, a method that (as Theorem 3 testifies) is strikingly in harmony with the proof-theoretical insights of labelled deduction (it is clear, for example, that there are interesting connections with the methods of Basin, Matthews and Viganò, as presented, for example, in [3]). The interplay of ideas from the hybrid logic and LDS communities promises to lead to a deeper understanding of what the modal enterprise is all about.

Acknowledgments

I am very grateful to David Basin and Luca Viganò for their encouragement and helpful remarks on the previous version of the paper, and for lots of stimulating discussion besides. Thanks also to the two referees, whose comments have helped me to clarify a number of points.

Patrick Blackburn
Computerlinguistik
Universität des Saarlandes
D-66041 Saarbrücken, Germany
patrick@coli.uni-sb.de

References

[1] J. Allen. Towards a general theory of action and time. *Artificial Intelligence*, 23(2):123–154, July 1984.

[2] C. Areces, P. Blackburn, and M. Marx. Hybrid logics. Characterization, interpolation and complexity. Technical Report CLAUS-Report 104, Computerlinguistik, Universität des Saarlandes, 1999. `http://www.coli.uni-sb.de/cl/claus`.

[3] D. Basin, S. Matthews, and L. Viganò. Labelled propositional modal logics: Theory and practice. *Journal of Logic and Computation*, 7:685–717, 1997.

[4] J. van Benthem. Correspondence theory. In D. Gabbay and F, Guenthner, editors, *Handbook of Philosophical Logic*, volume 2, pages 167–247. Reidel, Dordrecht, 1984.

[5] P. Blackburn. Fine grained theories of time. Manuscript, 1999.

[6] P. Blackburn. Modal logic and attribute value structures. In M. de Rijke, editor, *Diamonds and Defaults*, Synthese Language Library, pages 19–65. Kluwer Academic Publishers, Dordrecht, 1993.

[7] P. Blackburn. Tense, temporal reference, and tense logic. *Journal of Semantics*, 11:83–101, 1994.

[8] P. Blackburn. Internalizing labelled deduction. Technical Report CLAUS-Report 102, Computerlinguistik, Universität des Saarlandes, 1998. `http://www.coli.uni-sb.de/cl/claus`. To appear in the *Journal of Logic and Computation*.

[9] P. Blackburn and J. Seligman. Hybrid languages. *Journal of Logic, Language and Information*, 4(3):251–272, 1995.

[10] P. Blackburn and J. Seligman. What are hybrid languages? In M. Kracht, M. de Rijke, H. Wansing, and M. Zakharyaschev, editors, *Advances in Modal Logic*, volume 1, pages 41–62. CSLI Publications, Stanford University, 1998.

[11] P. Blackburn and E. Spaan. A modal perspective on the computational complexity of attribute value grammar. *Journal of Logic, Language and Information*, 2(2):129–169, 1993.

[12] P. Blackburn and M. Tzakova. Hybrid completeness. *Logic Journal of the IGPL*, 6:625–650, 1998.

[13] P. Blackburn and M. Tzakova. Hybridizing concept languages. *Annals of Mathematics and Artificial Intelligence*, 24:23–49, 1998.

[14] P. Blackburn and M. Tzakova. Hybrid languages and temporal logics. *Logic Journal of the IGPL*, 7(1):27–54, 1999.

[15] R. Bull. An approach to tense logic. *Theoria*, 36:282–300, 1970.

[16] S. Demri. Sequent calculi for nominal tense logics: a step towards mechanization? In N. Murray, editor, *Conference on Tableaux Calculi and Related Methods (TABLEAUX), Saratoga Springs, USA*, volume 1617 of *LNAI*, pages 140–154. Springer Verlag, 1999.

[17] S. Demri and R. Goré. Cut-free display calculi for nominal tense logics. In N. Murray, editor, *Conference on Tableaux Calculi and Related Methods (TABLEAUX), Saratoga Springs, USA*, volume 1617 of *LNAI*, pages 155–170. Springer Verlag, 1999.

[18] D. Gabbay. An irreflexivity lemma with applications to axiomatizations of conditions on linear frames. In U. Mönnich, editor, *Aspects of Philosophical Logic*, pages 67–89. Reidel, Dordrecht, 1981.

[19] D. Gabbay. *Labelled Deductive Systems*. Clarendon Press, Oxford, 1996.

[20] G. Gargov and V. Goranko. Modal logic with names. *Journal of Philosophical Logic*, 22:607–636, 1993.

[21] V. Goranko. Hierarchies of modal and temporal logics with reference pointers. *Journal of Logic, Language and Information*, 5(1):1–24, 1996.

[22] B. Konikowska. A logic for reasoning about relative similarity. *Studia Logica*, 58:185–226, 1997.

[23] R. Ladner. The computational complexity of provability in systems of modal propositional logic. *SIAM Journal of Computing*, 6(3):467–480, 1977.

[24] S. Passy and T. Tinchev. Quantifiers in combinatory PDL: completeness, definability, incompleteness. In *Fundamentals of Computation Theory FCT 85*, volume 199 of *LNCS*, pages 512–519. Springer, 1985.

[25] S. Passy and T. Tinchev. An essay in combinatory dynamic logic. *Information and Computation*, 93:263–332, 1991.

[26] A. Prior. *Past, Present and Future*. Oxford University Press, 1967.

[27] A. Prior and K. Fine. *Worlds, Times and Selves*. University of Massachusetts Press, 1977.

[28] M. Reape. A feature value logic. In C. Rupp, M. Rosner, and R. Johnson, editors, *Constraints, Language and Computation*, Synthese Language Library, pages 77–110. Academic Press, 1994.

[29] H. Sahlqvist. Completeness and correspondence in the first and second order semantics for modal logic. In S. Kanger, editor, *Proceedings of the Third Scandinavian Logic Symposium. Uppsala 1973.*, pages 110–143. North-Holland, Amsterdam, 1975.

[30] M. Schmidt-Schauß and G. Smolka. Attributive concept descriptions with complements. *Artificial Intelligence*, 48:1–26, 1991.

[31] J. Seligman. A cut-free sequent calculus for elementary situated reasoning. Technical Report HCRC-RP 22, HCRC, Edinburgh, 1991.

[32] J. Seligman. Situated consequence for elementary situation theory. Technical Report Logic Group Preprint IULG-92-16, Indiana University, 1992.

[33] J. Seligman. The logic of correct description. In M. de Rijke, editor, *Advances in Intensional Logic*, pages 107–135. Kluwer, 1997.

[34] R. Smullyan. *First-Order Logic*. Springer Verlag, 1968.

[35] M. Tzakova. Tableaux calculi for hybrid logics. In N. Murray, editor, *Conference on Tableaux Calculi and Related Methods (TABLEAUX), Saratoga Springs, USA*, volume 1617 of *LNAI*, pages 278–292. Springer Verlag, 1999.

LABELLED ABDUCTION

Compiled Labelled Abductive Systems

Krysia Broda and Dov M. Gabbay

Abstract In this paper it is explained how to use Gabbay's Labelled Deductive Systems (LDS) to provide a framework for abduction. The work builds upon the approach of compiled LDS (CLDS), developed by Broda and Russo, in which a proof procedure for the required LDS system is proved sound and complete with respect to a first order semantics. Previously, CLDS used natural deduction or tableaux style proof methods. This paper considers an example application yielding an abductive system for function free clausal logic. LDS resolution and reduction inference rules are given together with their first order semantics and shown to correspond to standard abduction.

1 OVERVIEW

A methodology for including abduction within LDS is described in this paper. The method is illustrated by means of a particular case study for a clausal logic without function symbols. The approach is general enough to encompass abduction for other logics.

Abduction is a reasoning technique that draws (hypothetical) conclusions from a given theory of background knowledge and a particular piece of input data for a given consequence relation. In particular, given background knowledge $\mathcal{T}$ and input data $\mathcal{O}$, sometimes called the *observation*, and a consequence relation $\vdash_{AP}$, an abduction procedure $\mathcal{AP}$ allows the reasoner to conclude Δ, where $\mathcal{T} \cup \Delta \models \mathcal{O}$ and usually also only when $\mathcal{T} \cup \Delta$ is consistent. Alternatively, it can be viewed as the process of finding missing data needed to derive the conclusion $\mathcal{O}$ from theory $\mathcal{T}$. Gabbay, in [14], described a concept of abduction (within the framework of LDS), based on the following intuitive principles:

***1** Abduction depends on proof procedures for $\vdash_{AP}$.

***2** Abductive principles are to be considered as part of the background knowledge.

D. Basin et al. (eds.), Labelled Deduction, 107–134.
© 2000 *Kluwer Academic Publishers. Printed in the Netherlands.*

The first principle assumes that the natural environment for abduction is a database of background knowledge $\mathcal{T}$ together with an abductive procedure $\mathcal{AP}$ for checking whether queries $\mathcal{O}$ follow from the data. In case $\mathcal{O}$ does follow, it is written as $\mathcal{T} \vdash_{AP} \mathcal{O}$. An abduction principle $\mathcal{AP}$ is a procedure, that incorporates rules $\mathcal{R}$, and that for any $\mathcal{T}$ and $\mathcal{O}$ can produce an (abduced) theory Δ, where $\mathcal{T} \cup \Delta \models \mathcal{O}$. Also, $\Delta = \mathcal{AP}(\mathcal{R}, \mathcal{T}, \mathcal{O})$. The abduction process $\mathcal{AP}$ may involve additional parameters and restrictions. For instance, Δ may have to be consistent with $\mathcal{T}$, or be "minimal" in some sense, or be "useful". An example of a "non-useful" Δ would be the choice of $\Delta = \mathcal{O}$. In many cases, the abduced information in Δ is often restricted to ground literals and sometimes further constrained to belong to a given set of *abducibles*. (See [17].) Applications and investigations of abduction have been made in the area of fault diagnosis [18, 16], but it has also been used in planning applications [21] and to explain negation as failure [15]. Much of the recent work in abduction has been within the realm of logic programming (see [13] for a good survey). An algorithm for abduction in Logic Programming, called SLDNFA, was given in [11]. That algorithm dealt with first order abducibles and assumed Logic Programming completion semantics.

This paper is concerned with refining the abduction concept by adding a logic component to control the abduced set Δ. To explain what this may involve, reconsider the basic situation $\Delta = \mathcal{AP}(\mathcal{R}, \mathcal{T}, \mathcal{O})$. Suppose that $\mathcal{AP}$ is a two-stage process, in which *Stage 1* is a rough algorithm giving several options for possible candidate sets of abducibles, Δ_1, Δ_2, Δ_3, and *Stage 2* employs an additional logic $\mathcal{L}_L$, to allow reasoning amongst the Δ_i to extract an acceptable Δ. The procedure $\mathcal{AP}$ is a bridge between the two logics. The logic $\mathcal{L}_L$ may be in a language and logic different from that of $\mathcal{T}$, $\mathcal{O}$ and $\mathcal{R}$, giving a third principle

***3** Abduction is a 2 stage process involving one logic in which the abduction is made and a second logic to reason about the abducibles.

A typical example is abductive explanation of a missing premise in a philosophical argument or an economic/political argument of a historical figure. There may be several options for the missing premise, but an external logic of the historical context may dictate which is the most plausible. For example, what were the implicit assumptions of Socrates, when he argued for "the good"? This paper offers this separation of logics for abduction in the context of clausal reasoning, by employing the framework of LDS. The advantages that may accrue from using LDS will derive from the ability to separate the logic of the non-abducibles from that of the abducibles. An interesting and wide-ranging discussion of abduction can be found in [1] and see, of course, [14] (chapter 11). The next example informally introduces the notion of consistency that sanctions the drawing of abductive consequences used in this paper.

Example 1 *Let $\mathcal{T}$ be the theory $\{P \leftarrow \neg a,\ \ Q \leftarrow P\}$ and suppose that the input data to be "explained" is Q. Let the set of abducibles be $\{a, b\}$. Then clearly Q can be deduced if $\neg a$ holds, or $\neg a$ is a candidate for Δ (Stage 1). A simple abductive procedure is to allow the conclusion $\neg a$ to be drawn if it is consistent with $\mathcal{T}$ (Stage 2).*

Now, suppose that the theory includes the additional data $\neg(\neg b \wedge \neg a)$. The previous 'solution' of $\neg a$ is still consistent with the theory, but now the abducible b can be deduced. To prevent an inconsistency arising from a subsequent abduction of $\neg b$ and to enable speedy rejection of such a step, b is abduced as well. The abduction of b could also be seen as being true to the idea of abducibles – if an abducible is derivable it should be abduced also. This is also sanctioned by the abductive procedure (Stage 2).

In case the extra data had been $a \vee b \vee c$ then either of b or c could be abduced to satisfy this stronger notion of "consistency". Data such as $\neg(\neg a \wedge \neg b)$ are often treated as constraints on a theory and dealing with the consequences of abduced literals with such data is known as the ramification problem. *The proposed syntax will, in fact, represent data involving only abducibles in the form $\neg(\neg a \wedge \neg b)$, so abducibles occur only as conditions of implications.*

The strong notion of consistency mentioned in Example 1 is necessary in some applications of abductive reasoning. For example, in planning using the event calculus [12], where abducibles are predicates indicating particular actions have happened, it is necessary to abduce actions that are implied by the theory.

This paper shows how an abductive reasoning process might be implemented in LDS. The particular restrictions chosen here can be varied in many different ways, without affecting the basic approach. The reasoning engine in this paper for dealing with negative literals is classical clausal logic. This is chosen because it is a familiar system and will therefore not obscure the ideas of the new LDS approach. There are other alternatives, related to logic programming, such as negation as failure (naf) [10]. Indeed, this flexibility is one of the reasons for using LDS, in that it enables the logic that controls selection of abduced formulas to be different from the main logic of the application. An interesting combination would be a logic programming procedure for the main logic of wffs and a classical procedure for dealing with the abducibles. An abductive procedure may also allow non-atomic conclusions to be drawn.

Labelled deductive systems (LDS) were introduced by Gabbay [14] as a uniform framework so that differences between logics of the same family can be presented uniformly, the differences being captured by a *labelling algebra*. Furthermore, in the LDS framework, different families of logics can be treated in a uniform way. Instead of dealing with wffs and various side conditions necessary to make valid deductions from them, the syntax deals with *declarative units* of the form $\alpha : \lambda$, where α is a wff from a particular language (for example, modal logic) and λ is a term, called a *label*, from a *labelling language*.

Depending on the logic family the term λ is interpreted differently. In modal logic, for instance, the declarative unit $\alpha : \lambda$ is interpreted as α is true at point, or possible world, λ. The structure and symbols of the labelling language differ for each family of logics, but remain the same within one family. In this paper, α is a formula from a clausal logic language $\mathcal{L}_P$, and λ is a term that represents a subset of the set of all abducibles, which are also ground literals from a restricted language. A ground declarative unit $\alpha : \lambda$ can be interpreted informally as "if abducibles in λ can be believed then α can be derived". Inference rules operate on the whole declarative unit.

In particular, this paper describes a development of the abduction framework within the Compiled Labelled Deductive System (CLDS) framework [20, 8]. Such a system compiles the two logics into a single logic, in this case classical first order logic, which is used to provide a common semantics. In some cases also, the compilation allows for a standard theorem prover to be employed for implementation. In the previous CLDS applications a semantics for a LDS was given in terms of first order schemas [8]. The various inference rules of the system were then proved sound and complete with respect to the semantics, using standard methods which were adapted to the syntax using declarative units. The first order schemas could be compiled into a first order theory by instantiating the axioms for any particular problem, which could then be submitted to a standard first order theorem prover. For any given CLDS for a particular logic, correspondence with a standard approach for that logic is proven, for which it is usually required to restrict the initial theory, showing that a CLDS theory for a particular logic generalises the standard system for that logic. The application to abduction can also follow this structure as is outlined in [7].

A CLDS theory consists of a *configuration*, which is a tuple $\mathcal{C} = \langle \mathcal{D}, \mathcal{T} \rangle$ (and possibly other elements), in which $\mathcal{D}$ is referred to as a *diagram* and $\mathcal{T}$ as a theory. A diagram in general consists of literals stating certain properties of labels, for instance, which label is related to which, and, in this application, what abducibles are believed. These things are all governed by a labelling algebra. The theory $\mathcal{T}$ relates hypothetical beliefs λ to clauses α, a membership statement having the form $\alpha : \lambda \in \mathcal{T}$ (or sometimes $\in \mathcal{C}$ when there should be no confusion). A CLDS is completed by defining a set of inference rules that derive one configuration from another and the labelling algebra. The presentation given here uses inference rules based on resolution since the syntax is based on clauses. Of course, the usual refinements for resolution could be imposed on top of the rules given here if required, or other systems such as tableaux could be used, without altering the basic idea. From these rules a first order semantics for abduction is derived, which is sound and complete with respect to the inference rules. The correspondence with the standard approach to abduction is also made. The CLDS system for abduction will be referred to as A_{CLDS}.

In order to be able to compare the CLDS framework for abduction with a standard framework, some definitions and algorithms used in the standard approach are given briefly in the next section. The reader should refer to [17, 13] for a fuller account. Section 3 details the CLDS framework for abduction and the semantics is described in Section 4 A worked example is given in Section 5, whilst some concluding remarks are made in Section 6

2 ABDUCTION FOR CLAUSES

In this section some terms related to abduction and used elsewhere in the paper are given. The presentation is restricted to a clausal language without function symbols.

Syntax. Given the signature $\mathcal{H} = \{CS, PS, AS\}$, where CS, PS and AS are disjoint finite sets of symbols and CS is a set of constant symbols, PS is a set of predicate symbols including $\perp$ representing falsehood, and AS is a set of abducible predicate symbols, then $\mathcal{L}_P$ is the usual clausal language based on $\mathcal{H}$. (Elements of PS and AS may have arity zero.) Elements in AS will usually begin with script letters, such as $a, b, c, \ldots$, drawn from the beginning of the alphabet. The variables $\mathcal{V}$ of the language are a countable set of names disjoint from all other symbols and whose initial letter is drawn from the set $\{u, v, w, x, y, z\}$. The set of ground atoms formed from predicate symbols and constants from $\mathcal{H}$ is called the *Base*, but more generally, an *atom* may also include variables from $\mathcal{V}$ as arguments. A *literal* is an atom or negated atom. The *complement* of an atom is its negation and the complement of a negated atom is the atom. An element of $\mathcal{L}_P$, a *clause*, is a universally quantified finite disjunction of literals. If a clause contains only positive/negative/abducible literals it is called *positive/negative/abducible.* A clause with exactly one literal is called a *fact* and a clause containing a literal and its negation is a *tautology.* A clause with no variables is called *ground.* The empty clause is written as $\perp$.

Resolution Rules of Inference. For clausal theories the standard inference rules are resolution, subsumption and factoring. These are defined below, together with substitutions; for more information the reader is referred to [19].

A *substitution* σ is a finite set of (*variable,term*) pairs, where each variable is *bound* to the corresponding term. Then $Vars(\sigma)$ is $\{v \mid (v,t) \in \sigma\}$ and if $v \in Vars(\sigma)$ then $v \notin t$ for any term t of a pair in σ. Given a substitution σ and a term, literal, or clause, E, $E\sigma$ is the result of *instantiating* E by σ and is the term, literal, or clause that results from simultaneously replacing each occurrence of $v \in Vars(\sigma)$ by the corresponding term t of the pair $(v,t) \in \sigma$.

The restriction on variables in $Vars(\sigma)$ leads to substitutions being maintained in an idempotent, or fully computed form, so that $(E\sigma)\sigma = E\sigma$. A *unifier* of two terms or literals t_1, t_2 is a substitution that, when applied to t_1 and t_2, gives the same result. A *most general unifier* θ of t_1 and t_2 , written m.g.u., is a unifier of t_1 and t_2 such that, for any other unifier σ of t_1 and t_2, there exists a substitution ρ satisfying $(t_1\theta)\rho = (t_2\theta)\rho = t_1\sigma = t_2\sigma$. Clauses are maintained in variable disjoint form, that is, no two clauses share a common variable. Given two clauses $G \vee L_1$ and $H \vee \neg L_2$, where G and H are clauses and L_i are atoms, their *resolvent* is $(G \vee H)\theta$, where θ is the m.g.u. of L_1 and L_2. L_1 and $\neg L_2$ are called *complementary*. G *subsumes* H if $G\sigma \subseteq H$, and G *factors* to G' if $G' = G\sigma \subset G$, both for some σ, treating the clauses as a set of literals, so that identical literals in $G\sigma$ are merged into one. A *refutation* is a clausal derivation of $\bot$.

The next definition is crucial to the approach taken in this paper and distinguishes between the standard (weak) notion of consistency, here called potential consistency, and a stronger notion that requires derivable abducibles to be explicitly abduced.

Definition 2 *Let $\mathcal{T}$ be a clausal theory and Δ be a set of ground abducible literals such that there is no abducible and its negation in Δ. Then $\mathcal{T} \cup \Delta$ is* potentially consistent *iff the empty clause cannot be derived from $\mathcal{T} \cup \Delta$.*

$\mathcal{T} \cup \Delta$ is consistent *iff it is potentially consistent and no abducible clause can be derived from $\mathcal{T} \cup \Delta$ other than those subsumed by an element of Δ. $\mathcal{T} \cup \Delta$ is* inconsistent *if it is not potentially consistent.*

This definition will be repeated in Section 3, but in terms of the LDS framework. However, it is noted here that one effect of the definition is to force a consistent theory (and abducibles) to include in Δ one abducible from each instance of a derivable abducible clause.

Example 3 *The theory $\{P \vee a, a \vee b, \neg a\}$, where a, b are abducible propositions, is potentially consistent but not consistent. The theory $\{P \vee a, a \vee b, \neg a\} \cup \Delta$, where $\Delta = \{b\}$, is consistent. The theory $\{a \vee b, \neg a, \neg b\}$ is inconsistent, as is $\{a \vee b, \neg a\} \cup \Delta$, where $\Delta = \{\neg b\}$.*

A Simple Abductive Procedure. To prove an observation, the clausal form of its negation is formed (and will here be denoted by $\mathcal{O}$ and also called the observation), and refutations of $\Delta \cup \mathcal{T} \cup \mathcal{O}$ are found by abducing Δ. The observation will not contain any occurrences of abducible predicates. A typical abductive procedure within a clausal framework, from a (potentially consistent) theory $\mathcal{T}$ and observation $\mathcal{O}$, might proceed as follows:

Stage 1: A derivation of the empty clause $\bot$ from $\mathcal{T} \cup \mathcal{O}$ is sought. In case $\bot$ cannot be derived, if there is a derivation of an abducible clause then the set of complements of the literals in this clause is called Ass. The elements

in this set, if assumed, would enable the empty clause to be derived. Only non-tautologous abducible clauses will be considered in this step, in order that the set Ass be internally consistent.
Stage 2: Ass is then checked for (strong) consistency with the theory $\mathcal{T}$. In case the empty clause cannot be derived all derivable abducible clauses are found and at least one literal from each ground instance is abduced. This is possible since Ass is potentially consistent with the theory. If Ass is not potentially consistent with $\mathcal{T}$ then the set is abandoned and a second possible set of assumed abducibles is found by returning to stage 1.

At the end of stage 1, in case the derived abducible clause is not ground, then it is first instantiated in all possible ways. Any one of the instances can be selected to form Ass. There are other ways to deal with non-ground abducible clauses, analogous to the introduction of free variables into tableaux systems [6], but these are not considered here. Alternatively, the theory could be instantiated to a ground theory prior to searching for the empty clause. In any case, it is assumed in this paper that the set Ass only consists of ground literals.

It is also possible to interleave the consistency check with the formation of Ass: whenever a clause is derived containing a ground abducible literal a, its complement is assumed and a consistency check is made immediately. As mentioned before, the strong consistency check for an abducible b, say, may require additional literals to be abduced. For example, if the clause $\neg c$ is derived during the consistency check for b, then the assumption of the abducible $\neg c$ should be made. This will ensure that c can never later be asserted as a belief (for if it were asserted, it would lead to inconsistency). In this case stage 2 must be repeated for $\neg c$. A set Ass is a *solution* if $\mathcal{T} \cup Ass$ is consistent and $\mathcal{T} \cup Ass \cup \mathcal{O} \models$.

In the next section the abductive procedure will be presented differently, in the framework of CLDS. The benefit of this approach is in the separation of the reasoning with abducibles from the ordinary reasoning with the theory, so that the reasoning for abducibles and the theory can be in different logics.

3 ABDUCTION IN A CLDS FRAMEWORK

Syntax. As mentioned in Section 1, CLDS operates within a framework involving two kinds of syntactic entities: declarative units and "belief literals" (literals taking labels as arguments, that are part of a diagram), which will henceforth be called B-literals. The B-literals involve a unary predicate B that allows for assumptions about abducibles to be accumulated separately from the theory. The inference rules will take account not only of the atomic structure of a declarative unit but also of its label and of the B-literals. A declarative unit $\alpha : \lambda$, which can be read as "if (conjunction of) beliefs in λ are held then

α holds", will correspond to the clause $\alpha \leftarrow \bigwedge(\lambda)$, or $\alpha \vee \neg(\lambda)$, where $\neg(\lambda)$ is the disjunction of the negation of each element of λ. As in the clausal framework of Section 2 the wffs α are drawn from a clausal first order language $\mathcal{L}_P$ without function symbols and based on a signature of constants CS, predicates PS and abducible predicates AS, with variables from $\mathcal{V}$. The labels λ belong to the labelling language $\mathcal{L}_L(A)$. The combined languages $\mathcal{L}_P$ and $\mathcal{L}_L(A)$ are called $Func(\mathcal{L}_P, \mathcal{L}_L(A))$.

Definition 4 *Let AS be a finite set of abducible predicates disjoint from PS. The set A of atoms and their negations formed from AS, CS and $\mathcal{V}$ are called* abducibles, *or* atomic labels *and $A_{gd} \subset A$ is the set of ground abducibles. Often, A is used also to refer to the set of atoms only, the inclusion of the negations being implicitly assumed. A set of atomic labels, a subset of A, is also called a label. The* Labelling language *$\mathcal{L}_L(A)$ includes elements of the following two forms:*

- *Self-consistent sets of abducibles (labels);*
- *B-literals $B(x)$, where $x \in A_{gd}$.*

A label in which the atomic labels λ and λ' are both members is called self-inconsistent. *A label is* self-consistent *if it is not self-inconsistent.*

The labels of $\mathcal{L}_L(A)$ are usually written without the set brackets for convenience. Thus $\{a, b, c\}$ is written as abc (or cba as the order doesn't matter). (The argument A of $\mathcal{L}_L(A)$ is also often omitted, when it is clear from the context.) $B(a)$ is true iff a is believed. Negated B-literals are not encountered in this application. The negation of an atomic label $\lambda \in A$ is called its *conjugate* and is written as λ', where $\lambda' \in A$ also. The atomic labels λ and λ' obviously correspond to the literals λ and $\neg\lambda$ respectively. (Remember that in a clause the elements of a label are negated, so these would actually appear in a disjunction as $\neg\lambda$ and λ.) Note that, if $a(x)$ and $\neg a(y)$ both belong to some label, that label is not self-inconsistent, since x and y need not be instantiated identically. A declarative unit $\alpha : \lambda$ is also called a *labelled clause* and satisfies the property that λ is self-consistent. In case λ is not a ground label and θ is a substitution, then $\lambda\theta$ is the result of applying θ to the atomic labels in λ.

Two combination operations are defined on self-consistent labels:

- The join of λ_1 and λ_2 is $\lambda_1 \cup \lambda_2$, unless this is self-inconsistent, in which case the join is not defined. It is written as $join(\lambda_1, \lambda_2)$.
- The combination of λ_1 and λ_2 with respect to c and d is $(\lambda_1 \cup \lambda_2)\theta - \{c, d\}\theta$, where $c \in \lambda_1$, $d \in \lambda_2$ and $d\theta = c'\theta$ for m.g.u. θ of c and d, unless the result is self-inconsistent, in which case the combination is not defined. It is written as $Res(\lambda_1, \lambda_2, c, d)$ because of its relationship with resolution. In case $d = c'$, then it is written $Res(\lambda_1, \lambda_2, c)$.

These two properties, together with some other useful properties, are collectively called an abductive labelling algebra and are formally defined in Definition 5.

Definition 5 *Let $\mathcal{L}_L(A)$ be a labelling language based on the set of abducibles A. An* abductive labelling algebra *($\mathcal{L}_A$) consists of the following properties on elements from $\mathcal{L}_L(A)$. Note that, in property (R1), the unifier θ is the most general unifier of λ_4 and λ'_3.*

(J) $\forall \lambda, \lambda_1, \lambda_2[\lambda = join(\lambda_1, \lambda_2) \leftrightarrow \lambda = \lambda_1 \cup \lambda_2 \wedge \neg\exists\lambda_3[\lambda_3 \in \lambda \wedge \lambda'_3 \in \lambda]]$

(R1) $$\forall \lambda, \lambda_1, \lambda_2, \lambda_3, \lambda_4 \left[\begin{array}{l} \lambda = Res(\lambda_1, \lambda_2, \lambda_3, \lambda_4) \leftrightarrow \\ \left[\begin{array}{l} \lambda_3 \in \lambda_1 \wedge \lambda_4 \in \lambda_2 \wedge \neg\exists\lambda_5(\lambda_5 \in \lambda \wedge \lambda'_5 \in \lambda) \wedge \\ \exists\theta(\lambda_4\theta = \lambda'_3\theta) \wedge \lambda = (\lambda_1 \cup \lambda_2)\theta - \{\lambda_3\theta, \lambda_4\theta\} \end{array} \right] \end{array} \right]$$

(I) $\forall \lambda . \neg\exists\lambda_1[\lambda_1 \in \lambda \wedge \lambda'_1 \in \lambda]$

(B) $\forall \lambda[B(\lambda) \rightarrow \exists a[a \in A_{gd} \wedge \lambda = a]]$

(Con) $\neg\exists\lambda[B(\lambda) \wedge B(\lambda')]$

Inference rules and derivations. A *labelled clause* is a declarative unit of the form $\alpha : \lambda$, where α is a first order clause from $\mathcal{L}_P$ and λ is a label from $\mathcal{L}_L$ satisfying (I). Notice that, whereas in clausal abduction a clause may have a positive abducible literal, for example $P \vee b$ (b abducible), as a labelled clause it is represented as $P : b'$. An abducible clause is always represented as $\perp : \lambda$ for some λ. For example, $\neg(\neg a \wedge \neg b)$ becomes $\perp : a'b'$.

A *configuration* of A_{CLDS} is the triple $(\mathcal{T}, \mathcal{D}, \langle \mathcal{T}_0, \mathcal{O} \rangle)$, where $\mathcal{D}$ is a *diagram* and consists of a set of B-literals satisfying (Con) and (B), $\mathcal{T}$ is a *labelled theory*, a set of labelled clauses, $\mathcal{T}_0$ is the initial theory and $\mathcal{O}$ is the set of labelled clauses derived from the negated initial observation. The labels of clauses in $\mathcal{O}$ are all empty. The initial configuration is defined as $\mathcal{C}_1 = (\mathcal{T}_1, \mathcal{D}_1, \langle \mathcal{T}_0, \mathcal{O} \rangle)$. The initial theory $\mathcal{T}_1$ is the union of $\mathcal{T}_0$ and $\mathcal{O}$, and the initial diagram $\mathcal{D}_1$ is the empty set. These assumptions will be made in the rest of the paper, although the initial emptyness restriction on $\mathcal{D}_1$ could be relaxed. Inference rules will be applied to derive new configurations, including the addition of beliefs to the diagram. Let $\mathcal{C} = (\mathcal{T}, \mathcal{D}, \langle \mathcal{T}_0, \mathcal{O} \rangle)$ be an A_{CLDS} configuration, then $\mathcal{C}^c = (\mathcal{T}_0, \mathcal{D}, \langle \mathcal{T}_0, \mathcal{O} \rangle)$ is called the *corresponding configuration* and is used to detect consistency of the current B-literals in $\mathcal{D}$ with the initial theory $\mathcal{T}_0$. A set of B-literals satisfying (Con) is also called *self-consistent* and a ground label λ is *self-consistent with a diagram D* iff $\{B(z) | z \in \lambda\} \cup \mathcal{D}$ is self-consistent.

Just as with clausal abduction there is a strong and a weak form of consistency for a configuration. For the strong form, the belief set must satisfy the property that no derivation of $\perp : \emptyset$ is possible from the configuration, whatever additional beliefs might be made. For the weaker form, called *potential consis-*

tency, there is a belief set B that, if added to the configuration, would enforce strong consistency. Of course, in order to show potential consistency, it is still necessary to find a set of beliefs which would enforce the strong consistency! Initial corresponding configurations can either be required just to be potentially consistent, leaving it to the computation to build up a suitable belief set B, or they can be pre-processed to derive a belief set B that enforces consistency. In that case, the initial diagram would be B. Because there may be many different consistent belief sets B, rather than choose between them it is easier to make the first assumption, which is adopted here.

Definition 6 *Let* $\mathcal{C} = (\mathcal{T}, \mathcal{D}, \langle \mathcal{T}_0, \mathcal{O}\rangle)$ *be an* A_{CLDS} *configuration.* $\mathcal{C}$ *is* consistent *iff the empty labelled clause* $\perp : \emptyset$ *cannot be derived from it using the rules Rs defined in Definition 8.*

$\mathcal{C}$ *is* potentially consistent *if there exists a diagram* $\mathcal{D}' \supseteq \mathcal{D}$ *such that* $(\mathcal{T}, \mathcal{D}', \langle \mathcal{T}_0, \mathcal{O}\rangle)$ *is consistent.*

$\mathcal{C}$ *is* inconsistent *if it is not potentially consistent.*

$\mathcal{C}$ *is* minimally consistent *iff* C *is consistent and for no subset of* $\mathcal{D}'$ *of* $\mathcal{D}$ *is* $(\mathcal{T}, \mathcal{D}', \langle \mathcal{T}_0, \mathcal{O}\rangle)$ *consistent. Note that proper subsets of the diagram in a minimally consistent* $\mathcal{C}$ *will still be potentially consistent, by definition.*

The definitions in 6 and 2 are not exactly the same, as the abduction procedure dynamically affects the configuration. However, in Section 4 it is shown that the two definitions are equivalent, in that they classify clausal theories and the corresponding configurations in the same way.

Example 7 $\mathcal{T}_0 = \{P : a', \perp : a'b'\}$, $\mathcal{D} = \{B(a')\}$ *is potentially consistent, since the additional belief* $B(b)$ *makes it impossible to derive* $\perp : \emptyset$*; it is not consistent since the additional belief* $B(b')$ *allows* $\perp : \emptyset$ *to be derived. However, if* $\mathcal{D} = \{B(a'), B(b)\}$ *then it is consistent too.*

Definition 8 *Let* A *be a set of abducibles,* $\mathcal{L}_L(A)$ *and* $\mathcal{L}_A$ *be the associated labelling language and labelling algebra and* $\mathcal{L}_P$ *be a clausal language, then the* Abductive LDS system $(\mathcal{L}_L(A), \mathcal{L}_P, \mathcal{L}_A, Rs)$ *is referred to as an* A_{CLDS}*. Rs is the set of inference rules on configurations, namely, rules (LR), (LC), (BP), (B-A), (B-C), (LI), (LS), (LF), defined below.*

Definition 9 *If* $A = (\mathcal{L}_L, \mathcal{L}_P, \mathcal{L}_A, Rs)$ *is an* A_{CLDS}*, then a* derivation *of* $\mathcal{C}' = (\mathcal{T}', \mathcal{D}', \langle \mathcal{T}_0, \mathcal{O}\rangle)$ *from the initial configuration* $\mathcal{C}_1 = (\mathcal{T}_1, \mathcal{D}_1, \langle \mathcal{T}_0, \mathcal{O}\rangle)$*, written as* $\mathcal{C}_1 \vdash_{AP} \mathcal{C}'$*, is a finite sequence of proof steps each using one of the rules in Rs. A derivation is a* labelled refutation *of* $\mathcal{C}_1$ *if* $\mathcal{T}'$ *includes the clause* $\perp : \emptyset$*. The beliefs in* $\mathcal{D}'$ *give rise to the abduced set* $\Delta = \{z : B(z) \in \mathcal{D}'\}$.

Example 1 can easily be recast into this framework as follows:

Example 10 *Let the set of atomic labels in A be $\{a, b\}$ and the theory be: $Q \vee \neg P : \emptyset$, $P : a'$, $\bot : a'b'$. The negation of the observation is $\neg Q : \emptyset$, ready to search for a refutation (i.e. a derivation of $\bot : \emptyset$). The labelled clause $\bot : a'$ can be derived. The belief $B(a')$ is then recorded and the data processed to account for this. In particular $\bot : b'$ is derivable, showing the beliefs are not consistent if $\neg b$ is provable. To avoid this outcome the belief $B(b)$ is assumed. The final result is the set of beliefs b and a' (i.e. b and $\neg a$) as before.*

The assumption of beliefs $B(a')$ and $B(b')$ would also lead to $\bot : \emptyset$ being derivable from the original theory, showing it was only potentially consistent. The initial theory will always be assumed to be potentially consistent and in this example it will be consistent if b is abduced.

A corresponding configuration is always potentially consistent, for $\mathcal{T}_0$ is potentially consistent and B-literals are only added to the initial and subsequent diagrams if they maintain consistency (and hence potentially consistency) with $\mathcal{T}_0$. (Since the values of $\mathcal{T}_0$ and $\mathcal{O}$ remain unchanged throughout a derivation, they will sometimes be omitted from the configuration triple.) A labelled clause or B-literal is said to be *dependent* on observation $\mathcal{O}$ if its derivation involves some clause in $\mathcal{O}$.

The next sequence of definitions defines the set of abductive inference rules used in this paper. The notation $\mathcal{C} + [\pi]$ denotes the addition of declarative unit or B-literal π to the appropriate part (theory or diagram) of configuration $\mathcal{C}$ as an assumption, later to be discharged. The notation $\mathcal{C} + \pi$ denotes the addition of declarative unit or B-literal π to the appropriate part (theory or diagram) of configuration $\mathcal{C}$.

Definition 11 *Let $G_i \vee L_i : \lambda_i$, $i = 1, 2$, be two labelled clauses such that L_1 and L_2 are complementary with m.g.u. θ and $\lambda_3\theta = join(\lambda_1\theta \cup \lambda_2\theta)$ is self-consistent, then the labelled clause $(G_1 \vee G_2)\theta : \lambda_3\theta$ is the result of* Labelled Resolution *(LR).*

Definition 12 *Let $G_i : \lambda_i$, $i = 1, 2$, be two labelled clauses such that $Res(\lambda_1, \lambda_2, c, d)$ is defined for some atomic labels c, d, then the* labelled combination *(LC) of G_1 and G_2 is $(G_1 \vee G_2)\theta : Res(\lambda_1, \lambda_2, c, d)$, where θ is the m.g.u. of c and d.*

In case there are more than one such pair of atomic labels c, d, there is more than one Labelled combination. It would be possible to restrict the use of (LC) to ground labels by prior use of the rule (LI) in the case when c and d were not ground.

Definition 13 *Let $G_i : \lambda_i$, $i = 1, 2$, be two labelled clauses, such that $G_1\theta \subseteq G_2\theta$ and $\lambda_1\theta \subseteq \lambda_2\theta$. Then $G_1 : \lambda_1$* labelled subsumes *(LS) $G_2 : \lambda_2$. The subsumed clause can be removed from $\mathcal{T}$.*

Definition 14 *Let $G : \lambda$ be a labelled clause, $G\theta$ factor to G^f and $\lambda\theta$ factor to λ^f. Then the clause $G^f : \lambda^f$ is a* labelled factor *(LF) of $G : \lambda$. The factored clause is added to $\mathcal{T}$.*

It is assumed that identical literals in either the wff or the label of a labelled clause are always merged; thus factoring must involve some instantiation in either G or λ. It is unnecessary in the propositional case.

Definition 15 *Let $G : \lambda$ be a labelled clause such that λ is not ground and $\lambda\theta = \lambda'$ be a ground instance of λ that is self-consistent, then $G\theta : \lambda'$ is a* labelled ground instance *of $G : \lambda$ formed by* labelled instantiation *(LI). All such labelled ground instances are added to $\mathcal{T}$.*

Note that it is not necessary to force G (as well as λ) to be ground by the above rule. The rule is needed as the rules (B-A) and (B-C) require ground labels in labelled clauses of the form $\perp : \lambda$. It is possible to add one labelled ground instance at a time as long as some record is kept of which instances have been added. Once all instances have been added the original clause can be removed since it is equivalent to their conjunction. The next three rules deal with B-literals; the reader is reminded that the argument of a B-literal is always ground.

Definition 16 *Let $\mathcal{C} = (\mathcal{T}, \mathcal{D})$ be an A_{CLDS} configuration and $B(a) \in \mathcal{D}$. Then $\mathcal{T}' = \{G : \lambda'' \mid G : \lambda \in \mathcal{T} \wedge a \in \lambda \wedge (\lambda'' = \lambda - \{a\})\} \cup \{G : \lambda'' \mid G : \lambda'' \in \mathcal{T} \wedge a' \notin \lambda'' \wedge a \notin \lambda''\}$ is the new theory formed by* belief propagation *(BP) of belief $B(a)$. The other parts of $\mathcal{C}$ are unchanged. Because a is ground, a' is also ground, so no substitution is involved.*

In other words, after applying (BP) using $B(a)$, declarative units with a label containing a' are removed (effectively by subsumption by a), and the labels of the others are reduced by a if it occurs in them. In case λ needs to be instantiated to force a and a' ground, the rule (LI) is first used. The rule could be made more general by considering propagation to labelled clauses $G : \lambda$ in which either a or a' occurred in $\lambda\theta$ for some θ. Such a generalisation would again require constraints to indicate which instances of labelled clauses had been utilised, to avoid a lot of subsumption checks.

Definition 17 *Let $C = (\mathcal{T}, \mathcal{D}, \langle \mathcal{T}_0, \mathcal{O} \rangle)$ be an A_{CLDS} configuration and $\mathcal{C}^c$ be the corresponding configuration. If $\perp : \lambda \in C$, $\lambda \neq \emptyset$ and λ is ground, then $\mathcal{C}' = (\mathcal{T}, \mathcal{D}', \langle \mathcal{T}_0, \mathcal{O} \rangle)$ is derived by (B-A), where $\mathcal{D}' \supseteq \mathcal{D} \cup B(\lambda)$, $B(\lambda) = \{B(x) | x \in \lambda\}$ and D' is self-consistent, and only in case $(\mathcal{T}_0, \mathcal{D}', \langle \mathcal{T}_0, \mathcal{O} \rangle)$ (called the (B-A) augmented corresponding configuration), is consistent. There may be several $\mathcal{D}'$ that can be chosen; they are not arbitrarily selected, but arise through application of the (B-C) rule, when applied to check consistency of the added B-literals in $B(\lambda)$.*

This rule is used to enable the subsequent derivation of $\bot : \emptyset$ in $\mathcal{C}'$. If λ is not ground then an application of (LI) must first be made. The rule is only applied if the new B-literals do not violate the consistency of $\mathcal{C}^c$. (It is also possible to vary this rule and only add $B(\lambda)$ to the diagram, leaving the computation to build-up the other beliefs required for consistency; in other words, to maintain potential consistency only.) The rule is also not specific about the new diagram $\mathcal{D}'$. To enforce a minimally inconsistent configuration might impose an unnecessary burden on the implementor, whereas to overconstrain $\mathcal{D}'$, making it larger than necessary, may also increase the search space. The reader may also wonder, rightly, if it is necessary to continually revert to $\mathcal{T}_0$ every time in order to check consistency, for surely, many of the deduction steps already taken to derive $\mathcal{T}$ could also have been made from $\mathcal{T}_0$, and some repetition could be saved. For presentation of the method it seems simpler to use $\mathcal{T}_0$ each time, but an implementation would no doubt use the updated theory and avoid recomputing those steps that were not dependent on the observation $\mathcal{O}$ (if any).

Definition 18 *Let $C = (\mathcal{T}, \mathcal{D}, \langle \mathcal{T}_0, \mathcal{O} \rangle)$ be an A_{CLDS} configuration and $\mathcal{C}^c$ be the corresponding configuration. If $\bot : \lambda \in C$, for ground $\lambda \neq \emptyset$, then $\mathcal{C}' = (\mathcal{T}, \mathcal{D}', \langle \mathcal{T}_0, \mathcal{O} \rangle)$ is derived by (B-C), where $\mathcal{D}' \supseteq \mathcal{D} \cup B(x')$ for some $x \in \lambda$ and $\mathcal{D}'$ is self-consistent, and only in case $(\mathcal{T}_0, \mathcal{D}', \langle \mathcal{T}_0, \mathcal{O} \rangle)$ (called (B-C) augmented corresponding configuration), is consistent.*

This rule is used in an attempt to prevent the derivation of $\bot : \lambda$ in $\mathcal{C}^c$ by causing it to be subsumed by later applications of the (BP) rule. The rule is usually initiated by an application of (B-A). The (B-C) rule is useful to show consistency. It implements the Stage 2 process referred to in the introduction.

These eight basic operations are adequate to build an abductive framework in LDS and will form the proof rules of the A_{CLDS} system. A particular algorithm using the above rules is given next, that follows the stages outlined in Section 2 The problem raised after Example 10 can now be clarified. The initial corresponding configuration in that example was only potentially consistent. There is a set of beliefs, possibly more than one, that would make it consistent, but rather than enforcing any particular set of such beliefs, it is left for them to evolve through the application of (B-A) and (B-C) rules. These rules *do* enforce consistency, and so at the first such application at least one of those belief sets will be found. But whereas, at the start, there may be many belief sets to enforce consistency, after some rule applications of (LR) and other "resolution" steps, only some of these belief sets may be possible as a result of the first (B-A) application.

Example 19 *Let the initial (labelled) theory $\mathcal{T}_1$ be $\{Q : \emptyset, \neg Q : b', \bot : ab\}$, where $Q : \emptyset$ is the observation. It is potentially consistent, but not consistent until the addition of some beliefs. If a computation is started from $\mathcal{T}_1$, then immediately there are two possibilities: The observation can be used in the* *initial step, as in set-of support refinements for resolution [9], and then the (B-A) rule is attempted with new belief $B(b')$. Consistency is achieved with $\mathcal{T}_0$ and then (BP) yields the desired result. If the observation is not immediately used, then (B-A) might be attempted with beliefs $B(a)$ and $B(b)$, which will not lead to consistency with $\mathcal{T}_0$, so the rule cannot finally be applied. On the other hand, if the initial configuration was made consistent by adding either $B(a')$ or $B(b')$, then the labelled clause $\bot : ab$ would be subsumed immediately, avoiding the problem. But only the second of the two consistent configurations leads to a solution. The theory $\mathcal{T}_0 = \{Q : a, \neg Q : b', \bot : a', \bot : b\}$ is not even potentially consistent as it cannot be made consistent by applying the (B-C) rule.*

It would appear, therefore, that a good strategy is to make use of the observation if possible, otherwise all possible ways of deriving inconsistency from $\mathcal{T}_0$ will be investigated to no gain. It may be a good plan to check consistency for one abducible at a time in rule (B-A), especially if the clauses to which (BP) applies have to be instantiated. If they do not have to be so, then it is probably better to perform the check on all abducibles in $B(\lambda)$ at once. It is clear that many refinements of the rules exist and the ones chosen here are just one possibility (the simplest).

A Simple Abductive Procedure in A_{CLDS}. Assume an initial configuration $C_1 = (\mathcal{T}_1, \mathcal{D}_1, \langle \mathcal{T}_0, \mathcal{O} \rangle)$, together with corresponding initial configuration $\mathcal{C}_1^c = (\mathcal{T}_0, \mathcal{D}_1, \langle \mathcal{T}_0, \mathcal{O} \rangle)$, which is assumed potentially consistent. (In fact, if it is required to know whether $\mathcal{C}_1^c$ *is* consistent, then the (B-C) rule can be used to find out. On the other hand, if $\mathcal{C}_1^c$ is not potentially consistent then, at the first attempt to check consistency, possible consistent extensions of $\mathcal{D}_1$ will be investigated and it will be disovered that there are none.) Steps similar to those outlined in Section 2 can now be applied using the A_{CLDS} inference rules in order to obtain a refutation. The set of belief literals give rise to Δ as in Definition 9. The rules (LR), (LC), (LS), (LF), (LI) and (BP) will collectively be referred to as *resolution rules*. The rules (BP), (B-A) and (B-C) can be applied in a controlled way, dividing an abductive derivation into three different phases: Phase 1 derives $\bot : \lambda$ using resolution rules only and then, by (LI), all ground instances of $\bot : \lambda$; Phase 2 tries to abduce $B(\lambda)$ by rule (B-A) and must check consistency of $B(\lambda)$ with $\mathcal{T}_0$ as a sub-computation. To this end rules (LI) and (BP) are next used, and then in Phase 3, any inconsistencies must be avoided, if possible, using rule (B-C). Phase 1 corresponds to the Stage 1 mentioned in the introduction, whereas Phase 3 corresponds to Stage 2. Phase 2 is intermediate, taking the result of Stage 1 and initiating Stage 2. The algorithm is given in step form. There is no consideration of efficiency, so that computations in phases 2 and 3 are always started using $\mathcal{T}_0$. This rule order is not the only possible one.

Phase 1: Derive configuration C_2 from C_1, where $\bot : \lambda_2 \in C_2$ using resolution rules. At least one of the clauses from $\mathcal{O}$ should be used in the derivation, for the belief set $B(\lambda_2)$ will never be potentially consistent with $\mathcal{C}_1^c$, by assumption.
Case 1: If derivation of $\bot : \lambda_2$ is not possible then no abductive inference can be made.
Case 2: If λ_2 is empty there is no need for further abduction; the configuration is already inconsistent.
Case 3: If $\lambda_2 \neq \emptyset$ then apply (LI) and move to Phase 2.
Phase 2: Attempt to apply rule (B-A) to one of the instances generated in Case 3: apply resolution rules to the (B-A) augmented configuration $\mathcal{C}_2^c$ to derive $\bot : \lambda_3$ in $\mathcal{C}_3$.
Case 4: If there is no such derivation then $\mathcal{C}_2^c$ is consistent, so generate the (B-A) configuration from $\mathcal{C}_2$ and repeat Phase 1 (case 2 will be next).
Case 5: If $\lambda_3 = \emptyset$ then fail, since $\mathcal{C}_2^c$ is inconsistent and so rule (B-A) cannot be applied. Return to Phase 1 to look for a different λ_2. If the application of (LI) in Case 3 yielded more than one instance of $\bot : \lambda_2$ then a different instance can be selected and Phase 2 re-attempted.
Case 6: If $\lambda_3 \neq \emptyset$ then apply (LI) move to Phase 3.
Phase 3: Attempt to apply rule (B-C) to one of the instances generated in Case 6: apply resolution rules to the (B-C) augmented configuration $\mathcal{C}_3^c$ to derive $\bot : \lambda_4$ in configuration $\mathcal{C}_4$.
Case 7: If there is no such derivation then $\mathcal{C}_3^c$ is still potentially consistent, so generate the (B-C) configuration from $\mathcal{D}_4$ and repeat Phase 2 using the augmented $\mathcal{C}_4^c$ in place of $\mathcal{C}_2^c$.
Case 8: If $\lambda_4 = \emptyset$ then fail, as $\mathcal{C}_3^c$ is inconsistent. Select a different atomic label from λ_3 and repeat Phase 3. If all selections result in failure then return to Phase 1 and select a different instance of λ_2 (derived in Case 3) or derive a new λ_2.
Case 9: If $\lambda_4 \neq \emptyset$ then treat this as in Case 6, using λ_4 in place of λ_3 and the current configuration, $\mathcal{C}_4^c$, in place of $\mathcal{C}_3^c$.

This method can very quickly yield a result as shown in Example 20.

Example 20 *The initial theory $\mathcal{T}_1$ is $\{\neg Q \vee P : \emptyset,\ Q : a,\ \neg Q : b',\ \bot : ab,\ \neg P : \emptyset\}$ in which $\neg P : \emptyset$ is the observation. The initial diagram is empty. The initial corresponding configuration $\mathcal{C}_1^c$ is* not *consistent, as the addition of beliefs $B(a)$ and $B(b)$ enable $\bot : \emptyset$ to be derived, as would the beliefs $B(a)$ and $B(b')$. However, it is potentially consistent as the belief $B(a')$ added to the initial diagram prevents a derivation of $\bot : \emptyset$. Consistency could therefore be enforced at the start by setting the initial diagram $\mathcal{D}_1 = \{B(a')\}$. It is also possible for the non-minimal $\mathcal{D}_1 = \{B(b'), B(a')\}$ to yield consistency of $\mathcal{C}_1^c$ in a different way. As this non-minimal belief set might be derived before the minimal one, it is for this reason the initial configuration is only required to be potentially consistent.*

From the initial configuration $\mathcal{C}_1$ there are 3 derivations in Phase 1: of $\bot : a$, $\bot : ab$ and $\bot : ab'$. The subsequent processing depends on the order in which these are taken. Since only the first one involves the observation it is taken first (Case 3). Rule (B-A) is attempted in Phase 2, after applying (BP) with $B(a)$. The (B-A) augmented configuration reduces to $\langle\{\neg Q \vee P : \emptyset, Q : \emptyset, \neg Q : b', \bot : b\},\quad B(a)\rangle$, which yields $\bot : b$ and also $\bot : b'$ (Case 6). Phase 3 is applied to the (B-C) augmented configuration (after applying (BP) with $B(b')$) $\langle\{\neg Q \vee P : \emptyset, Q : \emptyset, \neg Q : \emptyset\},\quad B(a), B(b')\rangle$. This is not consistent (Case 8), so the other possibility, $\bot : b'$, is processed. Phase 3 is applied to the (B-C) augmented configuration (after applying (BP) with $B(b)$) $\langle\{\neg Q \vee P : \emptyset, Q : \emptyset, \bot : \emptyset\},\quad B(a), B(b)\rangle$. This is again not consistent, so an alternative candidate from Phase 1 is found. Both alternatives lead immediately to failure since neither derivation involved the observation. So no abductive inference can be made here.

The reader can check that if the initial configuration was first made consistent the same result is obtained, although possibly more quickly. (It is also not difficult to show the following property: every abductive solution found if a non-minimal set of beliefs is used in the initial (consistent) diagram includes some solution found if a minimal initial diagram is taken.) The inference rules can be used in other sequences too. For example, the consistency test for the (B-A) augmented configuration could effectively be made one element of λ at a time by restricting application of (BP). In Phase 2, if all derivations of $\bot : \lambda_4$ are made before entering Phase 3, then a more informed selection of abduced beliefs might be possible, resulting in only minimal sets being selected. It is usual to require that only minimal Δ of abducibles are acceptable as solutions. Moreover, the use of rules (B-A) and (B-C) can be made in the opposite order, the same result eventually being found – in this case that there is no abductive derivation of $\bot : \emptyset$. Applying (B-C) before (B-A), both require a derivation of $\bot : \lambda$ as a prerequisite, amounts to anticipating possible applications of (B-A) before they have occurred. It is not particularly helpful to do this, since the wrong choice of λ' in the (B-C) application may be made. To be sure of trying all possibilities, a systematic search for derivations is desirable, as in the simple algorithm given, due to the different intentions of the (B-A) and (B-C) rules.

4 SEMANTICS OF A_{CLDS}

In order to give a semantics for the Abductive LDS proof system, a first order translation of an A_{CLDS} configuration $\mathcal{C}$ is sought such that its first order models can be defined as semantic structures for $\mathcal{C}$. A translated theory is derived from C, called FOT(C) from which a first order semantics for the A_{CLDS} system is obtained. The A_{CLDS} inference rules are then modified to deal in an appropriate and obvious way with FOT(C). This semantics will be seen, quite easily, to correspond with the clausal abductive system given in Section 2

Definition 21 *The translation of a declarative unit $E : \lambda$, FOT($E : \lambda$), is given by $E \vee \neg(\lambda)$, where $\lambda = \{z_1 \ldots z_k w_1' \ldots w_m'\}$ and $\neg(\lambda) = \neg z_1 \vee \ldots \vee \neg z_k \vee w_1 \vee \ldots \vee w_m$, $0 \leq k$, $0 \leq m$. The translation of a B-literal, FOT($B(x)$), is given by x. Together, these enable a first order translation of a configuration C, FOT(C).*

In other words, the $FOT(C)$ is given by the set of clauses from which the labelled clauses might have originally been derived! But please note that the definition of the semantic entailment that will be used is *not* the standard one, so the translated system is not an exact replica of the original. (It is also the case that the separation of abducibles from other parts of a formula is particularly simple for clausal form. Other normal forms could be used for theories in A_{CLDS}, for instance being derived from a tableau representation, although in the CLDS framework they would still be translated back into first order logic, but not necessarily into a form so apparently close to the original.)

The various inference rules of A_{CLDS} are reflected as follows: (LR) and (LC) become ordinary resolution between translated theory clauses, but with the proviso that the resolvent does not include two label literals z and $\neg z$. i.e. that they are not tautologies; (BP) using $B(z)$ (now just the atom z) becomes resolution with a B-fact or subsumption by a B-fact, depending on whether the atom z in $B(z)$ matches with z or z' in the label of a declarative unit; (B-A) and (B-C) add one or more B-facts such that the new set of B facts (abducible atoms) remains consistent with the original theory $\mathcal{T}_0$; (LS) and (LF) become standard subsumption and factoring. The rule (LI) is not a resolution rule, but simply forms instances of a clause and is incorporated here to ensure B-literals are ground.

The relation $\vdash_{AP}$ symbolises derivations between configurations. Semantic entailment is defined in terms of a first order entailment relation: $\mathcal{C} \models_{AP} \mathcal{C}'$ iff FOT($\mathcal{C}$) $\models_{AP(FOL)}$ FOT($\mathcal{C}'$), where the symbol $\models_{AP}$ is used for semantic entailment in CLDS and $\models_{AP(FOL)}$ as the first order relation it is mapped into. Also $\mathcal{C} \models_{AP} \alpha : \lambda$ and $\mathcal{C} \models_{AP} B(x)$ iff FOT($\mathcal{C}$) $\models_{AP(FOL)}$ FOT($\alpha : \lambda$) and FOT($\mathcal{C}$) $\models_{AP(FOL)}$ x, respectively. The definition of the relation $\models_{AP}$ takes into account the requirement for consistency with respect to the corresponding configuration.

Logical Implication Relation $\models_{AP}$. One of the sources of inspiration for this system was [5], in which inference rules are treated at the object-level. In this case, clearly $\models_{AP(FOL)}$ cannot be the standard first order entailment relation: consider the application of (B-A) to a configuration $\mathcal{C} = (\mathcal{T}, \mathcal{D}, \langle \mathcal{T}_0, \mathcal{O} \rangle)$. The pre-condition is that $\bot : \lambda \in \mathcal{T}$, for ground λ, and the new configuration is $\mathcal{C}' = (\mathcal{T}, \mathcal{D}', \langle \mathcal{T}_0, \mathcal{O} \rangle)$. The diagram $\mathcal{D}' = \mathcal{D} \cup B(z) \cup B(w)$, where $z \in \lambda$ and $w \in B_1$, B_1 being the set of additional abduced atoms to avoid inconsistency in the corresponding configuration of C. There is no immediate (standard) logical implication relation between FOT($\mathcal{C}$) and FOT($\mathcal{C}'$). Indeed, the standard definition is too strong in that it considers too many models. The definition is therefore weakened as defined below.

Definition 22 *$\mathcal{C} \models_{AP} \mathcal{C}'$ iff FOT($\mathcal{C}$) $\models^{FOT(\mathcal{C}^c)}$ FOT($\mathcal{C}'$), where*

$$T \models^{W} S \text{ iff } \exists \Pi [W \cup \Pi \text{ is potentially consistent } \wedge T \cup \Pi \models S]$$

Formally, the soundness and completeness of the inference rules with respect to this semantics can then be stated as follows:

Theorem 23 *Let $C = (\mathcal{T}, \mathcal{D}, \langle \mathcal{T}_0, \mathcal{O} \rangle)$ be an A_{CLDS} configuration derived from an initial configuration $(\mathcal{T}_0, \emptyset, \langle \mathcal{T}_0, \mathcal{O} \rangle)$, then $C \vdash_{AP} C'$ iff $C \models_{AP} C'$, or $FOT(C) \models_{AP(FOL)} FOT(C')$.*

The approach taken in this paper to prove soundness and completeness properties also follows that in [8], giving yet another instance of the standard approach for a CLDS.

The usual soundness properties of the rules that reduce to resolution are retained as this new definition of $\models^{W}$ is weaker than the standard one. In what follows, to avoid confusion between the various notions of consistency, when discussing configurations Definition 6 will be used, and when discussing clauses in FOL Definition 2 will be used. In case it is necessary to explicitly distinguish between them, either (A) or (C) will be appended. Recall also, that a consistent configuration or first order theory is also, respectively, a potentially consistent configuration or first order theory.

Soundness Property. The following property is to be shown: let $\mathcal{C} = (\mathcal{T}, \mathcal{D}, \langle \mathcal{T}_0, \mathcal{O} \rangle)$ be an A_{CLDS} configuration derived from an initial configuration $(\mathcal{T}_0, \emptyset, \langle \mathcal{T}_0, \mathcal{O} \rangle)$, then $\mathcal{C} \models_{AP} \mathcal{C}'$ if $\mathcal{C} \vdash_{AP} \mathcal{C}'$.

Consider the (LR) rule. Suppose $\mathcal{C}' = \mathcal{C} + [\alpha : \lambda]$, where $\alpha : \lambda$ is derived from $\mathcal{C}$ using the (LR) rule. Assume also that FOT($\mathcal{C}^c$) is (potentially) consistent, then it is sufficient to find a suitable Π to satisfy the two criteria of Definition 22. Choose $\Pi = \emptyset$, then FOT($\mathcal{C}^c$) $\cup\, \Pi$ is (potentially) consistent as FOT($\mathcal{C}^c$) is. Secondly, suppose H is a model of FOT($\mathcal{C}$) = FOT ($\mathcal{C}$) $\cup\, \emptyset$, then H is a model of $\alpha \vee \neg\lambda$ by the soundness of resolution. The cases for the (LC), (LF), (LI) and the increasing part of the (BP) rule are similar. For the subsumption rule (LS), since the FOT($\mathcal{C}$) is reduced by the rule there is nothing to show.

For the (B-A) rule, it is a prerequisite that $\perp : \lambda \in \mathcal{C}$ for ground λ, and then $\mathcal{C}' = \mathcal{C} + [B(\lambda)] + [B(\lambda_1)]$, where $B(\lambda) = \{B(z) | z \in \lambda\}$. It is sufficient to show that FOT($\mathcal{C}$) $\models^{FOT(\mathcal{C}^c)}$ FOT($B(\lambda) \cup B(B1)$), for the beliefs

$B(B1) = \{B(z)|z \in B_1\}$, where B_1 is the set of atomic labels sufficient to avoid inconsistency of the corresponding configuration for $\mathcal{C} \cup B(\lambda)$. Choose $\Pi = \{z|z \in \lambda\} \cup B_1$, then FOT($\mathcal{C}^c$) $\cup\Pi$ is consistent by the pre-condition of the rule, and if H is a model of FOT($\mathcal{C}$) $\cup$ Π then H is also a model of FOT($B(\lambda) \cup B(B_1)$). The argument for the (B-C) rule is very similar.

Completeness Property. The following property will be shown: let $\mathcal{C} = (\mathcal{T}, \mathcal{D}, \langle \mathcal{T}_0, \mathcal{O} \rangle)$ be an A_{CLDS} configuration, derived from an initial configuration $(\mathcal{T}_0, \emptyset, \langle \mathcal{T}_0, \mathcal{O} \rangle)$, then $C \vdash_{AP} \mathcal{C}'$ if $\mathcal{C} \models_{AP} \mathcal{C}'$.

In order to show completeness, the following useful and important property of abduction derivations (called the *res reduction* property) is used. The property allows the effects of applications of the (BP) rule to be "undone". In case a configuration $\mathcal{C}$ satisfies $\mathcal{C} \vdash_{AP} \alpha : \lambda$, and uses only resolution rules to derive $\alpha : \lambda$ from C, this will be written as $\mathcal{C} \vdash_{res} \alpha : \lambda$.

Lemma 24 *Let $\mathcal{C} = (\mathcal{T}, \mathcal{D}, \langle \mathcal{T}_0, \mathcal{O} \rangle)$ be a configuration. If there exists $\mathcal{D}'$ such that $\mathcal{D}' \supseteq \{B(z)|z \in \lambda\}$, $\mathcal{D}'$ is consistent with $\mathcal{C}^c$ and $\mathcal{C} + \mathcal{D}' \vdash_{res} \bot : \emptyset$, then it is also true that (1.) $\mathcal{C} \vdash_{AP} \bot : \lambda$ and (2.) $\mathcal{C} \vdash_{AP} \bot : \emptyset$.*

Proof Each of the resolution steps taken to derive $\bot : \emptyset$ from $\mathcal{C} + \mathcal{D}'$, apart from any (BP) steps, can be mimicked in the configuration $\mathcal{C}$; they will result in the derivation of $\bot : \lambda$ for some ground label λ instead of $\bot : \emptyset$. This gives conclusion (1.). For conclusion (2.), by steps of (B-A) and (B-C) (allowed because $\mathcal{D}'$ is consistent with C), and several (BP) steps, $\mathcal{C} \vdash_{AP} \bot : \emptyset$ can be derived. ■

As this abductive system is resolution based, the analogue of refutation completeness will be shown. The contrapositive of Lemma 24 is used in the proof of completeness, together with the notion of a maximally consistent configuration, obtained from a potentially consistent configuration $\mathcal{C}$, and from which a particular Herbrand model of FOT($\mathcal{C}$) can be obtained. The existence of such a configuration $\mathcal{C}_{mcc}$ is shown in Lemma 25. Informally, a maximally consistent configuration is a configuration to which all applicable rules have been applied and in all ways. It is also shown in Lemma 25 that $\mathcal{C}_{mcc}$ has the *max-belief* property, that is, for each abducible z, either $B(z)$ or $B(z')$ belongs to the maximal consistent configuration, but not both.

Lemma 25 *Let $\mathcal{C} = (\mathcal{T}, \mathcal{D}, \langle \mathcal{T}_0, \mathcal{O} \rangle)$ be a configuration such that $\mathcal{C}^c$ is potentially consistent and $\mathcal{C} \not\vdash_{res} \bot : \emptyset$.*

1. *$\mathcal{C}$ can be extended to a consistent configuration $\mathcal{C}_{mcc}$ such that $\mathcal{C}_{mcc} \not\vdash_{res} \bot : \emptyset$ and $\mathcal{C}_{mcc}$ satisfies the* max-belief *property.*

2. *A Herbrand model H of FOT($\mathcal{C}$) can be obtained from $\mathcal{C}_{mcc}$.*

Proof Part 1: The configuration $\mathcal{C}_{mcc}$= $(\mathcal{T}_{mcc}, \mathcal{D}_{mcc}, \langle \mathcal{T}_0, \mathcal{O} \rangle)$ is constructed inductively from $\mathcal{C} = (\mathcal{T}, \mathcal{D}, \langle \mathcal{T}_0, \mathcal{O} \rangle)$ according to (1) - (4) below:

1. $\mathcal{T} \subseteq \mathcal{T}_{mcc}$ and $\mathcal{D} \subseteq \mathcal{D}_{mcc}$;

2. For each of (BP), (LR), (LC), (LI), (LF), if $\alpha : \lambda$ can be derived by the rule from elements of $\mathcal{C}_{mcc}$, then $\alpha : \lambda$ is added to $\mathcal{C}_{mcc}$;

3. For each ground labelled clause $\alpha_1 \vee \ldots \vee \alpha_n : \lambda \in \mathcal{C}_{mcc}$, $n \geq 1$, there is at least one $\alpha_i : \lambda$ such that $\mathcal{C}_{mcc}+[\alpha_i : \lambda] \not\vdash_{res} \bot : \emptyset$, which is added to $\mathcal{C}_{mcc}$.

4. For each atomic abducible z, if $\mathcal{C}_{mcc}+ [B(z)] \not\vdash_{res} \bot : \emptyset$ and $\mathcal{D}_{mcc}+[B(z)]$ is self-consistent, then $B(z)$ is added to $\mathcal{D}_{mcc}$ and similarly for z'.

The construction of $\mathcal{C}_{mcc}$ from a configuration $\mathcal{C}$ that satisfies the conditions given in the Lemma and according to (1) -(4) yields a configuration $\mathcal{C}_{mcc}$ with self-consistent beliefs, such that $\mathcal{C}_{mcc} \not\vdash_{res} \bot : \emptyset$ and which has the *max-belief* property. This is shown next. It is shown that under each of the rules (2) - (4) the invariance of $\mathcal{C}_{mcc} \not\vdash_{res} \bot : \emptyset$ is maintained.

Property (2): Each rule (LC), (LR), (LF), (LI) and (BP) is considered and it is shown that the property $\mathcal{C}_{mcc} \not\vdash_{res} \bot : \emptyset$ is invariant under its application. For example, for the rule (LR): suppose $\beta_1 \vee \gamma_1 : \lambda_1 \in T_{mcc}$, $\neg\beta_1 \vee \gamma_2 : \lambda_2 \in T_{mcc}$, θ is the m.g.u. of β_1 and β_2 and $\mathcal{C}_{mcc} \not\vdash_{res} \bot : \emptyset$, then by (LR) and construction $(\gamma_1 \vee \gamma_2)\theta : join(\lambda_1, \lambda_2))\theta \in T_{mcc}$. If $\mathcal{C}_{mcc}+(\gamma_1 \vee \gamma_2)\theta : (join(\lambda_1, \lambda_2))\theta \vdash_{res} \bot : \emptyset$ then $\mathcal{C}_{mcc} \vdash_{res} \bot : \emptyset$, contradicting the assumption. The cases for the rules (BP), (LF), (LI) and (LC) are similar.

Property (3): The selection of some $\alpha_i : \lambda$ from a ground labelled clause $\alpha_1 \vee \ldots \vee \alpha_n : \lambda$ (not of the form $\bot : \lambda$) can also be made to maintain invariance of $\mathcal{C}_{mcc} \not\vdash_{res} \bot : \emptyset$: if $\mathcal{C}_{mcc}+\alpha_i : \lambda \vdash_{res} \bot : \emptyset$ for every i, then $\mathcal{C}_{mcc}+\alpha_1 \vee \ldots \vee \alpha_n : \lambda \vdash_{res} \bot : \emptyset$, contradicting the assumption. To see this, first apply the steps taken to derive $\bot : \emptyset$ from $\alpha_1 : \lambda$, which will now yield an instance of the labelled clause $\alpha_2 \vee \ldots \vee \alpha_n : \lambda$. Then apply the steps taken to derive $\bot : \emptyset$ from $\alpha_2 : \lambda$. Continue in this way until an instance of the clause $\alpha_n : \lambda$ is derived, and then derive $\bot : \emptyset$ from this.

Property(4): The case for the addition of elements $B(z)$ or $B(z')$ to D_{mcc} is similar also, by construction. To show the max-belief property, suppose that $B(z) \notin \mathcal{C}_{mcc}$ and $B(z') \notin \mathcal{C}_{mcc}$ for some atomic label z. Then $\mathcal{C}_{mcc}$+$[B(z)] \vdash_{res} \bot : \emptyset$ and $\mathcal{C}_{mcc}$+$[B(z')] \vdash_{res} \bot : \emptyset$. Hence $\mathcal{C}_{mcc} \vdash_{res} \bot : z$ and $\mathcal{C}_{mcc} \vdash_{res} \bot : z'$, since $\mathcal{C}_{mcc} \not\vdash_{res} \bot : \emptyset$. But then $\mathcal{C}_{mcc} \vdash \bot : \emptyset$ by (LC), a contradiction. Moreover, clearly $B(z)$ and $B(z')$ cannot both belong to $\mathcal{C}_{mcc}$ as then $\mathcal{C}_{mcc}$ is not self-consistent.

Part 2: A Herbrand model H_{mcc} can be constructed from $\mathcal{C}_{mcc}$ that satisfies FOT($\mathcal{C}_{mcc}$) and hence $FOT(\mathcal{C})$. The domain of the model is the Herbrand universe of FOT($\mathcal{C}_{mcc}$). Each atom β in the corresponding Herbrand base is interpreted true or false: if $\beta : \emptyset \in T_{mcc}$ then β is assigned true, otherwise it is assigned false. Notice that $\beta : \emptyset$ and $\neg\beta : \emptyset$ cannot both belong to T_{mcc}, else $\bot : \emptyset$ could be derived, a contradiction. Each atom z such that $B(z) \in D_{mcc}$ is assigned true. By the max-belief property, every abducible or its negation is assigned by this rule.

Suppose that some element of FOT($\mathcal{C}_{mcc}$) is false in H_{mcc}. First, by construction, the false element cannot have the form z derived from $B(z)$. Suppose the false element was a false ground instance of a (false) clause $\alpha \vee \neg(\lambda)$, then every atom in α and in $\neg(\lambda)$ must be false. In particular, this means that $\alpha_i : \emptyset \notin \mathcal{C}_{mcc}$ by construction, and λ is true so $B(z) \in \mathcal{C}_{mcc}$ for each $z \in \lambda$. But for at least one of the atoms $\alpha_i \in \alpha$ the labelled clause $\alpha_i : \lambda \in T_{mcc}$ and hence $\alpha_i : \emptyset \in \mathcal{C}_{mcc}$, a contradiction. Thus the false clause instance cannot be of the form $\alpha \vee \neg(\lambda)$. If the false clause instance was of the form $\neg(\lambda)$, then every atom in $\neg\lambda$ must be false, or every atom in λ must be true. That is, $B(\{z\}) \in D_{mcc}$ for every $z \in \lambda$. But this contradicts the fact that $\mathcal{C}_{mcc} \not\vdash_{res} \bot : \emptyset$. Since in all cases a contradiction is obtained, every element in FOT($\mathcal{C}_{mcc}$) is true in H_{mcc}. As $\mathcal{C}_{mcc} \supseteq FOT(C)$, H_{mcc} is also a model of $FOT(\mathcal{C})$. ∎

Now, the completeness property can be proved:

Theorem 26 *Let $\mathcal{C} = (\mathcal{T}, \mathcal{D}, \langle \mathcal{T}_0, \mathcal{O} \rangle)$ be a configuration. If $\mathcal{C} \models_{AP} \bot : \emptyset$, then $\mathcal{C} \vdash_{AP} \bot : \emptyset$.*

Proof The proof is by contradiction. Let $\mathcal{C} = (\mathcal{T}, \mathcal{D}, \langle \mathcal{T}_0, \mathcal{O} \rangle)$ be a configuration derived from an initial configuration containing $\mathcal{T}_0$. Assume that $\mathcal{C} \models_{AP} \bot : \emptyset$ and $\mathcal{C} \not\vdash_{AP} \bot : \emptyset$. A Herbrand model $\mathcal{H}$ will be exhibited that falsifies $\mathcal{C} \models_{AP} \bot : \emptyset$ and so leads to a contradiction, thus justifying that $\mathcal{C} \vdash_{AP} \bot : \emptyset$. The relation $\models_{AP}$ uses Weakened logical implication of Definition 22. By the property of Lemma 24, $\mathcal{C} \not\vdash_{AP} \bot : \emptyset$ implies that for every potential belief set Π (in addition to $\mathcal{D}$), either Π is inconsistent with $\mathcal{C}^c$, or $\mathcal{C} + [B(\Pi)] \not\vdash_{res} \bot : \emptyset$ (where $B(\Pi)$ is short for $\{B(z) | z \in \Pi\}$). This implies that either, for all such Π, Π is not consistent with $\mathcal{C}^c$, or, there is some such Π that *is* consistent with $\mathcal{C}^c$ and such that $\mathcal{C} + [B(\Pi_1)] \not\vdash_{res} \bot : \emptyset$. Considering the first disjunct, the particular $\Pi = \emptyset$ leads to $\mathcal{C}^c$ being inconsistent, a contradiction. Hence the second disjunct must hold. By Lemma 25, $\mathcal{C} + [B(\Pi_1)]$ can be extended to a maximal consistent configuration $\mathcal{C}_{mcc}$, and, again, by Lemma 25, a Herbrand model H_{mcc} of FOT($\mathcal{C} \cup B(\Pi_1)$) can be obtained. This again contradicts the assumption that $\mathcal{C} \models_{AP} \bot : \emptyset$, since there is a model H_{mcc} that satisfies FOT($\mathcal{C}^c$) $\cup\, \Pi_1$ (by the assumption on Π_1) but which does not satisfy $\bot$. ∎

Correspondence. The correspondence of the A_{CLDS} system with the standard abduction approach given in Section 2 is made as follows: The initial configuration $\mathcal{C}_1 = (\mathcal{T}_1, \emptyset, \langle \mathcal{T}_0, \mathcal{O} \rangle)$, where $\mathcal{T}_1$ is the translation into declarative units of each sentence in the initial theory, together with the clausal form of the negated observation. In case the derivation of $\bot : \emptyset$ is made from $\mathcal{C}_1$, then there is some π such that $B(\{z\})$, $z \in \pi$, is added consistently to $\mathcal{C}_1$. In other words, FOT($\mathcal{C}_1$)$\cup$ FOT($\{B(z) | z \in \pi\}$) $\vdash \bot$, and $FOT(\mathcal{T}_0) \cup$ FOT($\{B(z) | z \in \pi\}$) $\not\vdash \bot$ as required. So if a result is obtained in A_{CLDS} it is correct with respect to standard abduction. On the other hand, suppose $\mathcal{T}_1$ is a combined first order theory and observation, with $\mathcal{C}_1$ the translated initial configuration and $\mathcal{T}_0$ the initial theory without the observation, such that $\mathcal{T}_1 \cup \Pi \models \bot$ and $\mathcal{T}_0 \cup \Pi$ is consistent, for some Π. Therefore, $\mathcal{T}_1 \models_{AP} \bot : \emptyset$, by definition of labelled semantic entailment. By the completeness property applied to $\mathcal{C}_1$, standard abductive results can be obtained in A_{CLDS}.

Finally, the two definitions of consistency in Definitions 2 and 6 are also easily seen to be equivalent to each other. For instance, let $\mathcal{T}$ be a theory and Δ a set of ground abducibles, and suppose $\mathcal{T} \cup \Delta$ is potentially consistent(C). The corresponding situation in A_{CLDS} is a configuration $\mathcal{C} = (\mathcal{T}, \mathcal{D}, \langle \mathcal{T}_0, \mathcal{O} \rangle)$, where $\mathcal{D}$ is the diagram corresponding to Δ. If $\mathcal{C}$ were not potentially consistent(A), then for every $\mathcal{D}'$, such that $\mathcal{D}' \supseteq \mathcal{D}$, $\bot : \emptyset$ would be derivable. In particular, for $\mathcal{D}' = \emptyset$, or from the configuration $\mathcal{C}$, $\bot : \emptyset$ is derivable, contradicting the assumption. The other direction is similar. On the other hand, suppose that $\mathcal{T} \cup \Delta$ is consistent(C), then, additionally (to potential consistency), if $\mathcal{T} \cup \Delta \vdash \lambda$ for some abducible clause λ, then Δ subsumes λ; that is, some element $z \in \Delta$ occurs in λ. In the corresponding situation for A_{CLDS}, $\mathcal{C}$ must be potentially consistent(A). In case $\mathcal{C}$ were not consistent, then there is some $\mathcal{D}' \supset \mathcal{D}$ ($\mathcal{D}' \neq \mathcal{D}$ because $\mathcal{C}$ is potentially consistent), such that the new configuration $\mathcal{C}' = (\mathcal{T}, \mathcal{D}', \langle \mathcal{T}_0, \mathcal{O} \rangle)$ satisfies $\mathcal{C}' \vdash_{AP} \bot : \emptyset$. But this implies that some application of (B-A) has been made, due to a derivation of $\bot : \lambda$, where λ is not subsumed by the beliefs in Δ. But this implies $\mathcal{T} \cup \Delta$ can derive an abducible clause not subsumed by Δ, a contradiction. Again, the other direction is similar, as are the cases for inconsistency(C) and inconsistency(A).

5 DIAGNOSING BY ABDUCTION

This example is taken from [4] and shows the procedure operating on a larger theory. There are three components called $\{or_1, inv_1, inv_2\}$; the predicate $high(x, y)$, where x is a component and y is a port: either an input (i_i, i_2, i_3 or i_4) or an output (o_1, o_2 or o), reads as x has high voltage at port y. The observation is $\neg high(or_1, o)$, given $\neg high(inv_1, i_1)$ and $\neg high(inv_2, i_2)$. There are two kinds of abducibles, $ok(or_1)$,$ok(inv_1)$, $ok(inv_2)$ and $ok(or_1)'$, $ok(inv_1)'$, $ok(inv_2)'$, meaning, respectively, that the gate is normal (working properly), or abnormal (not working properly). (These are shortened to or_1, or'_1 etc, in the clauses below). The connections are shown in Figure 5.1. The aim is to abduce which of the gates is not working correctly; i.e. to find a diagnosis for the unexpected output. The labelled theory is given in the clauses below, in which the general clauses about the operation of or-gates and inverters have been instantiated for the particular situation depicted. This is so that the theory is propositional and does not need function symbols to represent inputs/outputs for the two inverters.

$[high(or_1, o) \leftrightarrow high(or_1, i_3) \vee high(or_1, i_4)] \leftrightarrow ok(or_1)$

(1) $high(or_1, o) \vee \neg high(or_1, i_3) : or_1$

(2) $high(or_1, o) \vee \neg high(or_1, i_4) : or_1$

(3) $high(or_1, i_3) \vee high(or_1, i_4) \vee \neg high(or_1, o) : or_1$

(4) $\neg high(or_1, i_3) \vee \neg high(or_1, o) : or'_1$

(5) $\neg high(or_1, i_4) \vee \neg high(or_1, o) : or'_1$

(6) $high(or_1, i_3) \vee high(or_1, o) \vee high(or_1, i_4) : or'_1$

$[high(inv, o) \leftrightarrow \neg high(inv, i)] \leftrightarrow ok(inv)$

(7)(8) $high(inv_j, o_j) \vee \neg high(inv_j, i_j) : inv'_j$

(9)(10) $high(inv_j, i_j) \vee \neg high(inv_j, o_j) : inv'_j$

(11)(12) $high(inv_j, o_j) \vee high(inv_j, i_j) : inv_j$

(13)(14) $\neg high(inv_j, o_j) \vee \neg high(inv_j, i_j) : inv_j$

(15) $\neg high(inv_1, i_1) : \emptyset$ (16) $\neg high(inv_2, i_2) : \emptyset$

(17) $high(or_1, i_3) \leftrightarrow high(inv_1, o_1) : \emptyset$ (19) $high(or_1, o) : \emptyset$

(18) $high(or_1, i_4) \leftrightarrow high(inv_2, o_2) : \emptyset$

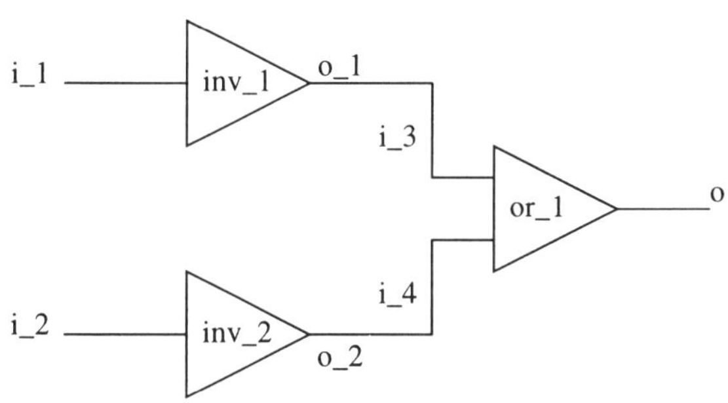

Figure 5.1

Clauses (1) - (3) detail the operation of a correctly operating or-gate and clauses (4)-(6) the operation of a faulty or-gate, both instantiated for the given gate or_1. Clauses (7)-(14) do the same for an inv-gate. Since there are two of these gates, (7) and (8), for example, would have inv_1/inv_2 in place of inv_j everywhere and similarly for o_j and i_j. Clauses (15) and (16) describe the inputs, and clauses (17) and (18) the connections between the gates. Clause (19) is the negation of the observed output, which was that the output voltage from the or-gate is low, or not high, as expected. Before applying the rules (LR) and (LC), notice that labelled subsumption is possible: (15) and (16) subsume (7),(13) and (8),(14) respectively, and (19) subsumes (1), (2) and (6). The following clauses : $\{3,4,5,9,10,11,12\}$ and $\{15,16,17,18,19\}$ may then be used in the computation. The corresponding initial configuration, $\mathcal{C}_1^c$, does not include the clauses (15), (16) or (19), so the subsumption steps are not possible from that configuration. Reasoning with unit clauses leads to the clauses (3') to (12') below.

(3')	(19,3)	$high(or_1, i_3) \vee high(or_1, i_4) : or_1$			
(4')	(19,4)	$\neg high(or_1, i_3) : or_1'$	(5')	(19,5)	$\neg high(or_1, i_4) : or_1'$
(9')	(15,9)	$\neg high(inv_1, o_1) : inv_1'$	(10')	(16,10)	$\neg high(inv_2, o_2) : inv_2'$
(11')	(15,11)	$high(inv_1, o_1) : inv_1$	(12')	(16,12)	$high(inv_2, o_2) : inv_2$

(4'), (11') and (17) now give $\bot : \{or_1', inv_1\}$ and similarly (5'), (12') and (18) yield $\bot : \{or_1', inv_2\}$. Alternatively, clauses (3'), (9') and (17) will yield $high(or_1, i_4) : \{inv_1', or_1\}$ and then (18) and (10') can be used to derive $\bot : \{inv_1', inv_2', or_1\}$. Other possibilities yield self-inconsistent labelled clauses and so are not made. Taking the first answer, $\{or_1', inv_1\}$, the consistency of $B(or_1')$ and $B(inv_1)$ with $\mathcal{C}_1^c$ is tested. The available clauses, after application of (BP) by the two new B-literals, are (4"), (5"), (6"), (11") and (13") given below, together with the original clauses (8), (10), (12), (14), (17) and (18).

(4")	$\neg high(or_1, i_3) \vee \neg high(or_1, o) : \emptyset$
(5")	$\neg high(or_1, i_4) \vee \neg high(or_1, o) : \emptyset$
(6")	$high(or_1, i_3) \vee high(or_1, o) \vee high(or_1, i_4) : \emptyset$
(11")	$high(inv_1, o_1) \vee high(inv_1, i_1) : \emptyset$
(13")	$\neg high(inv_1, o_1) \vee \neg high(inv_1, i_1) : \emptyset$

The empty clause cannot be derived from these clauses and the same is true for the clauses resulting from adding the appropriate B-literals derived from the second or third answers. (An easy way to see this is to apply the Davis-Putnam procedure [9] to the wff part of the declarative units.) Therefore, we conclude that there are two possible diagnoses: from the abductive solutions $\{or_1', inv_1\}$ and $\{or_1', inv_2\}$) it is obtained that the *or*-gate is working incorrectly, and from the solution $\{inv_1', inv_2', or_1\}$ it is obtained that the two inverters are working incorrectly.

It is clear from this example that the "consistency" test can be rather expensive, in that it can recompute many resolvents from the original clauses each time, sometimes with changed labels and sometimes not. A possible way to avoid this recomputation is mentioned in Section 6

6 CONCLUSIONS

This paper has shown how the abductive framework can be represented in Compiled Labelled Deductive Systems. In particular, it demonstrated that the non-monotonic reasoning inherent in abduction can be represented in a first order framework, A_{CLDS}. A translation of configurations into first order logic was given. By using a variant of classical logical implication (inspired by the nonmonotonic sequent rules in [5]) a soundness and completeness theorem for the translation was proved. The work demonstrates again the flexibility of the CLDS approach, since essentially the same methods were employed to show soundness and completeness as in previous applications.

The correspondence result shows that the A_{CLDS} system gives the same results as a standard classical abductive system would do, in which the basic rule was resolution. Although the A_{CLDS} rules can be applied in any order, to simulate the usual method for abduction requires a particular order. It is usually better to apply the (B-A) rule before (B-C), since the latter involves choice, where the former does not. There are various investigations/extensions to be made within this framework, which are very briefly discussed below.

Language Extension. The language $\mathcal{L}_{\mathcal{P}}$ could be extended to include clauses with function symbols and consideration of termination is the main interest for such an extension. Secondly, the maintenance of ground labels in clauses of the form $\bot : \lambda$ could be achieved by restricting the form of the clauses in some way, similar to restrictions on logic programs [13]. Finally, the restrictions of applying the (BP) rule to clauses with ground abducibles and on ground B literals could also be removed. Free variable methods employed in tableau prcedures [6] would be useful here.

Other Logics. The second, and more interesting investigation, is to consider different logics; for example, logic programming with negation as failure could be used for the logic of wffs, leaving the labelling language classical, as before. Another obvious area to investigate is the relation of the rules (B-A) and (B-C) to the operations in default logic [2]. The labelling language could also be time-dependent, perhaps including an interval as an argument. This would allow a form of temporal abduction, which could be used to perform model search in propositional linear temporal logic (PLTL) [3], in which a model for a temporal formula α is a Kripke model that forces α.

Pre-processing. In some presentations of abduction, especially when the theory is a logic program, there is an additional component called integrity constraints (IC), that constrain the theory in some way. The abduced literals could then be forced to be consistent with the ICs rather than with the rest of the theory, it being arranged that this also implies consistency with the rest of the theory. There are several different principles employed to effect the constraining influence of integrity constraints, any of which could be represented in the abductive procedure $\mathcal{AP}$; for example, the theory must be consistent with the ICs, or the ICs are implied by the theory, or the ICs impose some kind of meta-level constraint on the theory. In this paper the general theory is not enhanced in this way, in order to maintain simplicity. However, the rules introduced could be used to implement such constraining influences.

It is also noted that in a classical clausal theory, which is monotonic, the addition of abduced literals can only increase the possible derivations that can be made. Therefore, a clausal theory with no function symbols could be preprocessed in a finite time by resolution in order to derive a set of new clauses containing only abducibles that would enable more efficient checks to be made to determine consistency when new abduced data is to be added. In the inverter example, the empty clause cannot be derived in any way, so consistency could then always be detected immediately. George Valassopoulos investigated one such pre-procesing system in his MSc. thesis [22]. Such pre-processing might not always be useful; for example, for a very large theory it might be more expensive and involve far more clauses than processing for the particular set of observations. For another simple way to expedite the consistency check, the following method can be used. The corresponding configuration $\mathcal{C}_1^c$ is simply processed in tandem with C_1. Then, any steps that have already been made, that were applicable to $\mathcal{C}_1^c$, need not be repeated and any method of checking consistency can then be made.

Acknowledgement. The authors acknowledge the helpful comments suggested by the referees.

Krysia Broda
Department of Computing
Imperial College London, UK
kb@doc.ic.ac.uk

Dov M. Gabbay
Department of Computer Science
King's College London, UK
dg@dcs.kcl.ac.uk

References

[1] A. Aliseda-Llera. *Seeking Explanations: Abduction in Logic, Philosophy of Science and Artificial Intelligence*, ILLC-Publications, University of Amsterdam, 1997.

[2] G. Antoniou. *Nonmonotonic Reasoning*. MIT Press, 1997.

[3] H. Barringer, M. Fisher, G, Gough, D. Gabbay and R. Owens. *METATEM: A Framework for Programming in Temporal Logic*. In Stepwise Refinement of Distributed Systems: Models, Formalisms, Correctness,

[4] P. Baumgartner, P. Frohlich, U. Furbach, W. Nejdl. *Tableaux for Diagnosis Applications*. TABLEAUX97, Springer LNAI 1227, 1997.

[5] P. A. Bonatti and N. Olivetti. *A Sequent Calculus for Skeptical Default Logic*. Proc. TABLEAUX97, Springer LNAI 1227, 1997

[6] K. Broda. *The Application of Semantic Tableaux with Unification to Automated Theorem Proving*. PhD Thesis, Department of Computing (DOC), Imperial College, 1991.

[7] K.Broda and D. Gabbay. *Labelled Abduction - Compiled Labelled Abductive Systems*. Technical report. DOC, Imperial College, 1999.

[8] K. Broda and A. Russo. *A Unified Compilation Style Labelled Deductive System for Modal and Substructural Logic using Natural Deduction*. Technical Report 10/97, DOC, Imperial College, 1997.

[9] C. L. Chang and R. Lee. *Symbolic Logic and Mechanical Theorem Proving*. Academic Press 1973.

[10] K. Clark. *Negation as Failure*. In *Logic and Databases*, (Eds) H. Gallaire and J. Minker, Plenum Press, 1978.

[11] M. Deneker and D. De Schreye. *SLDNFA: an abductive procedure for normal abductive programs*, Proceedings of the Joint International Conference and Symposium on Logic Programming, 1992, MIT press.

[12] K. Eshghi. *Abductive Planning with Event Calculus*. Proc. ICLP88, MIT Press, 1988.

[13] T. S. Fung. *Abduction by Deduction*. PhD Thesis, Department of Computing, Imperial College,1995.

[14] D. Gabbay. *Labelled Deductive Systems, Volume I - Foundations*. Oxford University Press, 1996.

[15] A. C. Kakas, R. A. Kowalski and F. Toni. *Abductive Logic Programming*. Journal of Logic and Computation 2 (6). 1993.

[16] K. Konolige. *Abduction versus closure in causal theories*. Artificial Intelligence, Vol 53, 1992.

[17] D. Poole, A. Mackworth and R. Goebel. *Computational Intelligence.* Oxford University Press, 1997.

[18] R. Reiter. *A Theory of Diagnosis from First Principles.* Artificial Intelligence , 32, 1987.

[19] J.A. Robinson. *Logic, Form and Function.* Edinburgh Press, 1979.

[20] A. Russo. *Modal Logics as Labelled Deductive Systems.* PhD. Thesis, DOC, Imperial College, 1996.

[21] M. Shanahan. *Solving the Frame Problem.* MIT Press, 1997. Springer LNCS 430, 1990.

[22] G. Valassopoulos. *A Diagnostic System with the Use of Abduction.* MSc. Thesis, Department of Computing, Imperial College, 1998.

LABELLED TABLEAUX FOR PROPOSITIONAL LINEAR TIME LOGIC OVER FINITE FRAMES

Serenella Cerrito and Marta Cialdea Mayer

Abstract In this work we propose a proof-system for propositional Linear Time Temporal Logic over Finite Temporal Frames (LTL^{Fin}) in the tableau style, where formulae are equipped with labels in order to explicitly embed some semantic information in the inference rules. The labels mark formulae as true in a given time interval and linear inequality constraints (temporal constraints) are used to express ordering relations between time points. Branch closure is reduced to unsatisfiability over the integers of the set of temporal constraints in the branch. The proposed tableau calculus checks whether a finite set of formulae has a finite model whose cardinality is bounded by a constant given in input, and constructs such a model, if any. From a theoretical standpoint, the method can also be used to check finite satisfiability *tout court*. The following properties of the proposed proof-system are proved: termination, soundness and completeness w.r.t. both bounded and finite validity. This work is a revised and extended version of [6].

1 INTRODUCTION

Temporal logics are modal logics that allow one to reason about propositions whose truth values change over time. In the case of linear time temporal logic (LTL), the underlying model of time is a linear sequence of states. For many applications, e.g. the specification and verification of reactive systems, such a sequence of states is taken to be isomorphic to $\mathbb{N}$. This model of time is quite appropriate, in fact, to reason about nonterminating or continuously operating concurrent programs such as operating systems and network communication protocols.

However, there are problems in Computer Science and Artificial Intelligence where only a finite fragment of the time sequence is of interest. For instance, dynamical integrity constraints on databases evolving through time by means of updates may be expressed by (first order) temporal formulae. In particular, the so called *Anteriority Dynamic Dependencies* [4] state that if a given property of data holds at a state of the database history, then another property had been true at previous states (for instance: it is necessary to have been a student in order to be a teacher). Thus, at any given time point, only the finite time segment

D. Basin et al. (eds.), Labelled Deduction, 135–159.
© 2000 *Kluwer Academic Publishers. Printed in the Netherlands.*

going from the initial time point, corresponding to database creation, up to the current one is dealt with. Finite temporal frames are quite appropriate to study dynamical integrity constraints in databases [3]. Similarly, the specification of a planning problem may be expressed by means of a set of temporal logic formulae; since a plan is a finite sequence of actions, leading from the initial situation to the desired goal, it can be represented by a temporal model over a finite frame [6, 7]. In this work, we consider a version of linear temporal logic with a *finite* semantics, LTL^{fin}, where time is assumed to be finite both in the past and in the future.

Two different notions of validity for temporal formulae are considered:

finite validity : a formula A is finitely valid iff it is true in every interpretation over a finite time frame

bounded validity (or k-validity) : if k is a positive integer, A is k-valid iff A is true in every interpretation whose time frame is not greater than k.

Both notions are of interest. The first one corresponds to the usual notion of validity (in the framework of the finite semantics). The second one, or rather its dual notion of k-satisfiability, is useful, for instance, in applications to planning, where one may be interested in the existence of a plan of a given maximal duration.

When studying linear time logic, different sets of temporal operators may be considered: mainly, future time operators ($\Box$: always, $\Diamond$: eventually, $\bigcirc$: next, $\mathcal{U}$: until), possibly restricting to the fragment with $\Box$ and $\Diamond$ only, or both past and future time ones ("full" LTL). We define a labelled tableau calculus that allows one to check whether a propositional formula, in the language of full LTL, is either k-valid or finitely valid.

The work is organized as follows. Section 2 presents the syntax and semantics of LTL over finite and infinite temporal frames. In this section, we present also the main features of other tableau systems for LTL, "traditional" [17, 2] as well as "labelled"[13, 15], in order to highlight the use of *labels* in tableaux for linear temporal logic, and place our own contribution in context. Indeed, the tableau calculus that we propose carries strong similarities with [15]. In fact, the calculus is based on labels, that mark formulae as true in a given time interval, and linear inequality constraints on time points (temporal constraints). Branch closure is reduced to unsatisfiability over the integers of the set of temporal constraints in the branch. However, focusing on finite models makes it possible to give the "since" and "until" operators (which are not treated in [15]) an intuitive and simple treatment, without hindering termination. The tableau calculus proposed in this work is presented in Section 3 and its main properties (termination, soundness and completeness w.r.t. bounded and finite validity) are proved in Section 4 Section 5 concludes this work.

This work is a a revised and extended version of [6]. In the present paper, full proofs of the technical results are provided, and the relation of our approach with other systems is outlined.

2 LINEAR TIME TEMPORAL LOGIC

The language of LTL extends classical propositional logic by means of a set of modal operators, such as, for instance, $\Box$ (always in the future), $\blacksquare$ (always in the past), $\Diamond$ (eventually), $\blacklozenge$ (sometime in the past), $\circ$ (in the next state), $\bullet$ (in the last state), and the binary ones $\mathcal{S}$ (since) and $\mathcal{U}$ (until).

The semantics of the language is defined as follows. A temporal frame T is a (finite or infinite) denumerable sequence of states, that can be identified either with $\mathbb{N}$ or with a finite initial segment of the natural numbers. The version of LTL where the temporal frame is always assumed to be finite will be called LTL^{fin}.

The elements of a temporal frame are called *time points*. If $\mathcal{L}$ is an LTL temporal language and P the set of propositional letters in $\mathcal{L}$, an *$\mathcal{L}$-interpretation* $\mathcal{M}$ is a pair $\langle T, \sigma \rangle$, where:

- T is a temporal frame;
- $\sigma : T \to \mathcal{P}(P)$ is a function on time points, providing an interpretation to the propositional letters in P for any point in T. I.e. if $i \in T$, then $\sigma(i) \subseteq P$ is the set of propositional letters true at i.

The semantics of each temporal operator is defined by addition of a corresponding clause to the usual ones for the classical connectives, in the inductive definition of the satisfiability relation $\mathcal{M}_i \models A$, for $M = \langle T, \sigma \rangle$ and $i \in T$. Here we consider a very large set of operators, extending the classical language with $\neg$, $\vee$ and $\bot$ (the false):

1. Classical operators:
   (a) $\mathcal{M}_i \not\models \bot$
   (b) $\mathcal{M}_i \models p$ iff $p \in \sigma(i)$, for $p \in P$.
   (c) $\mathcal{M}_i \models \neg A$ iff $\mathcal{M}_i \not\models A$
   (d) $\mathcal{M}_i \models A \vee B$ iff either $\mathcal{M}_i \models A$ or $\mathcal{M}_i \models B$.
2. Future time operators:
   (a) $\mathcal{M}_i \models \Box A$ iff for all $j \geq i$, $\mathcal{M}_j \models A$. (*always in the future*)
   (b) $\mathcal{M}_i \models \Box^+ A$ iff for all $j > i$, $\mathcal{M}_j \models A$ (*strict always*).
   (c) $\mathcal{M}_i \models \Diamond A$ iff there exists $j \in T$ such that $j \geq i$ and $\mathcal{M}_j \models A$ (*eventually*).

(d) $\mathcal{M}_i \models \Diamond^+ A$ iff there exists $j \in T$ such that $j > i$ and $\mathcal{M}_j \models A$ (*strict eventually*).

(e) $M_i \models \circ A$ iff either $i+1 \notin T$ or $\mathcal{M}_{i+1} \models A$ (*weak next*).

(f) $M_i \models \odot A$ iff $i+1 \in T$ and $\mathcal{M}_{i+1} \models A$ (*strong next*).

(g) $\mathcal{M}_i \models A\mathcal{U}B$ iff there exists $j \in T$ such that $j \geq i$, $\mathcal{M}_j \models B$ and, for any k with $i \leq k < j$, $\mathcal{M}_k \models A$ (*until*).

(h) $\mathcal{M}_i \models A\mathcal{U}^+B$ iff there exists $j \in T$ such that $j > i$, $\mathcal{M}_j \models B$ and, for any k with $i < k < j$, $\mathcal{M}_k \models A$ (*strict until*).

(i) $\mathcal{M}_i \models A\mathcal{W}B$ iff either for all $k \geq i$, $\mathcal{M}_k \models A$ or there exists $j \in T$ such that $j \geq i$, $\mathcal{M}_j \models B$ and, for any k with $i \leq k < j$, $\mathcal{M}_k \models A$ (*weak until*).

3. Past time operators:

(a) $\mathcal{M}_i \models \blacksquare A$ iff for all $j < i$, $\mathcal{M}_j \models A$ (*always in the past*).

(b) $\mathcal{M}_i \models \Diamond\!\!\!\cdot\, A$ iff there exists $j \in T$ such that $j < i$ and $\mathcal{M}_j \models A$ (*sometime in the past*).

(c) $\mathcal{M}_i \models \bullet A$ iff either $i = 0$ or $\mathcal{M}_{i-1} \models A$ (*weak previous*).

(d) $\mathcal{M}_i \models \circledcirc A$ iff $i \neq 0$ and $\mathcal{M}_{i-1} \models A$ (*strong previous*).

(e) $\mathcal{M}_i \models A\mathcal{S}B$ iff there exists $j \in T$ such that $j < i$, $\mathcal{M}_j \models B$ and, for any k with $j < k < i$, $\mathcal{M}_k \models A$ (*since*).

(f) $\mathcal{M}_i \models A\mathcal{Z}B$ iff either for all $k < i$, $\mathcal{M}_k \models A$ or there exists $j \in T$ such that $j < i$, $\mathcal{M}_j \models B$ and, for any k with $j < k < i$, $\mathcal{M}_k \models A$ (*weak since*).

Note that future and past time operators are not perfectly symmetric: $\blacksquare$, $\Diamond\!\!\!\cdot$ and $\mathcal{S}$ are the mirror images of the strict operators $\Box^+$, $\Diamond^+$ and $\mathcal{U}^+$, respectively.

Different subsets of temporal operators can be taken as primitive, still allowing others to be definable. The smallest complete set of operators consists of the two binary operators $\mathcal{U}^+$ and $\mathcal{S}$. The strict "until" operator can alternatively be defined in terms of $\circ$ and $\mathcal{U}$. Here follow some useful equivalences, where $\top \equiv_{def} \neg\bot$ and $\wedge$ is defined as usual.

$\Diamond^+ A \equiv \top\mathcal{U}^+ A$	$\Box^+ A \equiv \neg\Diamond^+\neg A$
$\Diamond A \equiv A \vee \Diamond^+ A$	$\Box A \equiv A \wedge \Box^+ A$
$A\mathcal{U}B \equiv B \vee (A \wedge A\mathcal{U}^+B)$	$A\mathcal{W}B \equiv \Box A \vee A\mathcal{U}B$
$\odot A \equiv \bot\mathcal{U}^+ A$	$\circ A \equiv \neg \odot \neg A$
$\Diamond\!\!\!\cdot\, A \equiv \top\mathcal{S}A$	$\blacksquare A \equiv \neg \Diamond\!\!\!\cdot\, \neg A$
$\circledcirc A \equiv \bot\mathcal{S}A$	$\bullet A \equiv \neg\circledcirc\neg A$
$A\mathcal{Z}B \equiv \blacksquare A \vee A\mathcal{S}B$	

Truth is satisfiability in the initial state: a formula A is true in $\mathcal{M}$ (and $\mathcal{M}$ is a model of A) iff $\mathcal{M}_0 \models A$. Truth of sets of formulae is defined as usual. Satisfiability of a formula is truth in at least one model, validity is truth in all models.[1] A formula is finitely satisfiable (valid) iff it is true in some (every) interpretation over a finite temporal frame. If k is a positive integer, a formula is k-satisfiable (k-valid) if it is true in some (every) interpretation whose temporal frame $\langle 0, ..., m\rangle$ is such that $m \leq k$.

2.1 TRADITIONAL TABLEAUX FOR *LTL*

Different proof systems for linear temporal logic can be found in the literature, generally referring to the "infinite" semantics, i.e. the temporal frame is always assumed to be infinite in the future. In this work, we consider only tableau calculi. Most of them (for example, [17, 2]) are essentially based on the following equivalences:

$$\begin{aligned} \Diamond A &\equiv A \vee \circ\Diamond A \\ \Box A &\equiv A \wedge \circ\Box A \\ A\mathcal{U}B &\equiv B \vee (A \wedge \circ(A\mathcal{U}B)) \end{aligned}$$

that give the temporal operators a fixpoint characterisation.

The seminal work on tableau calculi for linear time temporal logic is [17]. In this work, a tableau is a directed graph with a unique root, where nodes are labelled by sets of formulae. Only future time operators are considered and $\Box$, $\Diamond$, $\circ$ and $\mathcal{U}$ are all taken as primitive. The rules dealing with temporal operators are the following (besides dual rules for negated formulae):

$$(\Box)\ \frac{\Box A, \Gamma}{A, \circ\Box A, \Gamma} \qquad\qquad \frac{\Diamond A, \Gamma}{A, \Gamma \quad\ \circ\Diamond A, \Gamma}\ (\Diamond)$$

$$(\mathcal{U})\ \frac{A\mathcal{U}B, \Gamma}{B, \Gamma \quad\ A, \circ(A\mathcal{U}B), \Gamma} \qquad\qquad \frac{\Lambda, \circ A_1, ..., \circ A_n}{A_1, ..., A_n}\ (\circ)$$

In the $\circ$-rule, Λ is a set of literals. The tableau construction proceeds by applying the classical rules and the rules for $\Box$, $\Diamond$ and $\mathcal{U}$ as far as possible, thus internally "analysing" a state of the searched model. When none of them is any longer applicable, the node has the form $\Lambda, \circ A_1, ..., \circ A_n$, where Λ is a set of literals. At this point the successor state can be created, by means of the $\circ$-rule: the next state inherits the formulae $A_1, ..., A_n$, it is internally analysed

[1] A different definition of validity can also be found in the literature: A is valid iff for any temporal interpretation $\mathcal{M}$ and for any time point i, $\mathcal{M}_i \models A$. However, it is easy to see that the two notions of validity are inter-definable.

in turn, and so on. Termination is guaranteed by loop checking: if a son of node n would be labelled by the set Γ and a node m already exists in the tableau with Γ, then no new node is created, but an edge is added from n to m.

When the construction of the tableau terminates, it is checked for closure. First of all, every path containing a node with a pair of complementary literals is marked as closed. But this is not sufficient to ensure satisfiability of open paths. In fact, it may happen that a still open path in the tableau contains a formula of the form $\Diamond B$ or $A\mathcal{U}B$ (an *eventuality*) and yet none of the subsequent nodes in the path contains B. This happens when the branch always contains the rightmost expansion of the application of the $\Diamond$ or $\mathcal{U}$ rule to the considered formula; in other terms, the choice of the point where B holds is postponed forever. In such cases, the eventuality is said not to be satisfied, and the branch is marked as unsatisfiable. Hence, the tableau closure test involves some mechanism to check whether all eventualities are sooner or later satisfied.

This mechanism can be extended to deal with full LTL by exploitation of a normal form result for LTL formulae, the Separation Theorem [11], like in the tableau system underlying propositional METATEM [2]. In fact, every LTL formula can be transformed into an equivalent one, that is a boolean combination of strict past time formulae, i.e. formulae whose main logical operator is a past time operator and do not contain any future time operator, and non-strict future time formulae, i.e. formulae that do not contain any past time operator. In other terms, pure past can be separated from present and future. The expansion of a strict past time formula P, in a leaf node of a tableau branch, is then performed by checking the truth of P in the "history" of the node, i.e. the branch from the root to the node. If it results to be true, then P is simply dropped, otherwise the branch is closed. The expansion of non-strict future time formulae is performed by use of the usual rules for classical and future time operators. Termination and test for tableau closure are as in [17]. It must be noted, however, that by the preliminary transformation of formulae required by this method, the length of a formula can grow exponentially.

2.2 LABELLED TABLEAUX FOR LTL WITH $\Box$ AND $\Diamond$

The systems introduced in [13] and [15] to which the present work is strongly indebted, are significantly different from the above ones. The main novelty of such approaches is the embedding of semantical information in the calculus itself, namely the fact that a temporal frame is isomorphic to $\mathbb{N}$, thereby making it possible to check satisfiability by use of specific algorithms for solving integer constraints. In both papers, only the LTL fragment containing just $\Box$ and $\Diamond$ as temporal operators is considered.

In [13] a labelled tableau is a tree whose nodes are pairs Γ, C, where Γ is a set of labelled formulae, of the form $s : A$, and C is a set of temporal constraints of the form $s \leq t$. The expressions s, t denote time points and have the form $k_0 + ... + k_n$. Intuitively, $s : A$ means that A holds at s. As a sample rule, consider the $\Diamond$ rule:

$$\frac{\{s : \Diamond A\} \cup \Gamma,\ C}{\{s + k : A\} \cup \Gamma,\ \{k \geq 0\} \cup C}$$

The expansion introduces a new label $s + k$, denoting the time point where A holds, and a constraint forcing such a point to either coincide with s or else follow s. It is apparent that, with the use of labels, an "eventuality" can be immediately satisfied, without the risk of postponing its *execution* indefinitely. The "always" operator is treated by means of two rules: the first one below takes reflexivity into account, while the second is called *synchronization* rule.

$$\frac{\{s : \Box A\} \cup \Gamma,\ C}{\{s : \Box A, s : A\} \cup \Gamma,\ C}$$

$$\frac{\{s : \Box A, t : \lambda\} \cup \Gamma,\ C}{\{s : \Box A, t : \lambda\} \cup \Gamma,\ \{t \leq s\} \cup C \qquad \{s : \Box A, t : \lambda, t : A\} \cup \Gamma,\ \{t > s\} \cup C}$$

Termination is ensured by the use of additional marks "freezing" labelled formulae, which must be re-activated in some cases not to loose completeness.

Additionally, a *conflict rule* establishes a link between formulae and constraints:

$$\frac{\{s : A, t : \neg A\} \cup \Gamma,\ C}{\{s : A, t : \neg A\} \cup \Gamma,\ \{s \neq t\} \cup C}$$

A node Γ, C in the tableau is contradictory iff C is unsolvable over $\mathbb{Z}$.

The main novelty of this work with respect to traditional tableaux is the use of linear constraints over world variables, so that satisfiability of a temporal formula (containing only the $\Box$ and $\Diamond$ temporal operators) is reduced to satisfiability of a set of constraints over $\mathbb{Z}$. This approach is carried on in [15], again considering LTL with only $\Box$ and $\Diamond$ as temporal operators. In this work, a tableau is a tree whose nodes are either temporal constraints or expressions of the form $[s, t]\Gamma$ or $[s, +\infty[\Gamma$ or $|\infty|\Gamma$ - where s and t denote time points and Γ is a finite multiset of formulae. The semantics of an expression $[s, t]\Gamma$ is the following: in each point of the time interval $[s, t]$, the disjunction of the formulae in Γ holds; $[s, +\infty[\Gamma$ can be read as stating that the disjunction of the formulae in Γ holds from s onwards, and $|\infty|\Gamma$ states that the *conjunction* of formulae in Γ holds infinitely often. Labelling sets of formulae, rather than single formulae, makes it possible to give a very simple formulation of the β rules; for instance, in the case of "closed" intervals $[s, t]$ the rule is:

$$\frac{[s, t]\Gamma, A \vee B}{[s, t]\Gamma, A, B}$$

The use of time intervals rather than single time points as labels makes it also possible to give a rather simple formulation of the $\Box$ rules, which, for instance, generate a node $[s, +\infty[A$ out of the node $[s, t]\Box A$. As a consequence, termination is easily ensured (formulae are eventually decomposed into literals).

New constraints can then be added to the branch, by means of a set of "resolution rules" (in fact, the disjunctions are now treated very much in the style of resolution), which play a role similar to the conflict rule in [13] and, just as it is done there, satisfiability is reduced to solvability of a set of integer constraints. By exploiting the fact that this can be checked in polynomial time, the calculus enables one to test satisfiability for the considered fragment of LTL in non-deterministic polynomial time (while the problem is $PSPACE$ complete for full LTL).

However, the introduction of the "since" and "until" operators in the same framework raises new difficulties. In [14], where full LTL is considered, the tableau expansion rules for such operators are mainly based on their fixpoint characterization, but an ingenious rewriting of the "contracted" formula into a propositional letter guarantees termination. After the terminating tableau expansion, branch closure is checked by reduction to a model checking problem in CTL with fairness constraints, in the style of [5]. Because of the special form of the set of active formulae in the branch, the model checking problem is somewhat simplified.

3 THE TABLEAU SYSTEM FOR $\boldsymbol{LTL^{FIN}}$

In [12], a decision procedure is defined for a temporal language whose semantics is given in terms of state sequences which may be either infinite or finite. It can easily be adapted to LTL^{fin}, where the state sequence is assumed to be always finite. The procedure is developed by means of a translation into LTL over infinite temporal frames.

In this work we propose a direct approach, defining a specific decision procedure for LTL^{fin}. In this section a tableau calculus for full LTL with the *finite* semantics is presented, that allows one to test whether a set of formulae admits models whose underlying temporal frame does not exceed a given size, hence to check k-validity. In Section 4.4 it is shown that the calculus can be used also to check finite validity *tout court*.

We take as primitive the operators $\neg$, $\wedge$, $\vee$, $\bot$, $\Box^+$, $\Diamond^+$, $\mathcal{U}^+$, $\blacksquare$, $\blacklozenge$ and $\mathcal{S}$. The calculus treats formulae in *negation normal form* (nnf), where no logical operator is in the scope of a negation. Two formulae A and B are equivalent iff for all $\mathcal{M}$ and i, $\mathcal{M}_i \models A$ iff $\mathcal{M}_i \models B$. It can easily be shown that, under this strong notion of equivalence, every formula can be transformed into an equivalent formula in nnf, by applying the usual equivalences for $\rightarrow, \wedge, \vee, \Box^+, \blacksquare, \Diamond^+, \blacklozenge$, and:

$$\neg(A\mathcal{U}^+B) \equiv (\neg B\mathcal{U}^+(\neg A \wedge \neg B)) \vee \Box^+\neg B$$
$$\neg(A\mathcal{S}B) \equiv (\neg B\mathcal{S}(\neg A \wedge \neg B)) \vee \blacksquare\neg B$$

The restriction to formulae in nnf allows us to give a more compact formulation of the rules and it could easily be replaced by a larger set of rules. Since formulae in nnf are considered, the operators $\Box^+$, $\mathcal{U}^+$, $\blacksquare$ and $\mathcal{S}$ are all to be taken as primitive. Although the $\Diamond^+$ and $\blacklozenge$ operators are definable, we prefer to give them a separate treatment in the tableau system.

Let $C = \{start, finish, d_1, d_2, d_3, ...\}$ be a set of *constants* (intuitively denoting time points, $start$ denoting the initial time point and $finish$ the final one of the searched model). A *state* is any expression of the form $c + n$, for $c \in C$ and $n \in Z$. The set of states is denoted by Σ. It is intended that $C \subset \Sigma$ (c can be rewritten as $c + 0$). If $s, t \in \Sigma$, then $s \leq t$ is a *temporal constraint*. A *labelled formula* is an expression of the form $[s, t]A$, where $s, t \in \Sigma$ and A is a formula in nnf. $[s, s]A$ will be abbreviated by $[s]A$, and $s \leq t - 1$ by $s < t$.

A tableau is a tree whose nodes are labelled either by temporal constraints or labelled formulae (that are called *logical nodes*). If S is a finite set of formulae in nnf and $K = \{finish \leq start + k\}$ for some integer $k \geq 0$ (representing the maximal size of the searched models), then tableaux for $S \cup K$ are initialized with the set $\{[start]A \mid A \in S\} \cup K$ and expanded by application of the rules in Table 1, where c denotes an element of C, and $s, t, s', t', ..$ elements of Σ. The set of nodes occurring above the line of a rule is called the *premise* of the rule, while the sets of nodes occurring below are the *expansions* of the premise.

A β-formula holding in a time interval is intuitively equivalent to a universal quantifier dominating a disjunction. It could be treated by means of a sort of "instantiation", in the style of the synchronization rule in [13], but this would raise a non trivial termination problem in the presence of both past and future time operators. The approach followed here is different. Note that a sort of contraction is implicit in the β-rule: the rightmost expansion of the rule contains a node with the same formula already occurring in the premise, even though the labels (intervals) of the nodes are different. As we shall see later on, this does not hinder termination.

The intuition behind the β-rule is the following: either A is true in the whole interval (leftmost branch), or there exists a smallest time point c in the interval where A is false, hence B is true; since c is chosen to be the first of such points, A is true in the (possibly empty) subinterval before it. The rule could also be formulated in a symmetric way, distinguishing two cases according to whether A is true at s or B is true at s. However, this would force the rule to be reapplied once for each point of the interval; on the contrary, with the proposed asymmetric formulation the interval may be cut into larger pieces at each application (obviously, the behaviour is the same in the worst case, where A and B are interleaved). The β-rule is a delicate point: note that it is indirectly charged to expand also $\mathcal{U}^+$- and $\mathcal{S}$-formulae. By contrast, the calculus defined in [15], where only the $\Box, \Diamond$ fragment of LTL is considered, has a very simple, non-branching β rule; however, it must be noted that having single formulae, instead of sets of formulae, labelled by time intervals allows us to give a conceptually simpler treatment of the $\mathcal{U}^+$- and $\mathcal{S}$-formulae than in [14].

Logical rules		
Propositional	*Future time*	*Past time*
α**-rule**	$\Box^+$**-rule**	$\blacksquare$**-rule**
$\dfrac{\begin{array}{c}[s,t]\,A\wedge B\\ s\le t\end{array}}{\begin{array}{c}[s,t]\,A\\ [s,t]\,B\end{array}}$	$\dfrac{\begin{array}{c}[s,t]\,\Box^+A\\ s\le t\end{array}}{[s+1,finish]\,A}$	$\dfrac{\begin{array}{c}[s,t]\,\blacksquare A\\ s\le t\end{array}}{[start,t-1]\,A}$
β**-rule**	$\mathcal{U}^+$**-rule**	$\mathcal{S}$**-rule**
$\dfrac{\begin{array}{c}[s,t]\,A\vee B\\ s\le t\end{array}}{[s,t]\,A \qquad \begin{array}{c}s\le c\\ c\le t\\ [s,c-1]\,A\\ [c]\,B\\ [c+1,t]\,A\vee B\\ c\in C \text{ fresh}\end{array}}$	$\dfrac{\begin{array}{c}[s,t]\,A\mathcal{U}^+B\\ s\le t\end{array}}{\begin{array}{c}[c]\,B\\ t<c\\ [t+1,c-1]\,A\\ [s+1,t]\,A\vee B\\ c\in C\ fresh\end{array}}$	$\dfrac{\begin{array}{c}[s,t]\,A\mathcal{S}B\\ s\le t\end{array}}{\begin{array}{c}[c]\,B\\ c<s\\ [c+1,s-1]\,A\\ [s,t-1]\,A\vee B\\ c\in C\ fresh\end{array}}$
	$\Diamond^+$**-rule**	$\blacklozenge$**-rule**
	$\dfrac{\begin{array}{c}[s,t]\,\Diamond^+A\\ s\le t\end{array}}{\begin{array}{c}[c]\,A\\ t<c\\ c\in C\ fresh\end{array}}$	$\dfrac{\begin{array}{c}[s,t]\,\blacklozenge A\\ s\le t\end{array}}{\begin{array}{c}[c]\,A\\ c<s\\ c\in C\ fresh\end{array}}$

Interval rule	Conflict resolution rules	
$\dfrac{[s,t]\,A}{t<s \qquad s\le t}$	$\dfrac{\begin{array}{c}s\le t\\ s'\le t'\\ [s,t]\,p\\ [s',t']\,\neg p\end{array}}{t<s' \qquad t'<s}$	$\dfrac{[s,t]\,\bot}{t<s}$

Table 1. Tableau expansion rules

When the interval rule is applied to expand $[s,t]A$, we say that it is applied to the interval $[s,t]$, independently of the formula A. This rule distinguishes the cases where an interval is empty or not. Its role is to provide the preconditions for the application of the logical and resolution rules. Intuitively, it is useless – and sometimes even incorrect – to expand a node $[s,t]A$ when the interval is empty and, given two nodes $[s,t]p$, $[s',t']\neg p$, there is no conflict to be solved if either $[s,t]$ or $[s',t']$ (or both) are empty. Note that such rule could be dispensed with, and a corresponding branching added to most of the other rules, handling the case where the considered interval is empty. For example, the $\Diamond^+$-rule would be

$$\frac{[s,t]\,\Diamond^+ A}{\begin{array}{cc} t < s & \begin{array}{c} s \leq t \\ [c]\,A \\ t < c \end{array} \end{array}}$$

Distinguishing the interval rule as a separate rule makes the formulation of the calculus more compact and clearer: obviously, the choice about the "emptiness" of an interval $[s,t]$ needs to be done just once in a branch, independently on the number of logical nodes labelled by $[s,t]$.

When the leftmost conflict resolution rule is applied, we say that it is applied to the intervals $[s,t]$ and $[s',t']$.

In the following, if $\mathcal{B}$ is a tableau branch, $const(\mathcal{B})$ denotes the set of constants occurring in $\mathcal{B}$, including $start$ and $finish$.

Definition 1 *Let C be a set of constants (including $start$ and $finish$) and $\mathcal{I}$ a mapping from C to the integers. The notation $\mathcal{I}*$ is used to denote the extension of $\mathcal{I}$ from states to the integers such that $\mathcal{I}^*(c+n) = \mathcal{I}(c) + n$ for every $c \in C, n \in \mathbb{Z}$.*

1. *Let $T = \langle 0, ..., k \rangle$ be a finite sequence of integers starting at 0. $\mathcal{I}$ is a* temporal mapping *for C with range T if $min\{\mathcal{I}(c) \mid c \in C\} = \mathcal{I}(start) = 0$ and $max\{\mathcal{I}(c) \mid c \in C\} = \mathcal{I}(finish) = k$. Hence, in particular, the range of a temporal mapping is always finite.*

2. *If K is a set of temporal constraints over C, then $\mathcal{I}$ is a solution to K iff:*

   (a) $\mathcal{I}$ is a temporal mapping for C;

   (b) if $s \leq t \in K$, then $\mathcal{I}^*(s) \leq \mathcal{I}^*(t)$.

3. *Let $\mathcal{B}$ be a tableau branch, $C = const(\mathcal{B})$ and $\mathcal{M}$ a temporal interpretation with frame T.*

   (a) If $\mathcal{I}$ is a temporal mapping for C with range T, then $\langle \mathcal{M}, \mathcal{I} \rangle$ satisfies $\mathcal{B}$ ($\langle \mathcal{M}, \mathcal{I} \rangle \models \mathcal{B}$) iff:

      i. $\mathcal{I}$ is a solution to the set of temporal constraints occurring in $\mathcal{B}$;

      ii. if $[s,t]A$ occurs in $\mathcal{B}$, then for every integer i, if $i \in T$ and $I^*(s) \leq i \leq I^*(t)$, then $\mathcal{M}_i \models A$.

(b) $\mathcal{B}$ *is* satisfiable *in* $\mathcal{M}$ *iff there exists a temporal mapping* $\mathcal{I}$ *for* C *such that* $\langle \mathcal{M}, \mathcal{I} \rangle \models \mathcal{B}$.

Definition 2 *Let* $\mathcal{B}$ *be a tableau branch and* K *the set of temporal constraints occurring in* $\mathcal{B}$. $\mathcal{B}$ *is* open *iff there exists a solution to* K. *Otherwise it is* closed.

Checking closure of a finite branch amounts to checking satisfiability of a finite set of integer constraints. This can be done by means of well known graph algorithms (see any standard textbook, e.g. [10]).

The following definition captures the intuitive idea of non redundant tableaux where no wasteful expansions are ever performed. In particular, closed branches are never expanded.

Definition 3 *A tableau branch* $\mathcal{B}$ *is canonical iff:*

- *The interval rule is applied at most once to each interval.*
- *Every labelled formula is expanded at most once by means of a logical rule.*
- *Each conflict resolution rule is applied at most once to each interval (second conflict resolution rule) or pair of intervals (first conflict resolution rule).*
- *No proper initial subsegment of* $\mathcal{B}$ *is closed.*

A tableau is canonical iff all its branches are canonical.

Definition 4 *If* $\mathcal{B}$ *is a tableau branch, then* $\mathcal{B}$ *is* complete *iff there exists no canonical expansion of* $\mathcal{B}$. *A tableau is complete if all its branches are complete.*

The following definition characterises tableaux proving that a formula A is true in all models bounded by a positive integer k (the adequacy between the syntactical notion defined below and the semantical notion of k-valid formula are proved in the next section). Here and in the rest of the paper, we use $\neg A$ as an abbreviation for the nnf of $\neg A$.

Definition 5 *Let* k *be a positive integer and* A *a formula. A* k-tableau proof *of* A *is a finite tableau for* $\{[start]\neg A, finish \leq start + k\}$ *such that each branch (going from the root to a leaf) is closed.*

Here follows a simple example, showing how to prove that $\Box^+\Diamond^+ p \rightarrow \Box^+\bot$ is valid (in fact $\Box^+\Diamond^+ p$ is true in a finite structure only if it consists of a single time point). Below, the application of the interval rule to intervals of the form $[c]$ is not shown, and neither are obvious premises of the form $c \leq c$. Moreover, for the sake of simplicity we abuse notation, identifying, for example, a state of the form $a + 1 - 1$ with a itself, or abbreviating $s + 1 \leq t$ by $s < t$ (i.e. we confuse syntax and semantics in the notation for constraints). The tableau is initialized with the nnf of the negation of the given formula, and no limit on the size of the temporal frame (see Section 4.4). The nodes are numbered in order to comment the tableau.

$$
\begin{array}{c}
1.\ [start]\,\Box^{+}\Diamond^{+}p \wedge \Diamond^{+}\neg\bot \\
\hline
2.\ [start]\,\Box^{+}\Diamond^{+}p \\
3.\ [start]\,\Diamond^{+}\neg\bot \\
\hline
4.\ [start+1, finish]\,\Diamond^{+}p \\
\hline
5.\ [c_1]\,\neg\bot \\
6.\ start < c_1 \\
\hline
\begin{array}{cc}
7.\ finish \leq start \qquad & \qquad 8.\ start < finish \\
 & 9.\ [c_2]\,p \\
 & 10.\ finish < c_2
\end{array}
\end{array}
$$

Nodes 2 and 3 derive from 1 by application of the α-rule. Node 4 results from 2 by the $\Box^{+}$-rule and nodes 5-6 from 3 by the $\Diamond^{+}$-rule. Nodes 7-8 result from the application of the interval rule to node 4. The branch ending at 7 is closed, because $\{start < c_1, finish \leq start\}$ is unsatisfiable (for any time constant c in the branch, the constraints $start \leq c \leq finish$ are implicitly present). Nodes 9-10 derive from 4 and 8 by the $\Diamond^{+}$-rule. This branch is also closed, since no time point can be greater than $finish$.

The next example shows the partial development of an open tableau for $A = \Diamond^{+}p \wedge \Box^{+}(\neg p \vee q)$, pointing out that every open branch corresponds to a model of the initial formula. We assume that the tableau is initialized with some limit k, for any $k \geq 1$.

Nodes 3 and 4 are the expansion of 2, and their expansions are 5, 6 and 7, respectively. Nodes 8 and 9 result from the application of the interval rule to 7; the branch with node 8 is closed, since, implicitly, $d_1 \leq finish$. Nodes 10 and 13–17 are the branches obtained from 7 and 9. The conflict resolution rule is applied to 5, 9 and 10, giving 11 and 12, that both close. Nodes 18–19 and 20–21 derive from 15 and 17, respectively, by the interval rule. The branches passing through 19 and 21 are not further developed. The open and complete branch ending at 20 represents the smallest (partial) model, with points $0 = \mathcal{I}(start) < 1 = \mathcal{I}(d_1) = \mathcal{I}(d_2) = \mathcal{I}(finish)$, with both p and q holding at 1. One of the further expansions of the branch passing through 21 would yield a more general description, with $0 = \mathcal{I}(start) < 1 = \mathcal{I}(d_2) \leq \mathcal{I}(d_1)$, where p holds at $\mathcal{I}(d_1)$ and q holds in the whole interval $[1, \mathcal{I}(finish)]$. The development of 19 explores models where $d_1, d_2 > 1$.

1. $\mathit{finish} \leq \mathit{start} + k$
2. $[\mathit{start}]\, A$

3. $[\mathit{start}]\, \Diamond^{+} p$
4. $[\mathit{start}]\, \Box^{+}(\neg p \vee q)$

5. $[d_1]\, p$
6. $\mathit{start} < d_1$

7. $[\mathit{start} + 1, \mathit{finish}]\, \neg p \vee q$

8. $\mathit{finish} \leq \mathit{start}$ | 9. $\mathit{start} < \mathit{finish}$

10. $[\mathit{start} + 1, \mathit{finish}]\, \neg p$ | 13. $\mathit{start} < d_2$

11. $d_1 \leq \mathit{start}$ | 12. $\mathit{finish} < d_1$

14. $d_2 \leq \mathit{finish}$
15. $[\mathit{start} + 1, d_2 - 1]\, \neg p$
16. $[d_2]\, q$
17. $[d_2 + 1, \mathit{finish}]\, \neg p \vee q$

18. $d_2 - 1 \leq \mathit{start}$ | 19. $\mathit{start} < d_2 - 1$

20. $\mathit{finish} \leq d_2$ | 21. $d_2 < \mathit{finish}$

4 TERMINATION, SOUNDNESS AND COMPLETENESS

In this section we show that the construction of any canonical tableau terminates and that the calculus is sound and complete w.r.t. to bounded validity, i.e., for any positive integer k and any formula A, if there is a k-tableau proof of A then A is k-valid (soundness), and, conversely, if A is k-valid, then there is a k-tableau proof of A (completeness). It is also shown that, from a theoretical standpoint, the calculus can be used to check finite validity *tout court*.

4.1 TERMINATION

Here we restrict our attention to canonical tableaux, thus ensuring, in particular, that the construction of a branch is abandoned as soon as it can be recognized to be closed. As already remarked, the existence of a solution for a given set of constraints can be tested by means of shortest-path graph algorithms.

The following definition captures the idea of a sequence of applications of the β-rule, each of which expands an expansion of the previous one.

Definition 6 *Let $\mathcal{B}$ be a tableau branch. A β-node in $\mathcal{B}$ is a node of the form $[s,t]A \vee B$. A β-chain in $\mathcal{B}$ is a sequence of β-nodes $X_0, X_1, \ldots$ such that, for every $i \geq 1$, X_{i-1} is expanded in $\mathcal{B}$ by application of the β-rule and X_i is* *the β-node in the corresponding rightmost expansion. A k-length-β-chain is a finite β-chain $X_0, X_1, ..., X_{k+1}$, constituted by $k+2$ nodes.*

A node X in $\mathcal{B}$ is the root node *of a β-chain if it is the first node in a β-chain and it is not itself obtained by an application of the β-rule, i.e. there exists a maximal length β-chain in $\mathcal{B}$ having X as its first node.*

The proof that the construction of canonical tableaux terminates uses the following lemmas.

Lemma 7 *Let $\mathcal{B}$ be a branch in a tableau and $\mathcal{I}$ a temporal mapping satisfying the constraints in $\mathcal{B}$. If $[s,t]A$ occurs in $\mathcal{B}$ and $\mathcal{I}^*(s) \leq \mathcal{I}^*(t)$, then $0 \leq I^*(s) \leq I^*(t) \leq \mathcal{I}(finish)$.*

Proof The proof is an easy induction on the length of $\mathcal{B}$. The statement is obviously satisfied by the logical nodes in the initial tableau. The inductive step considers different cases, according to the last expansion rule applied in the branch. What has to be shown is that, if every node of the form $[s,t]A$ in the premise satisfies $0 \leq \mathcal{I}^*(s) \leq \mathcal{I}^*(t) \leq \mathcal{I}(finish)$ whenever $\mathcal{I}^*(s) \leq \mathcal{I}^*(t)$, then every node in the expansion(s) having the the form $[s',t']B$ is such that, if $\mathcal{I}^*(s') \leq \mathcal{I}^*(t')$, then $0 \leq \mathcal{I}^*(s') \leq \mathcal{I}^*(t') \leq \mathcal{I}(finish)$. Here follow some cases of the inductive step.

1. Let us assume that the β-rule is applied to expand $[s,t]A \vee B$, and that $0 \leq \mathcal{I}^*(s) \leq \mathcal{I}^*(t) \leq \mathcal{I}(finish)$. Whichever the value of $\mathcal{I}(a)$, where a is the new constant introduced in the leftmost expansion, if $\mathcal{I}$ satisfies the constraints in $\mathcal{B}$, it holds that $s \leq a \leq t$, hence $0 \leq \mathcal{I}^*(s) \leq \mathcal{I}(a) \leq \mathcal{I}(finish)$. Moreover, $0 \leq \mathcal{I}^*(a+1)$; hence if $\mathcal{I}^*(a+1) \leq \mathcal{I}^*(t)$, then $0 \leq \mathcal{I}^*(a+1) \leq \mathcal{I}^*(t) \leq \mathcal{I}(finish)$ holds, too.

2. Let the $\mathcal{S}$-rule be applied to $[s,t]A\mathcal{S}B$, with $0 \leq \mathcal{I}^*(s) \leq \mathcal{I}^*(t) \leq \mathcal{I}(finish)$, producing $a < s$, $[a]B$, $[a+1, s-1]A$ and $[s, t-1]A \vee B$. The interval $[a]$ trivially satisfies the desired property.

   Moreover, $0 \leq \mathcal{I}^*(a+1)$ and $\mathcal{I}^*(s-1) \leq \mathcal{I}(finish)$, hence if also $\mathcal{I}(a+1) \leq \mathcal{I}^*(s-1)$, then $0 \leq \mathcal{I}^*(a+1) \leq \mathcal{I}^*(s-1) \leq \mathcal{I}(finish)$.

   Similarly, $0 \leq \mathcal{I}^*(s)$ and $\mathcal{I}^*(t-1) \leq \mathcal{I}(finish)$, hence if also $\mathcal{I}(s) \leq \mathcal{I}^*(t-1)$, then $0 \leq \mathcal{I}^*(s) \leq \mathcal{I}^*(t-1) \leq \mathcal{I}(finish)$.

■

Lemma 8

1. *If $\mathcal{B}$ is a canonical open branch in a tableau for $S \cup \{finish \leq start+n\}$, then $\mathcal{B}$ contains no k-length-β-chain with $k > n$.*

2. *Any canonical infinite branch contains an infinite β-chain.*

3. *Every open branch is finite.*

Proof

1. Let us assume that $\mathcal{B}$ does contain a k-length-β-chain with $k > n$. Then $\mathcal{B}$ contains a sequence of nodes having the form: $[s, t]\ A \vee B, [a_1 + 1, t]\ A \vee B, [a_2 + 1, t]\ A \vee B, ..., [a_{n+2} + 1, t]\ A \vee B$ and the sequence of constraints: $s \leq a_1, a_1 + 1 \leq a_2, ..., a_{n+1} + 1 \leq a_{n+2}$, and $a_{n+2} \leq t$. If $\mathcal{I}$ is is a solution for the constraints in $\mathcal{B}$, then, by Lemma 7, $0 \leq \mathcal{I}^*(s) \leq \mathcal{I}^*(t) \leq \mathcal{I}^*(finish)$ (in fact $s \leq t$ is in $\mathcal{B}$), so that $\mathcal{I}^*(a_{n+2}) \geq n + 1$. However, $\mathcal{I}^*(a_{n+2}) \leq \mathcal{I}^*(t) \leq \mathcal{I}^*(finish) \leq n$ holds too, that is absurd.

2. It is easily recognized that the calculus enjoys the subformula property, where $A \vee B$ is included among the subformulae of both $A\mathcal{U}B$ and ASB. Hence, if a canonical branch $\mathcal{B}$ is infinite, then there is an infinite number of nodes X_i in $\mathcal{B}$ having the form $[s_i, t_i]A$, for a given formula A. Clearly, A must be a disjunction. Let S be the set of such nodes. Now, every node in S belongs to some β-chain in $\mathcal{B}$. The set S can be partitioned into classes $C_1, C_2, ...$ such that X_i and X_j belong to the same class if and only if they belong to the same maximal β-chain.

   We consider two cases:

   - Every class C_i is finite, hence S is partitioned into an infinite number of classes. Therefore there is an infinite number of root nodes of β-chains in $\mathcal{B}$. With the exception of the (finite number of) nodes that occur in the initial tableau, each of such nodes R_i is the expansion of a node Y_i that is not a β-node. Hence, some rule other than the β-rule is applied infinitely often in $\mathcal{B}$. But this is absurd, since the initial tableau is always finite and $\mathcal{B}$ is canonical, so that the same formula cannot be expanded an infinite number of times. Therefore, only the second case is possible:
   - There is a class C_i containing an infinite number of nodes. In this case, an infinite number of β-nodes of $\mathcal{B}$, namely the elements of C_i, belong to the same maximal β-chain, that is therefore infinite.

3. Let $\mathcal{B}$ be a canonical open branch. Then the set of its temporal constraints has a solution $\mathcal{I}$, hence, by item 1, $\mathcal{B}$ contains no infinite β-chain and, by item 2, it is finite.

■

The next theorem follows immediately from the above lemma.

Theorem 9 *Any canonical tableau for* $S \cup \{finish \leq start + n\}$, *for any* $n \geq 0$, *is finite.*

4.2 SOUNDNESS WITH RESPECT TO BOUNDED VALIDITY

The proof that the tableau calculus is sound w.r.t. bounded validity uses the following lemma.

Lemma 10 *Let $\mathcal{B}$ be a branch in a tableau $\mathcal{T}$, $\mathcal{M}$ a temporal interpretation and $\mathcal{I}$ a temporal mapping for the constants occurring in $\mathcal{B}$. If $\langle \mathcal{M}, \mathcal{I} \rangle \models \mathcal{B}$ and $\mathcal{B}$ can be expanded, $\mathcal{B}_1$ (and $\mathcal{B}_2$) being its expansion(s), then for some $i = 1, 2$ there exists an extension $\mathcal{I}'$ of $\mathcal{I}$ to the fresh constants of $\mathcal{B}_i$ (if any) such that $\langle \mathcal{M}, \mathcal{I}' \rangle \models \mathcal{B}_i$.*

Proof Different cases must be considered according to the applied expansion rule. Let T be the temporal frame underlying $\mathcal{M}$ (and the range of $\mathcal{I}$). Here we show the treatment of the β-rule and the past time operators.

1. In the case of the β-rule, by hypothesis $\langle \mathcal{M}, \mathcal{I} \rangle \models [s,t]A \vee B$ and $\mathcal{I}^*(s) \leq \mathcal{I}^*(t)$, so that $0 \leq \mathcal{I}^*(s) \leq \mathcal{I}^*(t) \leq \mathcal{I}^*(finish)$, by Lemma 7. It follows that for every $i \in [\mathcal{I}^*(s), \mathcal{I}^*(t)]$, $\mathcal{M}_i \models A \vee B$. If for every such i, $\mathcal{M}_i \models A$, then the leftmost branch is clearly satisfied by $\mathcal{M}$ and $\mathcal{I}$. Otherwise, let k be the smallest element in the interval $[\mathcal{I}^*(s), \mathcal{I}^*(t)]$ such that $\mathcal{M}_k \not\models A$, hence $\mathcal{M}_k \models B$. The mapping $\mathcal{I}$ is extended to $\mathcal{I}'$ such that $\mathcal{I}'(a) = k$, where a is the new time constant produced by the rule. By the choice of k, clearly $\langle \mathcal{M}, \mathcal{I}' \rangle \models [s, a-1]A$ and $\langle \mathcal{M}, \mathcal{I}' \rangle \models [a]B$. If $k = \mathcal{I}^*(t)$, $\langle \mathcal{M}, \mathcal{I}' \rangle \models [a+1, t]A \vee B$ is voidly satisfied. Otherwise, the interval $[\mathcal{I}^*(a+1), \mathcal{I}^*(t)]$ is a subinterval of $[\mathcal{I}^*(s), \mathcal{I}^*(t)]$, hence again $\langle \mathcal{M}, \mathcal{I}' \rangle \models [a+1, t]A \vee B$.

2. In the case of the $\blacksquare$-rule, by hypothesis $\langle \mathcal{M}, \mathcal{I} \rangle \models [s,t]\blacksquare A$ and $\mathcal{I}^*(s) \leq \mathcal{I}^*(t)$. Hence for every $i \in T$ such that $\mathcal{I}^*(s) \leq i \leq \mathcal{I}^*(t)$, $\mathcal{M}_i \models \blacksquare A$. Since $0 \leq \mathcal{I}^*(s) \leq \mathcal{I}^*(t) \leq \mathcal{I}(finish)$ (by Lemma 7), $\mathcal{I}^*(t) \in T$ and $\mathcal{I}^*(t) \in [\mathcal{I}^*(s), \mathcal{I}^*(t)]$. Hence $\mathcal{M}_{\mathcal{I}^*(t)} \models \blacksquare A$, i.e. for every $i \in T$ such that $i < \mathcal{I}^*(t)$, $\mathcal{M}_i \models A$. This clearly implies that $\langle \mathcal{M}, \mathcal{I} \rangle \models [start, t-1]A$.

3. In the case of the $\blacklozenge$ rule, applied to a node $[s,t] \blacklozenge A$, by hypothesis we have that $\langle \mathcal{M}, \mathcal{I} \rangle \models [s,t] \blacklozenge A$ and $\mathcal{I}^*(s) \leq \mathcal{I}^*(t)$. By Lemma 7, $0 \leq \mathcal{I}^*(s) \leq \mathcal{I}^*(t) \leq \mathcal{I}(finish)$, hence $\mathcal{I}^*(s) \in T$. Therefore there exists $i \in T$ s.t. $i < \mathcal{I}^*(s)$ and $\mathcal{M}_i \models A$. If b is the new constant introduced by the rule, by setting $\mathcal{I}'(b) = i$ we get the desired conclusion.

4. Let the $\mathcal{S}$-rule be applied to $[s,t]A\mathcal{S}B$. By hypothesis, $\langle \mathcal{M}, \mathcal{I} \rangle \models A\mathcal{S}B$ and $\mathcal{I}(s) \leq \mathcal{I}^*(t)$. By Lemma 7, $0 \leq \mathcal{I}^*(s) \leq \mathcal{I}^*(t) \leq \mathcal{I}(finish)$. It follows that $\mathcal{I}^*(s) \in T$ and that there exists $i \in T$ such that $i < \mathcal{I}^*(s)$, $\mathcal{M}_i \models B$, and for any $j \in T$ such that $i < j < \mathcal{I}^*(s)$, $\mathcal{M}_j \models A$. Hence if we set $\mathcal{I}'(a) = i$, where a is the new constant introduced by the rule,

we get that $\langle \mathcal{M}, \mathcal{I}' \rangle$ satisfies $[a]B$, $a \leq s-1$ and $[a+1, s-1]A$. Let now consider any time point $j \in T$ such that $\mathcal{I}^*(s) \leq j \leq \mathcal{I}^*(t-1)$ and assume that $\mathcal{M}_j \not\models A \vee B$. Let k be the smallest of such points; then $\mathcal{I}^*(s) \leq k+1 \leq \mathcal{I}^*(t)$ but $\mathcal{M}_{k+1} \not\models A\mathcal{S}B$, a contradiction.

■

As a consequence of lemma 10 we have the following:

Theorem 11 (k-soundness) *Let k be a positive integer and A a formula in nnf. If there is k-tableau proof for A, then A is k-valid.*

Proof Suppose, by *reductio*, that $\mathcal{T}$ is a k-tableau proof of A, yet A is not k-valid. Hence $\neg A$ has at least a model $\mathcal{M}$ whose underlying temporal frame $T = \langle 0, .., n \rangle$ is such that $n \leq k$. Thus, there is a temporal mapping $\mathcal{I}$ for the constants $start, finish$ such that its range is T and $\langle \mathcal{M}, \mathcal{I} \rangle \models B_0$, where B_0 is the initial segment of $\mathcal{T}$ containing only the node $[start]\neg A$ and the constraint $finish \leq start + k$. By lemma 10 there are at least a branch $\mathcal{B}$ of $\mathcal{T}$, going from the root to a leaf, and an extension of $\mathcal{I}'$ of $\mathcal{I}$ such that $\langle M, \mathcal{I}' \rangle \models \mathcal{B}$. Hence the set of constraints in $\mathcal{B}$ is satisfiable. Yet, if $\mathcal{T}$ is closed, $\mathcal{B}$ is closed: a contradiction. ■

4.3 COMPLETENESS WITH RESPECT TO BOUNDED VALIDITY

The following lemma establishes the invertibility of the logical rules, with respect to their logical nodes.

Lemma 12 *Let $\mathcal{M} = \langle T, \sigma \rangle$ be a temporal interpretation and $\mathcal{I}$ a temporal mapping with range T. For every logical rule*

$$\frac{\begin{array}{c}[s,t]\,F\\ s \leq t\end{array}}{\mathcal{B}_1 \quad \mathcal{B}_2} \quad or \quad \frac{\begin{array}{c}[s,t]\,F\\ s \leq t\end{array}}{\mathcal{B}_1}$$

if $\langle \mathcal{M}, \mathcal{I} \rangle \models \mathcal{B}_i$ (for some $i = 1, 2$), then $\langle \mathcal{M}, \mathcal{I} \rangle \models [s,t]F$.

Proof If $\mathcal{I}^*(t) < \mathcal{I}^*(s)$ then $\langle \mathcal{M}, \mathcal{I} \rangle \models [s,t]F$ for any F. Hence we assume $\mathcal{I}^*(s) \leq \mathcal{I}^*(t)$. Here, we prove the result for the case of the β-rule and the future time operators.

1. If F is $A \vee B$, then the rule has two expansions. Let us assume that $\langle \mathcal{M}, \mathcal{I} \rangle$ satisfies the rightmost one (the other case is straightforward): $[s, a-1]A$, $[a]B$, $[a+1, t]A \vee B$ and $s \leq a \leq t$, where a is a fresh constant. Then obviously $\langle \mathcal{M}, \mathcal{I} \rangle \models [s, a]A \vee B$ and $\langle \mathcal{M}, \mathcal{I} \rangle \models [a+1, t]A \vee B$, hence (whichever the position of $\mathcal{I}(a)$ with respect to $\mathcal{I}^*(s)$ and $\mathcal{I}^*(t)$) $\langle \mathcal{M}, \mathcal{I} \rangle \models [s,t]A \vee B$.

2. If F is $\Box^+ A$, then $\langle \mathcal{M}, \mathcal{I} \rangle \models [s+1, finish]A$. Then clearly $\langle \mathcal{M}, \mathcal{I} \rangle \models [s,t]\Box^+ A$ (also in the case where $s = finish$).

3. If F is $\Diamond^+ A$, and $\langle \mathcal{M}, \mathcal{I} \rangle \models [a]A$ for some a such that $\mathcal{I}(a) > \mathcal{I}^*(t)$, then $\langle \mathcal{M}, \mathcal{I} \rangle \models [s,t]\Diamond^+ A$.

4. If F is $A\mathcal{U}^+ B$, then $\langle \mathcal{M}, \mathcal{I} \rangle$ satisfies $[a]B$, $[t+1, a-1]A$, $[s+1, t]A \vee B$, where a is a fresh constant such that $\mathcal{I}^*(t+1) \leq \mathcal{I}(a) = n$. Since $\langle \mathcal{M}, \mathcal{I} \rangle \models [t+1, a-1]A$, clearly $\mathcal{M}_{\mathcal{I}^*(t)} \models A\mathcal{U}^+ B$. Let now $i \in [\mathcal{I}^*(s), \mathcal{I}^*(t-1)]$. Since for all $j \in [\mathcal{I}^*(s+1), \mathcal{I}^*(t)]$, $\mathcal{M}_j \models A \vee B$, also for all $j \in [i+1, \mathcal{I}^*(t)]$, $\mathcal{M}_j \models A \vee B$. We distinguish two cases:

   - If for all $j \in [i+1, \mathcal{I}^*(t)]$, $\mathcal{M}_j \models A$, then n itself is the "witness" for i: $n > j$, $\mathcal{M}_n \models B$ and for all j, $i < j < n$, $\mathcal{M}_j \models A$. Hence in this case $\mathcal{M}_i \models A\mathcal{U}^+ B$.
   - Otherwise, let k be the smallest $j \in [i+1, \mathcal{I}^*(t)]$ such that $\mathcal{M}_k \not\models A$. Since $\mathcal{M}_k \models A \vee B$, $\mathcal{M}_k \models B$. Moreover $k > i$ and, by construction, for all j, $i < j < k$, $\mathcal{M}_j \models A$. Hence again $\mathcal{M}_i \models A\mathcal{U}^+ B$.

■

The lemma below essentially says that any complete open branch describes some model of its initial formula.

Lemma 13 *If $\mathcal{B}$ is a complete and open tableau branch and $\mathcal{I}$ is a solution of the set of temporal constraints occurring in $\mathcal{B}$, then there is an interpretation $\mathcal{M}$ such that its underlying temporal frame is the range of $\mathcal{I}$ and $\langle \mathcal{M}, \mathcal{I} \rangle \models \mathcal{B}$.*

Proof First of all, observe that by Lemma 1 the branch $\mathcal{B}$ is finite. Let T be the (finite) range of $\mathcal{I}$ and $\mathcal{M} = \langle T, \sigma \rangle$, where σ is such that:

> for each atom p and $i \in T$, $p \in \sigma(i)$ iff there is a node $[s,t]p$ in $\mathcal{B}$ such that $\mathcal{I}^*(s) \leq i \leq \mathcal{I}^*(t)$.

The interpretation function σ is well defined, since, if $[s,t]\bot$ is in $\mathcal{B}$, then K contains $t < s$, because $\mathcal{B}$ is complete, so that there are no elements $i \in T$ with $\mathcal{I}^*(s) \leq i \leq \mathcal{I}^*(t)$.

By hypothesis, $\mathcal{I}$ satisfies all the temporal constraints in $\mathcal{B}$, so it remains to be shown that for any logical node $[s,t]F$ in $\mathcal{B}$, $\langle \mathcal{M}, \mathcal{I} \rangle \models [s,t]F$. This is done by induction on the number of application of logical rules in $\mathcal{B}$. Note that if the constraint $t < s$ occurs in $\mathcal{B}$, then trivially $\langle \mathcal{M}, \mathcal{I} \rangle \models [s,t]F$. Hence, the only interesting cases arise when the constraint $s \leq t$ occurs in $\mathcal{B}$ (since $\mathcal{B}$ is complete, either $t < s$ or $s \leq t$ occurs in $\mathcal{B}$, for any $[s,t]F$ in $\mathcal{B}$). In the following, K denotes the set of constraints occurring in $\mathcal{B}$.

1. The base case is when there is no application of logical rules in $\mathcal{B}$, hence, since $\mathcal{B}$ is complete, the only logical nodes in $\mathcal{B}$ have the form $[s,t]F$, where F is a literal.
   - If F is an atom, the result holds by definition of σ.
   - If F is $\neg p$, assume that $\langle\mathcal{M},\mathcal{I}\rangle \not\models [s,t]\neg p$, i.e. there exists $i \in T$ such that $\mathcal{I}^*(s) \leq i \leq \mathcal{I}^*(t)$ and $p \in \sigma(i)$. Hence, by definition of σ, there is a node $[s',t']p$ in $\mathcal{B}$ where $\mathcal{I}^*(s') \leq i \leq \mathcal{I}^*(t')$. Since $\mathcal{B}$ is complete, both $s \leq t$ and $s' \leq t'$ are in K (otherwise, $t < s$ or $t' < s'$ would be in K, contradicting the fact that neither $[\mathcal{I}^*(s),\mathcal{I}^*(t)]$ nor $[\mathcal{I}^*(s'),\mathcal{I}^*(t')]$ are empty), and the conflict resolution rule is applied to $s \leq t$, $s' \leq t'$, $[s,t]\neg p$ and $[s',t']p$. Hence, either $t < s'$ or $t' < s$ is a node of $\mathcal{B}$. In both cases, the intersection of $[\mathcal{I}^*(s),\mathcal{I}^*(t)]$ and $[\mathcal{I}^*(s'),\mathcal{I}^*(t')]$ should be empty.
2. The inductive step follows from Lemma 12.

■

Hence we get the following:

Theorem 14 (k-completeness) *Let k be a positive integer and A a LTL formula in nnf. If A is k-valid, then there is a k-proof of A.*

Proof Let us assume that there is no k-proof of A. Hence, any canonical and complete tableau $\mathcal{T}$ for $\{[start]\neg A, start \leq finish + k\}$ is either infinite or else has at least a branch $\mathcal{B}$, from the root to a leaf, that is open. Theorem 9 rules out the possibility that $\mathcal{T}$ is infinite, thus let $\mathcal{I}$ be a solution of the constraints in $\mathcal{B}$. By Lemma 13, there is an interpretation $\mathcal{M}$ such that its underlying temporal frame is the range of $\mathcal{I}$ and $\langle M,\mathcal{I}\rangle \models \mathcal{B}$. Hence $\mathcal{M}$ is a model bounded by k such that $\langle M,\mathcal{I}\rangle \models [start]\neg A$, so that $\mathcal{M}_0 \models \neg A$. Therefore A is not k-valid. ■

4.4 COMPLEXITY AND FINITE VALIDITY

Since the rules of the calculus are invertible, the search for a model can be done by construction of a single (canonical) tableau, following any strategy to choose the nodes to be expanded. It can be shown that the length of a branch in a tableau for $S \cup \{finish \leq start + k\}$ is polynomial in $n \times k$, where n is the size of S. Since testing whether a branch is open or not can be done in polynomial time, a non deterministic algorithm to check k-satisfiability, implementing the proposed tableau system, is polynomial in $n \times k$. A deterministic algorithm to test k-satisfiability of a set of nnf formulae S is $O(2^{n \times k})$, since the number of branches must also be taken into account. However, the space required by depth first search is still polynomial in $n \times k$.

When tableaux are initialized without any constraint of the form $finish \leq start + k$, it is still true that, if $\mathcal{M}$ is a model of S, then any canonical tableau for S has a finite open branch that is satisfiable in $\mathcal{M}$, and, conversely, any complete and open tableau branch is satisfiable. This gives a semi-decision procedure for testing satisfiability in models of finite size.

However, the calculus also induces a decision procedure for checking LTL finite validity. It is already known that finite satisfiability in LTL is decidable; for instance, this can be proved by use of Lemma 4.5 in [16] (that can be easily adapted to our finite semantics). Roughly, the lemma says that, given any model $\mathcal{M}$ of an LTL formula F, if two time points satisfy exactly the same subformulae of F then they may collapse (and the interval between them disappear), so that a smaller model $\mathcal{M}'$ may be obtained. As a consequence of such a result, if F has a finite model, then it has a model whose size does not exceed the cardinality of the powerset of the set of its subformulae, i.e. $2^{|sub(F)|}$. In our framework, this implies that by setting the initial constraint in the tableau construction equal to $finish \leq start + 2^{|sub(F)|}$ we immediately get a system which is sound and complete with respect to finite validity in LTL.

This may be interesting from a theoretical standpoint, but clearly unpractical: if the bound k on the model size is exponential in n, a doubly exponential upper bound to the deterministic time complexity is obtained, and the space complexity is exponential in n.

5 CONCLUSIONS AND RELATED WORK

The contribution of this paper is the proposal of a tableau style proof system for propositional linear time temporal logic over finite frames.

The interest of focusing on finite structures comes from applications in Computer Science where computations or reasoning involving only a finite time interval are of interest. This is the case, for instance, of planning or databases applications. In the case of planning, a plan specification can be expressed by a set of LTL formulae and the search of a plan of a bounded duration can be implemented as the search of a bounded model for such a specification, i.e. it may be reduced to a bounded satisfiability problem [6, 7]. In the case of databases evolving through time by updates, the database "history", from its creation up to the current time, may be formalized as a finite temporal interpretation, and dynamical integrity constraints may be expressed by LTL formulae. The view that proof systems for finite and/or bounded validity are a useful tool in order to statically check, at compilation time, integrity constraint preservation by updates, is advocated in [3]. In that framework, proving that a given temporal formula (depending on the given constraint and the given update) is finitely valid guarantees that the update will never violate the constraint, while proving that it is k-valid guarantees that a given fixed number of updates will not cause violation of the constraint.

The tableau calculus we propose makes use of annotations on formulae, indicating the time intervals in which formulae are taken to hold, hence it is formulated in a labelled deduction style, similarly to [13, 15]. In the case of LTL, the use of labels makes it possible to encode the information that time points "behave" as natural numbers inside the proof system itself. With respect to non labelled calculi, such an approach presents, in principle, the following main advantages:

1. The possibility of "immediately executing" *eventualities* (i.e. formulae with an existential commitment), without the need to systematically choose the specific point where the considered formula must hold. This can prove particularly useful, for instance, in the search of a synthetic description of all the k-bounded models of a given formula, similarly to what happens when constraints are added to logic programming.

2. A symmetric treatment of past and future time operators, without the need of a preliminary transformation of formulae into a "separated" form.

3. Since the satisfiability of a formula is reduced to solvability of a set of integer constraints, satisfiability (validity) checking algorithms may take advantage of the existence of efficient techniques for integer constraints solving.

While our approach shares with both [13] and [15] the use of temporal labels and constraints, it is specifically indebted with the second work with respect to the use of time intervals, rather than time points, as labels for formulae. By referring to intervals, a simpler formulation of the inference rules can be given, so as to guarantee termination. Indeed, the use of intervals versus time points can be compared to the use of (bounded) quantification (over time points) versus enumeration.

Besides the focus on the finite semantics, a major difference between our approach and [15] is the formulation of the β-rules. Although our formulation may lead to inefficiency when only the "always" and "eventually" operators are considered, it makes it possible to give a conceptually very simple formulation of the "until" and "since" rules, that are not dealt with in [15] and are treated by means of model checking techniques in [14]. Note, however, that focusing on finite models could lead to simpler solutions also in the framework of [14]. In this case, in fact, some hard problems disappear, such as the problem of fairness conditions (in connection with formulae of the form $\Box\Diamond A$).

An open problem for future investigation is the possibility of using the same approach in the case of the infinite semantics. This is a hard task, mainly because the methods used here in order to ensure termination without recurring to loop checking are of no use when the temporal frames can be infinite.

In a different direction, a natural development of this work is addressing the case of first order LTL. It is well known that LTL over infinite temporal frames is not recursively axiomatizable [1], and the same happens when temporal frames are assumed to be finite [9]. However, the calculus proposed in this work can be lifted to first order in order to test bounded validity [8].

Serenella Cerrito
Université de Paris-Sud
LRI, Bât 490,
F-91405 Orsay Cedex, France
serena@lri.fr

Marta Cialdea Mayer
Dipartimento di Informatica e Automazione
Università Roma Tre
via Vasca Navale 79
I-00146 Roma, Italy
cialdea@dia.uniroma3.it

References

[1] M. Abadi. The power of temporal proofs. *Theoretical Computer Science*, 65:35–83, 1989.

[2] H. Barringer, M. Fisher, D. Gabbay, G. Gough, and R. Owens. METATEM: a framework for programming in temporal logic. In *Proc. of REX Workshop on Stepwise Refinement of Distributed Systems: Models, Formalisms, Correctness*, Springer, 1989.

[3] V. Benzaken, S. Cerrito, and S. Praud. Vérification statique de contraintes d'intégrité dynamiques. In *Proc. of 15èmes Journées Bases de Données Avancées (BDA99)*, Bordeaux, France, 1999.

[4] N. Bidoit and S. DeAmo. A first step towards implementing dynamic algebraic dependencies. In *Proceedings of 5th International Conference on Database Theory (ICDT'95)*, pages 308–319. Springer, 1995.

[5] J. R. Burch, E. M. Clarke, K. L. McMillan, D. L. Dill, and L. J. Hwang. Symbolic model checking: 10^{20} states and beyond. *Information and Computation*, 98(2):142–170, 1992.

[6] S. Cerrito and M. Cialdea Mayer. Bounded model search in linear temporal logic and its application to planning. In H. De Swart, editor, *International Conference on Automated Reasoning with Analytical Tableaux and Related Methods*, pages 124–140. Springer, 1998.

[7] S. Cerrito and M. Cialdea Mayer. Using linear temporal logic to model and solve planning problems. In F. Giunghiglia, editor, *Proceedings of the 8th International Conference on Artificial Intelligence: Methodology, Systems, Applications (AIMSA'98)*, pages 141–152. Springer, 1998.

[8] S. Cerrito, M. Cialdea Mayer, and S. Praud. First order linear temporal logic over finite time structures. In *Proc. of the 6th Int. Conf. on Logic for Programming and Automated Reasoning (LPAR'99)*. Springer, 1999.

[9] S. Cerrito, M. Cialdea Mayer, and S. Praud. First order linear time temporal logic over finite time frames is not semi-decidable. In *Proc. of the Workshop Methods for Modalities 1 (M4M)*, 1999.

[10] T. H. Cormen, C. E. Leiserson, and R. L. Rivest. *Introduction to Algorithms*. The MIT press, 1990.

[11] D. Gabbay. Declarative past and imperative future: Executable temporal logic for interactive systems. In B. Banieqbal, H. Barringer, and A. Pnueli, editors, *Proc. of Colloquium on Temporal Logic in Specification*, pages 409–448. Springer, 1989.

[12] G. D. Gough. Decision procedures for temporal logic. Technical Report UMCS-89-10-1, Department of Computer Science, University of Manchester, 1984.

[13] R. Hähnle and O. Ibens. Improving temporal logic tableaux using integer constraints. In D. M. Gabbay and H. J. Ohlbach, editors, *Proceedings of the First International Conference on Temporal Logic (ICTL 94)*, pages 535–539. Springer, 1994.

[14] P.H. Schmitt and J. Goubault-Larrecq. A tableau system for full linear temporal logic. Draft, available at: http://www.dyade.fr/fr/actions/vip/jgl/ltl2.ps.gz.

[15] P.H. Schmitt and J. Goubault-Larrecq. A tableau system for linear-time temporal logic. In E. Brinksma, editor, *3rd Workshop on Tools and Algorithms for the Construction and Analysis of Systems (TACAS'97)*. Springer, 1997.

[16] A. P. Sistla and E. M. Clarke. The complexity of propositional linear temporal logics. *Journal of the ACM*, 32(3):733–749, 1985.

[17] P. Wolper. The tableau method for temporal logic: an overview. *Logique et Analyse*, 28:119–152, 1985.

FIBRED MODAL TABLEAUX

Dov M. Gabbay and Guido Governatori

Abstract We describe a general and uniform tableau methodology for multi-modal logics arising from Gabbay's methodology of fibring and Governatori's labelled tableau system KEM.

1 INTRODUCTION

Gabbay [11, 14] presented a general methodology for combining logical systems called fibring. Beckert and Gabbay [5] have studied under which conditions proof systems are well-suited for fibring, and they have argued that labelled tableaux are calculi of this kind. Moreover D'Agostino et al. [7, 8] have shown how to extend classical proof system KE [9], a combination of tableaux and natural deduction inference rules which allows for a restricted ('analytic') use of the cut rule, to deal with fibred logics, in particular for a combination of modal and substructural logics and the combination of different fragment of substructural implications. In [1, 18] a labelled extension of KE, called KEM, has been proven to be able to cope with a wide variety of (normal) modal logics, and in [3, 16, 17, 18] it has been showed how such a system can be extended to deal with multi-modal logics. In the present work we show how to adapt KEM in order to obtain a general and uniform tableau-like proof method for the fibred (dovetailed) combination of modal logics. In particular we show how the label formalism used in KEM matches Gabbay's semantic based fibring for (multi-) modal logics. The resulting system enjoys several interesting properties: it is modular, in the sense that the proof system for each logic is developed on its own and it is reused in the combination; it is uniform, in the sense that each system has its own individual features, but the framework remains constant among the combined systems; it seems to be flexible enough to deal with other logics, as well as their combinations; finally, although we use some logical machinery in describing the system, our presentation is natural and the idea behind it is very simple and easy to grasp.

Let L_1 and L_2 be two modal logics for which a tableau system exists. We start a tableau for a formula A of L_1, which has formulas of L_2 embedded in it. In the course of the proof for A, when we have to process a formula of L_2 we

D. Basin et al. (eds.), Labelled Deduction, 161–191.
© 2000 *Kluwer Academic Publishers. Printed in the Netherlands.*

begin a new proof for it in the appropriate system. It turns out that we do not really need to start a separate tableau, but we graft it into the original one. The labels will have a structure which will store nicely all the information about the grafting operation. This is possible because the key feature of KEM, besides its being based neither on resolution nor on standard sequent/tableau inference techniques, is that it generates models and checks them using a label scheme during the bookkeeping for the fibred model. The mechanism which KEM uses in manipulating labels is close to semantic fibring (dovetailing).

In this essay we shall stick ourselves with KEM based proof method, nevertheless the label formalism, and the combining methodology, can be used with almost whatever proof system allowing: 1) atomic proofs, and 2) composition of deductions.

The paper is organized as follows: in section 2 we shall introduce two basic methods for combining logical systems, namely fibring (2.1) and dovetailing (2.2); in section 4 we recall the basic tableau system KEM for modal logics; we shall relate the techniques from the previous sections in order to provide a general and uniform tableau methodology for multi-modal logics.

2 COMBINING MODAL LOGICS

The methodology of fibring allows us to combine arbitrary logical systems to form a new system in a uniform way "fibring" their models (for detailed expositions see [11, 12, 13, 14]). The main idea of fibring is very simple, and provides a new concept of possible world semantics. Let us suppose we want to combine two modalities $\Box_1$ and $\Box_2$ characterised, respectively, by the classes of models $\mathcal{K}_1$ and $\mathcal{K}_2$. We know how to evaluate $\Box_1 A$ in $\mathcal{K}_1$, $\Box_2 A$ in $\mathcal{K}_2$ and propositional formulas in both. All we need is a method for evaluating $\Box_1$ (resp. $\Box_2$) w.r.t. $\mathcal{K}_2$ (resp. $\mathcal{K}_1$). Each time we have to evaluate a formula $\mathcal{A}$ of the form $\Box_2 A$ in a world in a model of $\mathcal{K}_1$ we associate, via the fibring function $\mathbf{F}$, to the world a model in $\mathcal{K}_2$, or its actual world, where we calculate the truth value of the formula. Formally

$$w \models_{\mathbf{m}\in\mathcal{K}_1} \Box_2 A \iff \mathbf{F}_{\mathbf{m}}(w) \models_{\mathbf{m}'\in\mathcal{K}_2} \Box_2 A$$

$\mathcal{A}$ holds in w iff it holds in the model associated to w through the fibring function $\mathbf{F}$ (i.e., $\mathbf{m}'$); more precisely $\mathcal{A}$ holds in the actual world (i.e., $\mathbf{F}_{\mathbf{m}}(w)$) of $\mathbf{m}'$. But now we are in an appropriate model for evaluating $\mathcal{A}$.

In the next two sections we define two ways, fibring and dovetailing, in which semantics can be combined. We first explain the concepts by taking a simple example. Suppose we want to combine two modal logics L_1 and L_2. Let $\mathcal{K}_1, \mathcal{K}_2$ be the respective Kripke semantics of the logic. Let $\mathbf{m}$ be a model in $\mathcal{K}_1$ and let t be a possible world of $\mathbf{m}$. The semantic construction which combines the logics associates to t a model $\mathbf{n}$ in $\mathcal{K}_2$. The different methodologies of combination differ on the kind of model $\mathbf{n}$ we use. For *fibring* logics, we require that $\mathbf{n}$ be any model in $\mathcal{K}_2$. For *dovetailing*, a special case of fibring, we require that $\mathbf{n}$ be a model of $\mathcal{K}_2$ such that for any *atomic* p

$$t \vDash p \text{ iff } \mathbf{n} \vDash p$$

(i.e., the fibred model must agree with the values t gives to atoms).

First of all we have to introduce the language of the combined logic. Let I be a set of indices, and let $L_i, i \in I$ be modal logics in the respective language $\mathbb{L}_i$, with $\Box_i, i \in I$, respectively. The expressions E^i of the language $\mathbb{L}_i$ are built up using $\mathbb{L}_i$-constructors (connectives and operators) from a set of atomic units p^i. We schematically write $E^i(p_1^i, \ldots, p_n^i)$ to indicate that E^i is built up from the atoms $p_1^i, \ldots, p_n^i \in p^i$.

Definition 1 *The fibred language* $\mathbb{L}_{(x_1,\ldots,x_n)}$ *is defined as follows:*

- *Let* $\mathbb{L}_{(i)}$ *be* $\mathbb{L}_i$, $i \in I$;
- *Let* $\bar{y}$ *be* $(j, y_1, \ldots, y_k)$, $j, y_1, \ldots, y_k \in I$ *and* $i \neq j$. *Let* $\mathbb{L}_{(i)*\bar{y}}$ *be the family of all expressions of the form* $\alpha \in \mathbb{L}_{(y_1,\ldots,y_k)}$ *or* $\alpha = E^i(p_1/A_1, \ldots, p_n/A_n)$ *where* $E^i(p_1, \ldots, p_n) \in \mathbb{L}_i$ *and* $A_1, \ldots, A_n$ *are in* $\mathbb{L}_{\bar{y}}$, *and* p_k/A_k *indicates the substitution of* A_k *to* P_k *in* E^i;
- $\mathbb{L}_I = \bigcup_{\bar{y}} \mathbb{L}_{\bar{y}}$.

To clarify this notion let us consider two modal languages $\mathbb{L}_1$ and $\mathbb{L}_2$. The fibred language $\mathbb{L}_{(2,1,y_1,\ldots,y_k)}$ is the set of all expressions with outer constructor from $\mathbb{L}_2$ and with no more than $k+2$ nested alternation of constructors from $\mathbb{L}_1$ and $\mathbb{L}_2$.

2.1 FIBRING MODAL LOGICS

Let $\mathcal{K}_i$ be a class of models $\{\mathbf{m}_1^i, \mathbf{m}_2^i, \ldots\}$ for which L_i is complete. Each model $\mathbf{m}_n^i$ has the form (S, R, a, h) where S is the set of possible worlds, $a \in S$ is the actual world and $R \subseteq S^2$ is the accessibility relation. h is the assignment function, a binary function, giving a value $v(t,p) \in \{1,0\}$ for any $t \in S$ and atomic p. The actual world a plays a role in the semantic evaluation in the model, in so far as satisfaction in the model is defined as satisfaction at a. We can assume that the models satisfy the following condition:

$$S = \{x \mid \exists n \ aR^n x\}.$$

This assumption does not affect satisfaction in models because points not accessible from a by any power R^n of R do not affect truth values at a. Moreover we assume that all sets of possible worlds in any $\mathcal{K}_i$ are all pairwise disjoint, and that there are infinitely many isomorphic (but disjoint) copies of each model in $\mathcal{K}_i$. We use the notation $\mathbf{m}$ for a model and present it as

$\mathbf{m} = (S^{\mathbf{m}}, R^{\mathbf{m}}, a^{\mathbf{m}}, h^{\mathbf{m}})$ and write $\mathbf{m} \in \mathcal{K}_i$, when the model $\mathbf{m}$ is in the semantics $\mathcal{K}_i$. Thus our assumption boils down to $\mathbf{m} \neq \mathbf{n} \Rightarrow S^{\mathbf{m}} \cap S^{\mathbf{n}} = \varnothing$. In fact a model can be identified by its actual world, i.e., $\mathbf{m} = \mathbf{n}$ iff $a^{\mathbf{m}} = a^{\mathbf{n}}$.

Definition 2 *A fibred model is a structure*

$$(W, W_{i,i\in I}, W_a, R, w_0, h, \mathbf{F})$$

where $W = \bigcup_{\mathbf{m}\in\bigcup_i\mathcal{K}_i} S^{\mathbf{m}}$; $W_i = \{a^{\mathbf{m}} \mid \mathbf{m} \in \mathcal{K}_i\}$; $W_a = \bigcup_i W_i$; $R = \bigcup_{\mathbf{m}\in\bigcup_i\mathcal{K}_i} R^{\mathbf{m}}$; $w_0 \in W_a$ *is the actual world;* $h(t,p) = h^{\mathbf{m}}(t,p)$, *for the unique* $\mathbf{m}$ *such that* $t \in S^{\mathbf{m}}$; $\mathbf{F} : I \times W \mapsto W_i$, *is the fibring function. The fibring function* $\mathbf{F}$ *is a function giving for each* $i \in I$ *and each* $w \in W$ *another point (actual world) in* W_i *as follows:*

$$\mathbf{F}_i(w) = \begin{cases} w & \text{if } w \in S^{\mathbf{m}} \text{ and } \mathbf{m} \in \mathcal{K}_i \\ \text{a value in } W_i, & \text{otherwise} \end{cases}$$

such that if $x \neq y$, *then* $\mathbf{F}_i(x) \neq \mathbf{F}_i(y)$. *Satisfaction is defined as follows with the usual truth tables for boolean connectives:*

$$t \vDash p \quad \text{iff} \quad v(t,p) = 1$$

$$t \vDash \Box_i A \quad \text{iff} \quad \begin{cases} t \in \mathbf{m}^i \text{ and } \forall s(tRs \to s \vDash A) \\ t \in \mathbf{m}^j,\ i \neq j \text{ and } \mathbf{F}_i(t) \vDash \Box_i A \end{cases}$$

We say the model satisfies A *iff* $w_0 \vDash A$.

Theorem 3 (Completeness theorem for the fibred logic L_I^F) *Let* L_i, $i \in I$ *be modal logics in the respective language* $\mathbb{L}_i$ *with classes of structures* $\mathcal{K}_i$ *and set of theorems* $\mathbf{T}_i$ *(i.e.,* $\mathbf{T}_i = \{A \text{ of } \mathbb{L}_i \mid A \text{ is valid in all } \mathcal{K}_i \text{ models}\}$*). Let* $\mathbf{T}_I^F$ *be the following set of wffs of* L_I^F.

1. $\mathbf{T}_i \subseteq \mathbf{T}_I^F$, *for every* $i \in I$.

   1a. *If* A *is a Boolean combination of atoms and* $\alpha_n^{\bar{y}}$ *is in the* $\bar{y}$*-th fibred language, then* $A \to \alpha_n^{\bar{y}} \in \mathbf{T}_{\bar{y}}$ *implies* $A \to \bigvee_n \alpha_n^{\bar{y}} \in \mathbf{T}_I^F$, *where every atom in* $\alpha_n^{\bar{y}}$ *is in the scope of a modality.*

   1b. *If* $A(x_k) \in \mathbf{T}_i$ *then* $A(x_k/\Box_j\alpha_k) \in \mathbf{T}_I^F$, *for any* $\Box_j\alpha_k \in \mathbb{L}_j$, $j \in I$.

2. **Modal Fibring Rule:**
   If $\Box_i$ *is the modality of* L_i *and* $\Box_j$ *of* L_j, *where* i, j *are arbitrary with* $i \neq j$ *and*

$$C = \bigwedge_{k=1}^{n} \Box_i A_k \to \bigvee_{k=1}^{m} \Box_i B_k \in \mathbf{T}_I^F$$

then for all $d \in \mathbb{N}$, $\Box_j^d C \in \mathbf{T}_I^F$.[1]

3. *$\mathbf{T}_I^F$ is the smallest set closed under 1, 2, modus ponens and substitution.*

Then $\mathbf{T}_I^F$ is the set of all wffs of L_I^F valid in all the fibred structures of L_I^F.

Proof See [11, 14]. Notice that [11] deals with the case that the logics $L_i, i \in I$ have no atoms in common. ■

2.2 DOVETAILING MODAL LOGICS

Dovetailing arises in many applications where the fibred model at world t has the world t itself as its actual world. The notions of dovetailing results from that of fibring when for all $i \in I$ and for all atomic p

$$v(t, p) = v(\mathbf{F}_i(t), p) \ .$$

In such a case we can identify the actual world of the model fibred at t, $\mathbf{F}_i(t)$, with t. The fibring function $\mathbf{F}$ is no longer needed, since we identified t with $\mathbf{F}_i(t)$.

Let $L_i, i \in I$ be modal logics with $\mathcal{K}_i$ the class of models for L_i. Let L_I^D (the dovetailing combination of $L_i, i \in I$) be defined semantically through the class of all (dovetailed) models of the form (W, R, a, v), where W is a set of worlds, $a \in W$, v is an assignment as before, and for each $i \in I, R(i) \subseteq W \times W$. We require that for each i $(W, R(i), a, v)$ is a model in $\mathcal{K}_i$. We further require the following: Let $t \in W$ be such that there exist $n_1, \ldots, n_k$ and $i_1, \ldots, i_k$ such that $aR^{n_1}(i_1) \circ R^{n_2}(i_2) \ldots \circ R^{n_k}(i_k)t$ holds.

We define the notion of $w \models A$ by induction.

- $w \models p$ iff $v(w, p) = 1$ for p atomic.
- $w \models \Box_i A$ if for all $y \in W$, such that $wR(i)y$ we have $y \models A$.
- $\models A$ iff for all models and actual worlds $a \models A$.

Theorem 4 (Completeness theorem for the dovetailed logic L_I^D) *Let $L_i, i \in I$ be modal logics with semantical classes of structures $\mathcal{K}_i$ and set of theorems $\mathbf{T}_i$. Let $\mathbf{T}_I^D$ be the following set of wffs of L_I^D.*

1. *$\mathbf{T}_i \subseteq \mathbf{T}_I^D$, for every $i \in I$.*

[1] As a notation we use $\Box_i^0 A$ for A and $\Box_i^{n+1} A$ for $\Box_i \Box_i^n A$.

2. **Modal Dovetailing Rule:**
If $\Box_i$ is the modality of L_i and $\Box_j$ that of L_j, where i, j are arbitrary with $i \neq j$, and

$$C = \bigwedge_{k=1}^{n} \Box_i A_k \wedge \bigwedge_{k=1}^{m} \Diamond_i \neg B_k \to \bigvee_{k=1}^{r} p_k \in \mathbf{T}_I^D,$$

then for all $d \in \mathbb{N}$, $\Box_j^d C \in \mathbf{T}_I^D$. Where p_k are atoms or their negations, and $p_1, \ldots, p_r$ list all the atoms or their negations appearing in any A_k or B_k, $k = 1, 2, \ldots$.

3. *$\mathbf{T}_I^D$ is the smallest set closed under 1, 2, modus ponens and substitution.*

Then $\mathbf{T}_I^D$ is the set of all wffs of L_I^D valid in all the dovetailed structures of L_I^D.

Proof See [11]. ■

3 FROM FIBRING TO LABELLED TABLEAUX

Let us explain the fibring idea by looking at an example. This example will be treated in Section 4.3 below.

Consider two logics L_1 and L_2 with modalities $\{\Box_1, \Diamond_1\}$ and $\{\Box_2, \Diamond_2\}$ respectively. Assume L_i is an extension of **K** for its modality. Consider:

$$A = \Box_1 \Diamond_2 p \to \Diamond_1 \Box_2 \Diamond_2 p$$

This formula is in the combined language. We know how to dovetail these two logics to form a logic for the combined language. We know that if L_1 and L_2 are complete for classes of Kripke model $\mathcal{K}_1$ and $\mathcal{K}_2$ respectively, then we can fibre (dovetail) the models and get a class $\mathcal{K}$ of models of the form $\mathbf{m} = (S, R_1, R_2, w_0, v)$, with two accessibility relations. The class $\mathcal{K}$ is constructed from $\mathcal{K}_1$ and $\mathcal{K}_2$. It contains some specific models of the form above and completeness holds.

Let us see whether we can find an $\mathbf{m}$ which is a counter-model for A. Let us check what is needed to be in $\mathbf{m}$ and then we can check whether such an $\mathbf{m}$ can be in the class $\mathcal{K}$. We want to have

1. $w_0 \models \Box_1 \Diamond_2 p$ 2. $w_0 \not\models \Diamond_1 \Box_2 \Diamond_2 p$

From (1) we get that at an arbitrary point t such that $w_0 R_1 t$ we must have $t \models \Diamond_2 p$. Let $t = W_1^1$ be such an arbitrary point. We use capital letters to indicate that the point t is *arbitrary* and we use the superscript '1' to indicate that $w_0 R_1 t$ holds (i.e. we are choosing t because we are evaluating a modality of L_1). The subscript just counts that this is the first 'W^1' we are using. Hence we have

$$W_1^1 \vDash \Diamond_2 p.$$

Now we are evaluating the second modality. Therefore there exists a specific point s such that $W_1^1 R_2 s$ and $s \vDash p$. Again we use the suggestive notation $s = w_1^2$ and write $w_1^2 \vDash p$. The lower case w indicates that w_1^2 is not arbitrary but specific and the superscript '2' indicates that we are dealing with an L_2 modality. Similarly, since

$$w_0 \nvDash \Diamond_1 \Box_2 \Diamond_2 p$$

we must have that for an arbitrary x such that $w_0 R_1 x$ we have $x \nvDash \Box_2 \Diamond_w p$. According to our agreed notation, we represent such an arbitrary x by

$$W_2^1 \nvDash \Box_2 \Diamond_2 p.$$

The superscript '1' indicates that this point arose because of an L_1 modality. The subscript '2' indicates that this is a second such point used so far and the capital W (as opposed to a lower case w) indicates that it is an arbitrary point (accessible to w_0). We can continue, since

$$W_2^1 \nvDash \Box_2 \Diamond_2 p$$

there exists a specific point y, $W_2^1 R_2 y$ such that $y \nvDash \Diamond_2 p$. Using our conventions, we represent this point by

$$w_2^2 \nvDash \Diamond_2 p.$$

We can continue and represent an arbitrary point W_1^2 such that $w_2^2 R_2 W_1^2$ and $W_1^2 \nvDash p$. We can represent what we got so far in a tree in Figure 7.1(a).

The T, F indicate whether the formula is supposed to hold or not in the node. The uppercase/lowercase distinction tells us whether the point is arbitrary or not and the superscript tells us whether the point is R_1 or R_2 accessible to the point below it. We are relying on the geometrical drawing of a tree to represent the situation. If we do not want to draw trees, we can represent each point as the *sequence* of points leading up to it. Thus we have the table in Figure 7.1(b):

We now address the question of whether the situation in Figure 7.1(a) does provide a counter-model for A. Of course, if both $\Box_1$ and $\Box_2$ are **K** modalities, then we can provide a counter-model for A, since in this case, the dovetailed semantics $\mathcal{K}$ contains any model $\mathbf{m}$ with arbitrary v, R_1 and R_2. We know this fact from the completeness theorem of dovetailed logic (Theorem 4). If we do not want to rely on semantics, we can ask what would be the corresponding condition on the trees obtained (as we see in the example of Figure 7.1(a)) that

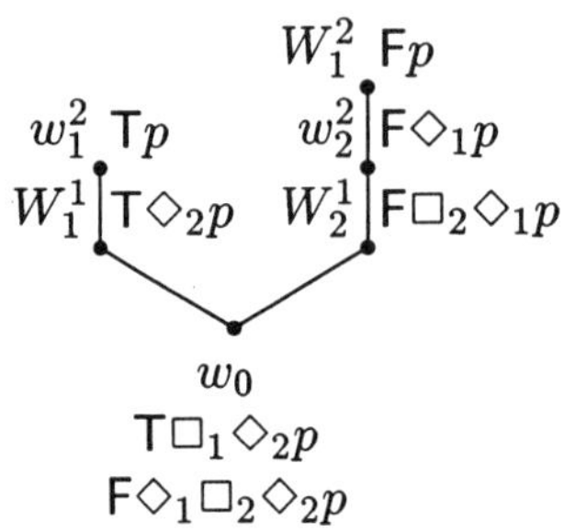

(a)

Point in tree	Sequence Representation
w_0	w_0
W_2^1	(W_2^1, w_0)
w_2^2	$(w_2^2, (W_2^1, w_0))$
W_1^2	$(W_1^2, (w_2^2, (W_1^1, w_0)))$
W_1^1	(W_1^1, w_0)
w_1^2	$(w_1^2, (W_1^1, w_0))$.

(b)

Figure 7.1

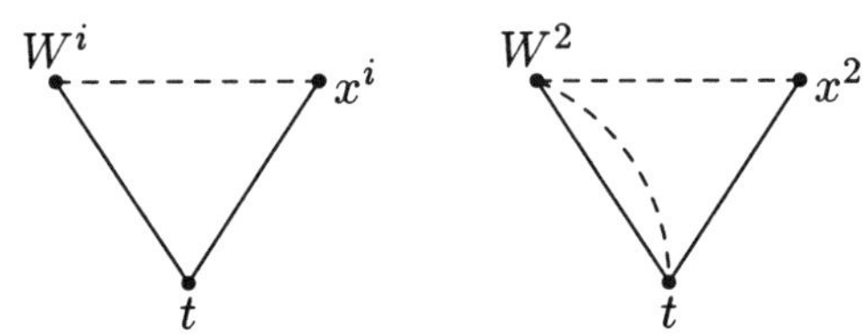

(a) (b)

Figure 7.2

will tell us whether we can get a counter-model or not. In other words, is the tree 'closed' therefore there is no counter-model, i.e. A is a theorem?

The following are the kind of rules we want:

In the subtree of Figure 7.2(a), W^i can unify with x^i. By repeated applications of the rule we try to unify points and get a contradiction. For example, if in Figure 7.2(a) we have W^i Tp and x^i Fp we get a contradiction.

Different logics will have different unification rules. For example, if the logic L_2 is a $\mathbf{T}$ modality, then in this case we use the rule depicted in Figure 7.2(b), where W^2 can unify with either x^2 or t.

A tableaux system for a dovetailed logic with $\{\Box_1, \Box_2\}$ would need the following components.

- Rules for building trees as we did in Figure 7.1(a) or using sequences as in Figure 7.1(b).
- Unification rules that unify labels in order to identify two labels; for example, one with Fq and the other with Tq.

- Completeness theorem for the resulting tableaux system for the dovetailed logic.

The next sections do all that in detail. It is a messy business and is done best by a theorem prover. Work on functional translation of this kind has been done by Ohlbach [20], and optimisation have been proposed in [21] and by Gasquet [15].

4 LABELLED TABLEAUX FOR MODAL LOGICS

In [1, 18] a tableau-like proof system, called KEM, has been presented, and it has been proven to be able to cope with a wide variety of (normal) modal logics. KEM is based on D'Agostino and Mondadori's [9] classical proof system KE, a combination of tableau and natural deduction inference rules which allows for a restricted ("analytic") use of the cut rule. The key feature of KEM, besides its being based neither on resolution nor on standard sequent/tableau inference techniques, is that it generates models and checks them using a label scheme for bookkeeping fibred models. In [3, 16, 17, 18] it has been shown how this formalism can be extended to handle various systems of multi-modal logic with interaction axioms. The mechanism KEM uses in manipulating labels is close to semantic fibring (dovetailing).

4.1 LABEL FORMALISM

In this section we introduce the label formalism we shall use in the course of the chapter. KEM uses *Labelled Signed Formulas* (*LS*-formula for short), where an *LS*-formula is an expression of the form $SA : t$, where A is a wff of the logic, S (the truth sign) is in $\{\mathsf{T}, \mathsf{F}\}$, and t is a label. Notice that the label we are referring in this section are not the labels of I. In the case of modal logic we have a type of labels corresponding to various modalities, and each set of atomic labels is partitioned into the set of variables and the set of constants. Formally

$$\Phi^{i\in I} = \Phi_C^{i\in I} \cup \Phi_V^{i\in I}$$

where $\Phi_C^{i\in I} = \{w_1^i, w_2^i, \dots\}$ and $\Phi_V^{i\in I} = \{W_1^i, W_2^i, \dots\}$. The set of constant world symbols and variable world symbols are respectively

$$\Phi_C = \bigcup_{i\in I} \Phi_C^{i\in I} \qquad \Phi_V = \bigcup_{i\in I} \Phi_V^{i\in I}$$

The set of labels $\Im$ is then defined inductively as follows:

$$\Im = \bigcup_{1\leq k} \Im_k \text{ where } \Im_k, k \in \mathbb{N}:$$

$$\Im_1 = \Phi_C \cup \Phi_V;$$
$$\Im_2 = \Im_1 \times \Phi_C;$$
$$\Im_{n+1} = \Im_1 \times \Im_n, (n > 1).$$

According to the above definition a label is either a (i) an element of the set Φ_C, or (ii) an element of the set Φ_V, or (iii) a path term (s', s) where (iiia) $s' \in \Phi_C \cup \Phi_V$ and (iiib) $s \in \Phi_C$ or $s = (t', t)$ where (t', t) is a label. From now on we shall use $t, s, r, \ldots$ to denote arbitrary labels.

As an intuitive explanation, we may think of a label $t \in \Phi_C$ as denoting a world (a *given* one), and a label $t \in \Phi_V$ as denoting a set of worlds (*any* world) in some Kripke model. A label $s = (t', t)$ may be viewed as representing a path from t to a (set of) world(s) t' accessible from t (i.e., from the world(s) denoted by t).

Definition 5 *For any label $t = (s', s)$ we shall call s' the* head *of t, s the* body *of t, and denote them by $h(t)$ and $b(t)$ respectively.*

Notice that these notions are recursive (they correspond to projection functions): if $b(t)$ denotes the body of t, then $b(b(t))$ will denote the body of $b(t)$, $b(b(b(t)))$ will denote the body of $b(b(t))$; and so on.

Definition 6 *We call each of $b(t)$, $b(b(t))$, etc., a* segment *of t. Let $b^m(t)$, be any segment of t (obviously, by definition every segment $s(t)$ of a label t is a label); then $h(b(t))$ will denote the head of $b^m(t)$. We shall call a label t* restricted *if $h(t) \in \Phi_C$, otherwise* unrestricted.

We shall think of the $\Im_1$ labels as possible worlds in a fibred (dovetailed) model, and labels in $\Im_n$ as paths leading to them starting from the actual world. For example a label such $t = (W_2^i, (w_2^j, w_0))$ denotes, with respect to a fibred (dovetailed) model, the set of worlds in a model for L_i accessible from the actual world $a^i = \mathbf{F}_i((w_2^j, w_0))$. Since dovetailed models can be reduced to structures of the form $(W, R(i), h)$, the label t turns out to represents the set of worlds accessible via $R(i)$ from a world accessible through $R(j)$ from the actual world w_0.

Definition 7 *For any label t, we define the length of t, $\ell(t)$, as the number of world-symbols in t, i.e., $\ell(t) = n \Leftrightarrow t \in \Im_n$. $s^n(t)$ will denote the segment of t of length n, i.e., $s^n(t) = s(t)$ such that $\ell(s(t)) = n$. We shall use $h^n(t)$ as an abbreviations for $h(s^n(t))$. Notice that $h(t) = h^{\ell(t)}(t)$.*

Definition 8 *For any label $t, \ell(t) > n$, we define the* counter-segment-n *of t, as follows:*

$$c^n(t) = h(t) \times (\cdots \times (h^k(t) \times (\cdots \times (h^{n+1}(t), w_0)))) \ (n < k < \ell(t))$$

where w_0 is a dummy label, i.e., a label not appearing in t (the context in which such a notion occurs will tell us what w_0 stands for. In most cases it will denote the actual world).

The counter-segment-n defines what remains of a given label after having identified the segment of length n with a 'dummy' label w_0. The appropriate dummy label will be specified in the applications where such a notion is used. However, it can be viewed also as an independent atomic label. In the contest of fibring w_0 can be thought as denoting the actual world obtained via the fibring function from the world denoted by $s^n(t)$.

Example 9 Given the label $t = (w_4, (W_3, (w_3, (W_2, w_1))))$, according to the above definitions its length $\ell(t)$ is 5, the head $h(t)$ is w_4, the body $b(t)$ is $(W_3, (w_3, (W_2, w_1)))$, the segment of length 3 is $s^3(t) = (w_3, (W_2, w_1))$, and the relative counter-segment-3 is $c^3(t) = (w_4, (W_3, w_0))$, where $w_0 = s^3(t) = (w_3, (W_2, w_1))$.

To clarify the notion of counter-segment, which will be used frequently in the course of the present work, we present, in the following table the list of the segments of t in the left-hand column and the relative counter-segments in the right-hand column.

$$\begin{array}{ll} s^1(t) = w_1 & c^1(t) = (w_4, (W_3, (w_3, (W_2, w_0)))) \\ s^2(t) = (W_2, w_1) & c^2(t) = (w_4, (W_3, (w_3, w_0))) \\ s^3(t) = (w_3, (W_2, w_1)) & c^3(t) = (w_4, (W_3, w_0)) \\ s^4(t) = (W_3, (w_3, (W_2, w_1))) & c^4(t) = (w_4, w_0) \\ s^5(t) = t & c^5(t) = w_0 \end{array}$$

So far we have provided definitions about the structure of the labels without regard of the elements they are made of. The following definitions will be concerned with the type of world symbols occurring in a label.

Let t be a label and t' an atomic label, in what follows we shall use $(t'; t)$ as a notation for the label (t', t) if $t' \neq h(t)$, or for t otherwise

Definition 10 *We say that a label t is i-*preferred *iff $h(t) \in \Phi^i$.*

Definition 11 *We say that a label t is i-*pure *iff each segment of t of length $n > 1$ is i-preferred, and we shall use $\Im^i$ to denote the set of i-pure labels.*

4.2 UNIFICATIONS

In the course of proofs labels are manipulated in a way closely related to the semantic of the logics under analysis. Labels are confronted and matched using a specialised logic dependent unification mechanism. The notion of two labels t and s unifying means that the intersection of their denotations is not empty and that we can move to such a set of worlds, i.e., to the result of their unification.

According to the semantics each modality is evaluated in an appropriate model corresponding to a model in the class of models characterising the logic the modality corresponds to. Similarly we provide an unification for each logic, the unification characterising such a logic in KEM formalism, then we graft them into a single unification for the whole L_I^δ.

4.2.1 Basic Unifications (Axiom Unifications). We add a set of auxiliary unindexed atomic labels $\Phi^A = \{w_0, w_0', \dots\}$, that will be used in unifications and proofs. Intuitively they stand for distinguished worlds in the various models. We define two substitutions, $\sigma^{\mathbf{D}}$ and $\sigma^{\mathbf{F}}$ resp. dovetailing and fibring substitution, in the usual way as a mapping

$$\begin{aligned} \sigma^\delta &= \mathbf{1}_{\Phi^A \cup \Phi_C} \\ \sigma^{\mathbf{D}} &: \Phi_V^i \longrightarrow \Im^i \cup \Phi^A \\ \sigma^{\mathbf{F}} &: \Phi_V^i \longrightarrow \Im^i \end{aligned}$$

i.e., identity for constants and auxiliary labels; and a mapping of variables onto i-pure labels in the case of fibring, and either onto i-pure labels or auxiliary ones for the other. The only difference between fibring and dovetailing substitutions is that, in the former, a variable cannot be mapped onto an auxiliary label. Henceforth we use σ to mean indifferently, unless specified, either $\sigma^{\mathbf{D}}$ or $\sigma^{\mathbf{F}}$.

The substitution for composite labels is as follows: if $t = (s', s)$, then

$$\sigma(t) = (\sigma(s'), \sigma(s))$$

For two labels t and s, and a substitution σ, if σ is a unifier of t and s then we shall say that t, s are *σ-unifiable*. We shall (somewhat unconventionally) use $[t|s]\sigma$ to denote both that t and s are σ-unifiable and the result of their unification. In particular

$$\forall t, s, r \in \Im, [t|s]\sigma = r \text{ iff } \exists \sigma(\sigma(t) = \sigma(s) \text{ and } \sigma(t) = r)$$

On this basis we may define several specialised, logic-dependent notions of σ-unification. As a case study we choose the normal modal logics arising from the combination of the axioms K, D, T, 4, B, and 5. Notice that the unifications listed below mimic the conditions on the accessibility relation corresponding to the appropriate axiom (see the accompanying examples for explanations).

$$[t|s]\sigma^K = [t|s]\sigma \quad \text{if at least one of } t \text{ and } s \text{ is restricted, and} \\ \forall n \leq \ell(t), [s^n(t)|s^n(s)]\sigma^K$$

$$[t|s]\sigma^D = [t|s]\sigma$$

Example 12 To exemplify the difference between σ^K and σ^D, let us consider first the labels

$$(w_3, (W_1, w_1)) \qquad\qquad (W_2, (w_2, w_1))$$

Obviously t and s σ^K- and σ^D-unify on $(w_3, (w_2, w_1))$ with the substitution

$$\begin{aligned} \sigma :& W_1 \mapsto w_2 \\ & W_2 \mapsto w_3 \end{aligned}$$

On the other hand the labels

$$t = (w_2, (W_1, w_1)) \qquad\qquad s = (W_2, (W_1, w_1))$$

σ^D- but not σ^K-unify. This is due to the fact that both $s^2(t)$ and $s^2(s)$ are variables, while in the definition of σ^K it is required that at least one of them is a constant. The reason for this condition on σ^K is that the interpretation of W_1 is the set of worlds accessible from w_1, but such a set may be empty so the denotation of W_1 would be empty; this is not the case with σ^D since the corresponding accessibility relation is serial, so W_1 cannot be empty.

$$[t|s]\sigma^T = \begin{cases} [s^{\ell(s)}(t)|s]\sigma & \text{if } \ell(t) > \ell(s), \text{ and} \\ & \forall n \geq \ell(s), [h^n(t)|h(s))]\sigma = [h(t)|h(s)]\sigma \\ [t|s^{\ell(t)}(s)]\sigma & \text{if } \ell(s) > \ell(t), \text{ and} \\ & \forall n \geq \ell(t), [h(t)|h^n(s)]\sigma = [h(t)|h(s)]\sigma \end{cases}$$

Example 13 For the notion of σ^T-unification, take for example the labels

$$t = (w_3, (W_1, w_1)) \qquad\qquad s = (w_3, (W_2, (w_2, w_1)))$$

Here $[W_2|w_3]\sigma = [w_3|w_3]\sigma$. Then the two labels σ^T-unify to $(w_3, (w_2, w_1))$. This intuitively means that the world w_3, accessible from a sub-path $s(s) = (W_2, (w_2, w_1))$, after the deletion of W_2 from s, is accessible from any path t which turns out to denote the same world(s) as $s(s)$; in fact the step from w_2 to W_2 is irrelevant because of the reflexivity relation of the model.

$$[t|s]\sigma^4 = \begin{cases} c^{\ell(t)}(s) & \ell(s) > \ell(t), h(t) \in \Phi_V \text{ and} \\ & w_0 = [t|s^{\ell(t)}(s)]\sigma \\ c^{\ell(s)}(t) & \ell(t) > \ell(s), h(s) \in \Phi_V \text{ and} \\ & w_0 = [s^{\ell(s)}(t)|s]\sigma \end{cases}$$

Example 14 For the notion of σ^4-unification, take for example the labels

$$t = (W_3, (w_2, w_1)) \qquad s = (w_5, (w_4, (w_3, (W_2, w_1))))$$

Here $s^{\ell(t)}(s) = (w_3, (W_2, w_1))$. Then t and s σ^4-unify to

$$(w_5, (w_4, (w_3, (w_2, w_1))))$$

since

$$[t|s^{\ell(t)}(s)]\sigma = [(W_3, (w_2, w_1))|(w_3, (W_2, w_1))]\sigma \, .$$

This intuitively means that all the worlds accessible from a sub-path $s^{\ell(t)}(s)$ of s are accessible from any path t which leads to the same world(s) denoted by $s^{\ell(t)}(s)$. Here W_3 stands for the set of worlds accessible from w_2; Then w_3, after the unification of (w_2, w_1) and (W_2, w_1), is one of such worlds. w_4 is accessible from w_3 and, via transitivity, from w_2. The same for w_5.

$$[t|s]\sigma^B = \begin{cases} [s^{\ell(t)-2n}(t)|s]\sigma & \text{if } h(t) \in \Phi_V \text{ and} \\ & [h(t)|h(s)]\sigma = [h^{\ell(t)-2n}(t)|h(s)]\sigma \\ [t|s^{\ell(s)-2n}(s)]\sigma & \text{if } h(s) \in \Phi_V \text{ and} \\ & [h(t)|h(s)]\sigma = [h(t)|h^{\ell(s)-2n}(s)]\sigma \end{cases}$$

Where $1 \leq n \leq V$, and $V = \ell(t) - m$, with m such that $\forall x, m \leq x \leq \ell(t), h^x(t) \in \Phi_V$.

Example 15 For σ^B we consider the labels

$$t = (W_3, (W_2, (w_2, (W_1, w_1)))) \qquad s = (W_4, (w_3, w_1)) \tag{7.1}$$

The labels t and s σ^B-unify since t has two variables, so we have two chances of going back: one steps from $b(t)$, or two steps from $b(b(t))$. In the first case we have to see whether $(w_2, (W_1, w_1)) = s^{\ell(t)-2n}(t), n = 1$ and s σ-unify. In the second case the label that have to σ-unify with s is $w_1 = s^{\ell(t)-2n}(t), n = 2$. But in this case the unification fails. The key idea of σ^B-unification is to match world symbols laying an even number of steps apart. The number of steps is given by the number of consecutive variables present in the labels. If the head of a label is a variable we can go back by two steps. In general we are allowed to return back of two steps for each variable. Labels like $(W_1, (w_2, w_1))$ and w_1 are a simple instance of such an unification. W_1 denotes the set of worlds accessible from w_2, but, since w_2 is accessible from w_1; so, by symmetry, w_1 is one of the world accessible from w_2.

$$[t|s]\sigma^5 = \begin{cases} ([h(t)|h(s)]\sigma; c^1(s^2(t))) & \text{if } \ell(t) > 2, \ell(s) > 1, h(t) \in \Phi_V, \text{ or} \\ & h(t) = h(s) \in \Phi_C \\ [t|s]\sigma & \text{if } \ell(t) = \ell(s) = 2 \\ ([t|h(s)]\sigma; c^1(s^2(s))) & \text{if } \ell(s) > 2, \ell(t) > 1, h(s) \in \Phi_V, \text{ or} \\ & h(t) = h(s) \in \Phi_C \end{cases}$$

where $w_0 = [s^1(t)|s^1(s)]\sigma$.

We exemplify how unifications corresponding to axioms obtained from the axioms listed above by prefixing $\Box^n$, $n \in \mathbb{N}$, to them can be defined.

$$[t|s]\sigma^O = [c^2(t)|c^2(s)]\sigma^D$$

where $w_0 = [s^2(t)|s^2(s)]\sigma^K$, and $O = \Box(\Box A \to \Diamond A)$.

4.2.2 High unifications (combined unifications). We are now able to combine the above unifications corresponding to the axiom characterising a logic into a single 'high' unification which will be used for defining the unifications characterising the logic we are concerned with.

$$[t|s]\sigma^{A_1^i \cdots A_n^i} = \begin{cases} [t|s]\sigma^{A_1^i} & \text{if } C_1^i \\ \vdots & \vdots \\ [t|s]\sigma^{A_n^i} & \text{if } C_n^i \end{cases} \qquad i \in I \tag{7.2}$$

where $A_1^i \cdots A_n^i$ stand for the axioms characterising L_i and $C_i, 1 \leq i \leq n$ are conditions varying from logic to logic. For example the high unification for **T**, which is characterized by the axioms D and T, is

$$[t|s]\sigma^{DT} = \begin{cases} [t|s]\sigma^T & \text{if } \ell(t) \neq \ell(s) \\ [t|s]\sigma^D & \text{otherwise} \end{cases}$$

and for **OM** which is **K** plus $M = \Box(\Box A \to A)$, the deontic version of **T**, the corresponding high unifications is

$$[t|s]\sigma^{OM} = \begin{cases} [t|s]\sigma^T & \text{if } [s^2(t)|s^2(s)]\sigma^O \\ [t|s]\sigma^O & \end{cases}$$

We then provide the definition of σ^{DT4} which is used in defining the logic unification for **S4**.

$$[t|s]\sigma^{DT4} = \begin{cases} [t|s]\sigma^D & \text{if } \ell(t) = \ell(s) \\ [t|s]\sigma^T & \text{if } \ell(t) < \ell(s), h(t) \in \Phi_C \\ [t|s]\sigma^4 & \text{if } \ell(t) < \ell(s), h(t) \in \Phi_V \end{cases} \tag{7.3}$$

It is worth noting that the conditions on axiom unifications are needed in order to provide a deterministic unification procedure. In general, if axiom unifications are given for each axiom characterising a logic, such conditions are not necessary.

4.2.3 Low unification (logic unifications). Combining recursively each high unification we obtain the σ_{L_i} unification for L_i as follows

$$[t|s]\sigma_{L_i} = \begin{cases} [c^n(t)|c^m(s)]\sigma^{A_1^i \cdots A_n^i} \\ [t|s]\sigma^{A_1^i \cdots A_n^i} \end{cases} \tag{7.4}$$

where $w_0 = [s^n(t)|s^m(s)]\sigma_{L_i}$.

Although the above definition provides a general method for obtaining the unification characterising the appropriate logic, for particular systems it may be preferred to define a different unification reflecting peculiar properties; for example, for **S5** which is characterised by the class of frames where the accessibility relation is an equivalence relation, it is convenient to define $\sigma_{\mathbf{S5}}$ as follows:[2]

$$[t|s]\sigma_{\mathbf{S5}} = \begin{cases} [h(t)|h(s)]\sigma & \min\{\ell(t), \ell(s)\} = 1 \\ ([h(t)|h(s)]\sigma, [s^1(t)|s^1(s)]\sigma) & \text{otherwise} \end{cases}$$

The format of the unifications for the logics containing axioms obtained by prefixing $\Box^n$ to one of the five axiom listed above is slightly more complex. We exemplify it by providing the unification for **OM**.

$$[t|s]\sigma_{\mathbf{OM}} = \begin{cases} [c^n(t)|c^m(s)]\sigma^{OM} & \\ [c^n(t)|c^m(s)]\sigma^{T} & n, m > 2 \\ [t|s]\sigma^{OM} & \end{cases}$$

where $w_0 = [s^n(t)|s^m(s)]\sigma_{\mathbf{OM}}$.

Example 16 According to (7.4) the logic unification for **S4** is

$$[t|s]\sigma_{\mathbf{S4}} = \begin{cases} [c^n(t)|c^m(s)]\sigma^{DT4} \\ [t|s]\sigma^{DT4} \end{cases}$$

where $w_0 = [s^n(t)|s^m(s)]\sigma_{\mathbf{S4}}$.

We provide a pair of labels t, and s that $\sigma_{\mathbf{S4}}$-unify.

$$t = (w_6, (w_5, (W_2, (w_1, (W_1, w_0))))) \quad s = (W_3, (w_4, (w_3, (w_2, (w_1, w_0)))))$$

The two labels do not σ^D-, nor σ^T-, nor σ^4-unify. We have then split them into appropriate counter-segments.

Graphically we can represent the steps of above unification as follows:

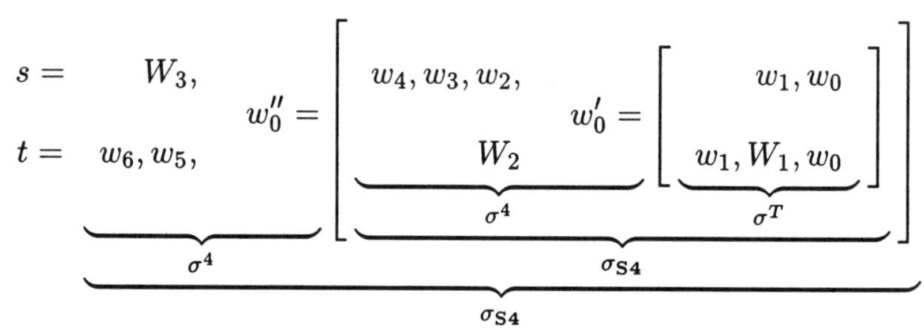

[2] Similar considerations hold for **K45** and **D45**. So specialised unifications for such logics have been provided in [16]. However in [18] equivalence between these specialised unifications and unifications obtained from the schema 7.4 has been proved.

In the first line we show the elements of the label s; the second line contains the elements of t. Notice that here w_0' and w_0'' are dummy labels and are common to t and s.

First we consider

$$s^3(t) = (w_1, (W_1, w_0)) \qquad \text{and} \qquad s^2(s) = (w_1, w_0) .$$

Such labels σ^T-unify on (w_1, w_0), see example 13. We identify their unification with w_0'; then

$$c^3(t) = (w_6, (w_5, (W_2, w_0'))) \quad \text{and} \quad c^2(s) = (W_3, (w_4, (w_3, (w_2, w_0')))) .$$

At this point we notice that

$$s^2(c^3(t)) = (W_2, w_0') \qquad s^4(c^2(s)) = (w_4, (w_3, (w_2, (w_1, w_0))))$$

σ^4-unify: $[w_0'|w_0']\sigma$ and $W_2 \in \Phi_V$. The result of their unification, w_0'', is $(w_4, (w_3, (w_2, (w_1, w_0))))$. Finally

$$c^4(t) = (w_6, (w_5, w_0'')) \qquad c^5(s) = (W_3, w_0'')$$

σ^4-unify. According to the definition of logic unification 7.4, and the previous considerations t and s $\sigma_{\mathbf{S4}}$-unify.

Example 17 In (7.1) we have seen a pair of labels that σ^B. Here we provide two labels that unify by applying recursively σ^B.

$$i = (W_3, (w_4, (W_2, (w_3, (W_1, (w_2, w_1)))))) \qquad k = w_1 \tag{7.5}$$

The unification is obtained according to the following decomposition.

$$[(W_3, (w_4, w_0))|w_1]\sigma^{KB}$$

where

$$w_0 = [(W_2, (w_2, w_0'))|w_1]\sigma_{\mathbf{KB}}$$

and

$$w_0' = [(W_1, (w_2, w_1))|w_1]\sigma_{\mathbf{KB}}$$

since

$$[(W_1, (w_2, w_1))|w_1]\sigma^{KB} = w_1$$

by an immediate application of σ^B.

For a wide class of unifications and a detailed account of the labelling algebra of KEM see [18].

4.2.4 *Fibred unification.* As L_I^δ is obtained by combining the logic L_i, $i \in I$ the corresponding unification is the combination of the σ_{L_i}-unifications characterising L_i.

$$[t|s]\sigma_{L_I^\delta} = \begin{cases} \bigcup_{i\in I}[c^n(t)|c^m(s)]\sigma_{L_i} \\ \bigcup_{i\in I}[t|s]\sigma_{L_i} \end{cases} \tag{7.6}$$

where $w_0 = [s^n(t)|s^m(s)]\sigma_{L_I^\delta}$, if $c^n(t), c^n(s)$ are i-pure, $i \in I$.

It is worth noting that the mechanism of the fibred unification is, essentially, the same mechanism governing the logic unification.

Example 18 We want to combine a deontic **D** modality $\Box_1$ with an epistemic **S5** modality $\Box_2$, let us call the resulting logic $\mathbf{ED}^\delta$. The unifications for $\mathbf{ED}^\delta$ are provided by the schema

$$[t|s]\sigma_{\mathbf{ED}^\delta} = \begin{cases} [c^n(t)|c^m(s)]\sigma_{\mathbf{S5}^2} \\ [t|s]\sigma_{\mathbf{S5}^2} \\ [c^n(t)|c^m(s)]\sigma_{\mathbf{D}^1} \\ [t|s]\sigma_{\mathbf{D}^1} \end{cases}$$

where $w_0 = [s^n(t)|s^m(s)]\sigma_{\mathbf{ED}^\delta}$, if $c^n(t)$ and $c^n(s)$ are i-pure, $i \in \{1, 2\}$.

According to the above definition, the labels

$$t = (W_2^2, (w_1^2, (W_1^1, w_0))) \qquad s = (w_2^2, (W_2^1, w_0))$$

$\sigma_{\mathbf{ED}^F}$-unify, and then $\sigma_{\mathbf{ED}^D}$-unify. In fact

$$c^2(t) = (W_2^2, (w_1^2, w_0')) \qquad c^2(s) = (w_2^2, w_0')$$

$\sigma_{\mathbf{S5}^2}$-unify, and $w_0' = [(W_1^1, w_0)|(w_1^1, w_0)]\sigma_{\mathbf{D}^1}$. On the contrary, if t had been $t' = (W_2^2, (w_2^1, (w_1^2, (W_1^1, w_0))))$, it would not have unified with s: $c^2(t')$ is not 2-pure. Let us consider now the labels

$$t = (W_1^1, w_0) \qquad s = (W_1^2, (w_1^2, (W_2^1, w_0)))$$

t and s do not $\sigma_{\mathbf{ED}^F}$-unify, but $\sigma_{\mathbf{ED}^D}$-unify, in so far as we can map W_2^2 onto w_0', where w_0' is as before.

We prove now an interesting property of $\sigma_{L_I^\delta}$-unifications.

Lemma 19 *$\forall t, s \in \Im$, if $[t|s]\sigma_{L_I^\delta} = r$, then $[t|r]\sigma_{L_I^\delta}$ and $[r|s]\sigma_{L_I^\delta}$.*

Proof Since $\sigma_{L_I^\delta}$ has been defined as the combination of σ_{L_i}-unifications, but each of them works on i-pure labels and there are no interactions among them, we just have to prove the property for each σ_{L_i}.

The proof will be by induction on the number of applications of $\sigma^{A_1^i \cdots A_n^i}$ in a σ_{L_i}-unification. Let n be the number of such applications. We give the proof only for **S4** and **S5**. The proof for the others systems is similar and can be found in [1, 18].

If $n = 1$ then we have to prove the property for $\sigma^{A_1^i \cdots A_n^i}$.[3] We prove it by induction on the length of labels.

If $\min\{\ell(t), \ell(s)\} = 1$ then we assume that $\ell(t) = 1$ (the proof for $\ell(s) = 1$ is similar). 1) $t \in \Phi_C$. If also $\ell(s) = 1$, we apply σ^D; in every case, by obvious considerations about σ, $r = [t|s]\sigma^D = t$, but $[t|t]\sigma^D$ and $[t|s]\sigma^D$. Therefore $[t|r]\sigma^{DT4}$ and $[r|s]\sigma^{DT4}$. If $\ell(s) > 1$ then $[t|s]\sigma^{DT4} = [t|s]\sigma^T$. Therefore $r = [t|s]\sigma^T = [t|s^1(s)]\sigma_T = t$, hence $[t|t]\sigma^D$ and $[t|s]\sigma^T$, and so, according to 7.3, $[t|r]\sigma^{DT4}$ and $[r|s]\sigma^{DT4}$.

2) $t \in \Phi_V$ then by the definition of σ it unifies with any label, in particular $[t|s]\sigma^D = s = r$, whence $[t|s]\sigma^D$ and $[s|s]\sigma^D$, then $[t|r]\sigma^{DT4}$ and $[r|s]\sigma^{DT4}$.

Let us suppose now that $\min\{\ell(t), \ell(s)\} = n > 1$, and that the property holds up to n for $\sigma^{A_1^i \cdots A_n^i}$. Thus we have the following cases.

If $\ell(t) = \ell(s)$ then $[t|s]\sigma^D = s$; by the inductive hypothesis $[b(t)|b(l)]\sigma^D$, $[b(s)|b(r)]\sigma^D$, $[h(t)|h(s)]\sigma^D$ and $[h(s)|h(r)]\sigma^D$; hence $[t|r]\sigma^D$ and $[s|r]\sigma^D$. Consequently $[t|r]\sigma^{DT4}$ and $(r, s)^{DT4}$.

If $\ell(t) < \ell(s)$ and $h(t) \in \Phi_C$, then $[t|s]\sigma^T = r$, where, by the inductive hypothesis $[b(t)|b(l)]\sigma^D$, $[s^{\ell(b(t))}(s)|b(r)]\sigma^D$. By the definition of σ^T, we know that, for $n \leq \ell(t)$ $h^n(r) = [h(t)|h(s)]\sigma = [h(t)|h^{\ell(t)}(s)]\sigma$; therefore $[t|r]\sigma^D$ and $[r|s]\sigma^T$. We can conclude that $[t|r]\sigma^{DT4}$ and $[r|s]\sigma_{DT4}$.

If $\ell(t) < \ell(s)$ and $h(t) \in \Phi_V$, then $[t|s]\sigma^4 = c^{\ell(t)}(s)$ where $w_0' = [t|s^{\ell(t)}(s)]\sigma$. By the inductive hypothesis and the definition of σ we have $[t|s^{\ell(t)}(r)]\sigma$ and $[s^{\ell(t)}(t)|s^{\ell(t)}(r)]\sigma$ and therefore $[t|r]\sigma^4$ and $[r|s]\sigma^D$, which means $[t|r]\sigma^{DT4}$ and $[r|s]\sigma^{DT4}$.

We have thus proved the inductive base for the lemma.

We can now assume that the lemma holds up to the n-th application of $\sigma^{A_1^i \cdots A_n^i}$. By the definition of σ_{L_i}, $[s^n(t)|s^m(s)]\sigma_{L_i} = w_0' = s^r(r)$ and $[c^n(t)|c^m(s)]\sigma^{A_1^i \cdots A_n^i} = c^r(r)$; but, by the inductive hypothesis, we know $[s^n(t)|s^r(r)]\sigma_{L_i}$ and $[s^m(s)|s^r(r)]\sigma_{L_i}$. By the property we have just proved for $\sigma^{A_1^i \cdots A_n^i}$ we obtain $[c^n(t)|c^r(r)]\sigma^{A_1^i \cdots A_n^i}$ and $[c^m(s)|c^r(r)]\sigma^{A_1^i \cdots A_n^i}$, which implies $[t|r]\sigma_{L_i}$ and $[r|s]\sigma_{L_i}$.

For **S5** if $\min\{\ell(i), \ell(s)\} = 1$, we have $[t|s]\sigma_{\mathbf{S5}}$ iff $[h(t)|h(s)]\sigma$, whence, if t is restricted, then $[t|s]\sigma_{\mathbf{S5}} = h(t) = r$ and thus $[t|s]\sigma_{\mathbf{S5}}$, i.e., $[h(t)|h(t)]\sigma$, and similarly for s; otherwise $[t|s]\sigma_{\mathbf{S5}} = h(s) = r$, therefore for the same

[3]Hereafter, in order to shorten proofs, when we have to consider labels of different lengths, we shall assume, unless specified, the first to be the shorter. Obviously proofs for the other cases can be carried out in the same way.

reason as in the previous case $[r|s]\sigma_{\mathbf{S5}}$ and $[t|r]\sigma_{\mathbf{S5}}$. If $\min\{\ell(t), \ell(s)\} > 1$ we can repeat the same argument of the other case with the difference that $r = ([h(t)|h(s)]\sigma, [s^1(t)|s^1(s)]\sigma)$. ■

Remark 20 *It is worth noting that we have no constraints on the component logics; they may be combined logics themselves.*

4.3 INFERENCE RULES

In displaying the rules of KEM we shall use Smullyan-Fitting [10] α, β, ν_i, π_i, $i \in I$ unifying notation as exposed in the following tables:

α	α_1	α_2
$\mathsf{T}A \wedge B$	$\mathsf{T}A$	$\mathsf{T}B$
$\mathsf{F}A \vee B$	$\mathsf{F}A$	$\mathsf{F}B$
$\mathsf{F}A \rightarrow B$	$\mathsf{T}A$	$\mathsf{F}B$
$\mathsf{F}\neg A$	$\mathsf{T}A$	$\mathsf{T}A$

β	β_1	β_2
$\mathsf{F}A \wedge B$	$\mathsf{F}A$	$\mathsf{F}B$
$\mathsf{T}A \vee B$	$\mathsf{T}A$	$\mathsf{T}B$
$\mathsf{T}A \rightarrow B$	$\mathsf{F}A$	$\mathsf{T}B$
$\mathsf{T}\neg A$	$\mathsf{F}A$	$\mathsf{F}A$

The relationships between α- and β-formula are

$$\begin{array}{lll} \alpha = \beta^C & \alpha_1 = \beta_1^C & \alpha_2 = \beta_2^C \\ \beta = \alpha^C & \beta_1 = \alpha_1^C & \beta_2 = \alpha_2^C. \end{array}$$

Formulas of type π_i and ν_i are classified as follows

ν_i	ν_0
$\mathsf{T}\Box_i A$	$\mathsf{T}A$
$\mathsf{F}\Diamond_i A$	$\mathsf{F}A$

π_i	π_0
$\mathsf{F}\Box_i A$	$\mathsf{F}A$
$\mathsf{T}\Diamond_i A$	$\mathsf{T}A$

Similarly we provide the relationships between ν_i- and π_i-formulas.

$$\begin{array}{ll} \nu_i = \pi_i^C & \nu_0 = \pi_0^C \\ \pi_i = \nu_i^C & \pi_0 = \nu_0^C. \end{array}$$

Given a signed formula X, X^C denotes the *conjugate* of X, i.e., the result of changing the sign of X to its opposite; two LS-formulas $X : t$ and $X^C : s$ such that $[t|s]\sigma_{L_I^\delta}$ and will be called $\sigma_{L_I^\delta}$*-complementary*.

$$\frac{\alpha : t}{\alpha_1 : t} \qquad \frac{\alpha : t}{\alpha_2 : t} \tag{α}$$

$$\frac{\begin{array}{c}\beta : t \\ \beta_1^C : s\end{array}}{\beta_2 : [t|s]\sigma_{L_I^\delta}} \; [t|s]\sigma_{L_I^\delta} \qquad \frac{\begin{array}{c}\beta : t \\ \beta_2^C : s\end{array}}{\beta_1 : [t|s]\sigma_{L_I^\delta}} \; [t|s]\sigma_{L_I^\delta} \tag{β}$$

$$\frac{\nu_i : t}{\nu_0 : (s,t)}\ s \in \Phi^i_V \text{ and new} \qquad (\nu_i)$$

$$\frac{\pi_i : t}{\pi_0 : (s,t)}\ s \in \Phi^i_C \text{ and new} \qquad (\pi_i)$$

$$\frac{}{X : t \qquad\qquad X^C : t}\ t \text{ restricted} \qquad (PB)$$

$$\frac{\begin{array}{c} X : t \\ X^C : s \end{array}}{\times [t|s]\sigma_{L^\delta_I}}\ [t|s]\sigma_{L^\delta_I} \qquad (PNC)$$

Here the α-rules are just the familiar linear branch-expansion rules of the tableau method, while the β-rules correspond to such common natural inference patterns as *modus ponens*, *modus tollens*, etc. The rules for the modal operators are as usual. 's new' in the proviso for the ν_i- and π_i-rule means: s must not have occurred in any label yet used. Notice that in all inferences via an α-rule the label of the premise carries over unchanged to the conclusion, and in all inferences via a β-rule the labels of the premises must be $\sigma_{L^\delta_I}$-unifiable, so that the conclusion inherits their unification. *PB* (the 'Principle of Bivalence') represents the (*LS*-version of the) semantic counterpart of the cut rule of the sequent calculus (intuitive meaning: a formula A is either true or false in any *given* world, whence the requirement that t should be restricted). *PNC* (the 'Principle of Non-Contradiction') corresponds to the familiar branch-closure rule of the tableau method, saying that from the occurrence of a pair of $\sigma_{L^\delta_I}$-complementary formulas on a branch we may infer the closure ('$\times$') of the branch. The $[t|s]\sigma_{L^\delta_I}$ in the 'conclusion' of *PNC* means that the contradiction holds 'in the same world'. Other logics might require additional rules in order to capture the full power of their semantics. See for example [17, 18]. As usual with refutation methods a KEM-proof of A consists of a successful attempt to construct a counter model for A by assuming that A is false in some arbitrary model, which means that we assume that A is false in the actual world of the model. So a KEM-proof in L for A ($\vdash_{\mathrm{KEM}(L)} A$)is a closed tree starting with $\mathsf{F}A$. A tree is closed iff all its branch are closed; a branch is closed iff it contains an application of *PNC*.

Let us consider the combined systems $\mathbf{ED}^\delta$ of example 18 and the formulas $A = \Box_1\Diamond_2 p \to \Diamond_1\Box_2\Diamond_2 p$ and $B = \Box_1 p \to \Diamond_1\Box_2\Diamond_2 p$. It is easy to see that A is valid in $\mathbf{ED}^\delta$. On the other hand, B holds in $\mathbf{ED}^D$, but fails in $\mathbf{ED}^F$. We provide now their KEM-proofs, and, from them, the fibred model where B

fails can be easily obtained.

1.	$\mathsf{F}\Box_1\Diamond_2 p \to \Diamond_1\Box_2\Diamond_2 p$	w_0
2.	$\mathsf{T}\Box_1\Diamond_2 p$	w_0
3.	$\mathsf{F}\Diamond_1\Box_2\Diamond_2 p$	w_0
4.	$\mathsf{T}\Diamond_2 p$	(W_1^1, w_0)
5.	$\mathsf{F}\Box_2\Diamond_2 p$	(W_2^1, w_0)
6.	$\mathsf{T}p$	$(w_1^2, (W_1^1, w_0))$
7.	$\mathsf{F}\Diamond_2 p$	$(w_2^2, (W_2^1, w_0))$
8.	$\mathsf{F}p$	$(W_1^2, (w_2^2, (W_2^1, w_0)))$

1.	$\mathsf{F}\Box_1 p \to \Diamond_1\Box_2\Diamond_2 p$	w_0
2.	$\mathsf{T}\Box_1$	w_0
3.	$\mathsf{F}\Diamond_1\Box_2\Diamond_2 p$	w_0
4.	$\mathsf{T}p$	(W_1^1, w_0)
5.	$\mathsf{F}\Box_2\Diamond_2 p$	(W_2^1, w_0)
6.	$\mathsf{F}\Diamond_2 p$	$(w_2^2, (W_2^1, w_0))$
7.	$\mathsf{F}p$	$(W_1^2, (w_2^2, (W_2^1, w_0)))$

Steps 2, and 3 are obtained by an application of an α-rule on 1, the other steps have been obtained by straightforward applications of ν_i- and π_i-rules, until we reach atomic formulas. We notice that in both trees we have complementary formulas, i.e., 6, 8 in the tree for A and 4, 7 in the tree for B. All that remains to do is verifying whether they are $\sigma_{\mathbf{ED}^\delta}$-complementary; i.e., we have to check if their labels $\sigma_{\mathbf{ED}^\delta}$-unify. In example 18 we have seen that the labels of 6 and 8 $\sigma_{\mathbf{ED}^\delta}$-unify. However these of 4 and 7 $\sigma_{\mathbf{ED}^D}$- but not $\sigma_{\mathbf{ED}^F}$-unify.

4.4 SOUNDNESS AND COMPLETENESS

In this section we prove soundness and completeness for KEM-based fibred modal tableaux. There are very similarity with the results of [5], however the material presented in this section points out close connections between unifications and modal fibring.

Theorem 21 *If a* KEM *tree closes it closes atomically.*

Proof A closed KEM tree means that each branch is closed, i.e., it contains two $\sigma_{L_I^\delta}$-complementary formulas $A : t$ and $A^C : s$.

$$\begin{array}{c} \vdots \\ A : t \\ \vdots \\ A^C : s \\ \times \end{array}$$

We prove the theorem by induction of the complexity of the complementary formulas. If they are literals then the branch closes atomically.

If they are not literals let us examine their form: if A is of type α then A^C is of type β; moreover $\alpha_1 = \beta_1^C$ and $\alpha_2 = \beta_2^C$. We apply an α-rule on $A : t$, obtaining $\alpha_1 : t$ and $\alpha_2 : t$. Since the relations just mentioned we can apply a β-rule w.r.t. $A^C : s$ and α_n $(n = 1, 2)$, from which we derive $\beta_{3-n} : [t|s]\sigma_{L_I^\delta}$. At this point the branch contains $\alpha_n : t$ and $\beta_n : [t|s]\sigma_{L_I^\delta}$, which are $\sigma_{L_I^\delta}$-complementary, in so far as $[t|[t|s]\sigma_{L_I^\delta}]\sigma_{L_I^\delta}$, see Lemma 19. If A is of type β we repeat the above reasoning applying the α-rule on A^C instead of A.

If A is of type ν_i, then A^C is of type π_i and $\nu_0 = \pi_0^C$. We apply a ν_i-rule on $A : t$ and a π_i-rule on $A^C : s$ obtaining $\nu_0 : (W_n, t)$ and $\pi_0 : (w_m, s)$, where W_n and w_m are new in the branch. The resulting formulas are $\sigma_{L_I^\delta}$-complementary due to the relationship between ν_i and π_i formulas and the fact that the labels obviously $\sigma_{L_I^\delta}$-unify. If A is of type π_i, then A^C is of type ν_i and we can repeat the same argument. ■

In the course of KEM-proofs labels are used to build appropriate models. Since the structure of the labels and unifications follows closely that of dovetailed and fibred models, we can repeat the same construction 'grafting' the models for each L_i through $\mathbf{F}$ into fibred and dovetailed models obtaining models for L_I^δ.

Theorem 22 *Let $L_i, i \in I$ be modal logics and let L_I^δ the resulting combined logic. If $\models_{K_i} A \iff \vdash_{\mathrm{KEM}(L_i)} A$ then*

1. $\models_{L_I^F} A \iff \vdash_{\mathrm{KEM}(L_I)} A$ *using* $\sigma^{\mathbf{F}}$;

2. $\models_{L_I^D} A \iff \vdash_{\mathrm{KEM}(L_I)} A$ *using* $\sigma^{\mathbf{D}}$.

Proof We prove the theorem by showing a) the set $\mathbf{T}_I^\delta$ is a subset of set of formulas provable in KEM and b) KEM rules are sound with respect to fibred and dovetailed models.

We start proving a). According to Theorems 3 and 4 a formula is a fibred (dovetailed) theorem if either it is a theorem of a component, or has been obtained by one of the axioms, or has been derived from an application of modal fibring (dovetailing) rule or modus ponens. We have then to prove that such rules and axioms are derived in KEM.

By hypothesis $\mathbf{T}_i$ coincides with the set of formulas provable in KEM for L_i.

For axiom 1b1 (Theorem 3), by hypothesis $A \to \alpha_n$ is a theorem of a fibred language, therefore $\mathsf{T}A : w_0$ and $\mathsf{F}\alpha_n : w_0$ lead to a closed KEM-tree. Let us start now a KEM-tree for $A \to \bigvee_n \alpha_n : w_0$, we obtain

$$\begin{array}{c} \mathsf{F}A \to \bigvee_n \alpha_n : w_0 \\ \mathsf{T}A : w_0 \\ \mathsf{F}\bigvee_n \alpha_n : w_0 \\ \mathsf{F}\alpha_1 : w_0 \\ \vdots \\ \mathsf{F}\alpha_n : w_0 \end{array}$$

At this point we can graft the proof for $A \to \alpha_n$, closing thus the tree.

For axiom 1b2 (Theorem 3), by hypothesis $A(x_j)$ has a closed KEM-tree, which means that each branch τ is closed; Theorem 21 implies that each branch is atomically closed, therefore each branch contains two σ_{L_i}-complementary labelled signed literals, let us say $x_\tau : t_\tau$ and $x^C_\tau : s_\tau$. We can now replace $\Box_j\alpha_j$ to x_τ obtaining $S\Box_j\alpha_j : t_\tau$ and $S^C\Box_j\alpha_j : s_\tau$. But the last two formulas are $\sigma_{L_I^F}$-complementary, then also in this case the tree is closed.

For Modus Ponens. By hypothesis A and $A \to B$ have closed KEM trees, let us call them $\mathcal{T}_1$ and $\mathcal{T}_2$.

$$\begin{array}{ccc} & \mathsf{T}B : w_0 & \\ \swarrow & & \searrow \\ \mathsf{T}A \to B : w_0 & & \mathsf{F}A \to B : w_0 \\ \mathsf{F}A : w_0 & & \mathcal{T}_2 : w_0 \\ \mathcal{T}_1 : w_0 & & \times \\ \times & & \end{array}$$

For modal and dovetailing rule, let us assume that C is a formula satisfying the conditions of the rules, then C has a closed KEM-tree, let us call it $\mathcal{T}$. We show now that also the tree for $\Box^d_j C$ is closed.

$$\begin{array}{c} \mathsf{F}\Box^d_j C : w_0 \\ \mathsf{F}\Box^{d-1}_j C : (w^j_1, w_0) \\ \vdots \\ \mathsf{F}C : (w^j_d, \dots (w^j_1, w_0)) \\ \mathcal{T} : w'_0 \\ \times \end{array}$$

All we have to do is to identify w'_0 with $(w^j_d, \dots (w^j_1, w_0))$ and we can repeat the proof for $C : w'_0$, in so far as $[(w^j_d, \dots (w^j_1, w_0))|(w^j_d, \dots (w^j_1, w_0))]\sigma_{L_I^\delta}$.

For b) in the course of KEM-proofs, we generate labels according to the structure of the formulas involved, but, as we have already said, they also generate (counter)-models. The labels are intended to denote possible worlds and relations among them. Remember (see section 4.1, and example 13): all the relevant information are recorded in the labels. So, to extract such information, we have to map labelled signed formulas to elements of fibred and dovetailed models. This is achieved with the help of three functions, namely g, r, and f. The function g will map labels to sets of possible worlds: a singleton for constants, a set of worlds (possibly empty) for variables, and an actual world for auxiliary labels. The accessibility relation R is assumed to be closed under specific conditions; but, we want to reconstruct it, through r, from the labels: path labels are intended to represent not only worlds, but also the chain of possible worlds leading to them. Finally, f, given an LS-formula, returns the evaluation of the formula with respect to the world(s) denoted by its label.

Let $\mathbf{m}^i = (S^i, R^i, a^i, h^i)$ be a model in $\mathcal{K}_i$ where: $S^i = \Phi^i_C$; R^i is a binary relation on S^i; $a^i \in \Phi^A$, and v^i is an evaluation function.

Let g be a function from the set of i-pure labels occurring in a KEM-proof to $\wp(S^i)$ thus defined:

$$g(t) = \begin{cases} \{h(t)_t\} & \text{if } h(t) \in \Phi^i_C \\ \{w_n \in S^i \mid g(b(t))R^i w_n\} & \text{if } h(t) \in \Phi^i_V \\ S^i & \text{if } t \in \Phi_V \\ \{a^i\} & \text{if } t \in \Phi^A \end{cases}$$

It may be possible that two labels have the same head, but they denote different worlds, this is way we have indexed $h(t)$ with the label itsself. However we shall drop the subscript, when this is harmless.

Let r be a function from the set of i-pure labels occurring in a KEM-proof to R^i thus defined:

$$r(t) = \begin{cases} \emptyset & \text{if } \ell(t) = 1 \\ g(t^1)R^i g(t^2), \dots, g(t^{n-1})R^i g(h(t)) & \text{if } \ell(t) = n > 1 \end{cases}$$

Let f be a function from LS-formulas occurring in a KEM-proof to v thus defined:

$$\begin{aligned} f(SA : t) &=_{def} v(w^i_n, A) = 1 \text{ if } S = \mathsf{T} \\ f(SA : t) &=_{def} v(w^i_n, A) = 0 \text{ if } S = \mathsf{F} \end{aligned}$$

for all $w^i_n \in g(t)$.

Until now we have examined i-pure labels. Let t be a not i-pure label. It can be decomposed into i-pure labels as follows: let $n \in \mathbb{N}$ such that $\forall m > n$, $s^m(t)$ is i-preferred. The label $c^n(t)$ is i-pure. A not i-pure label can be conceived as a recursive fibring of i-pure sub-labels.

Let $\mathbf{m}$ be a fibred structure, where the fibred function $\mathbf{F}$ is defined as follows:

$$\mathbf{F}_i(w_j) = g(w_0) \tag{7.7}$$

for each $w_n \in g(s^n(t))$, where $w_0 = h^1(c^n(t))$ and $c^n(t)$ is i-pure. Moreover we require that if $w_j \neq w_k$ then $\mathbf{F}_i(w_j) \neq \mathbf{F}_i(w_k)$. In the case of dovetailing we impose $\mathbf{F}_i(w_j) = w_j$

It is easy to see that $\mathfrak{m}$ is a fibred or a dovetailed model for L_I^δ.

Lemma 23 *For any $t, s \in \Im$, if $[t|s]\sigma_{L_I^\delta}$ then $g(t) \cap g(s) \neq \emptyset$.*

Proof The proof is by induction on the number of applications of σ_{L_i} in $\sigma_{L_I^\delta}$. First we have to prove the property for σ_{L_i} and therefore for $\sigma^{A_1^i \cdots A_n^i}$. For a detailed proof see [1, 18].

The proof is by induction on the number of applications of $\sigma^{A_1^i \cdots A_n^i}$ in σ_{L_i}. We need first to prove the following:

Lemma 24 *For any $t, s \in \Im$, if $[t|s]\sigma^{A_1^i \cdots A_n^i}$ then $g(t) \cap g(s) \neq \emptyset$.*

Proof We prove only the case for $DT4$, the other cases are similar and can be found in [1, 18]. The proof is by induction on the length of labels. If $\min\{\ell(t), \ell(s)\} = 1$, then at least one of t and s is either a constant or a variable, so that five cases will be present. By the definition of unifications t, s are either: i) two constants, or ii) a variable and a constant, or iii) two variables, or iv) a variable and a label, or v) a constant and a label.[4]

Case i) Two constants unify if and only if they are the same constant, and so $t = s$; therefore from the definition of g, $g(t) = g(s)$ and so $g(t) \cap g(s) \neq \emptyset$.

Case ii) If t (resp. s) is a variable and s (resp. t) is a constant, then $g(t) = S^i$ and $g(s) \in \wp(S^i)$ therefore also in this case $g(t) \cap g(s) \neq \emptyset$.

Case iii) and iv) These cases are identical to the previous ones because: 1) S^i is not empty, and 2) the variable is mapped to S^i and the label to some world(s) in it.

Case v) This case implies that $[t|s]\sigma^T$. Let us assume, for the sake of economy, that $\ell(t) = 1$ and $\ell(s) = n > 1$. If $[t|s]\sigma^T$, then for each $h(s(s))$ such that $\ell(s(s)) > 1$ either $h(s(s)) \in \Phi_V$, or $h(s(s)) = t$; therefore $r(s) = tR^ih^2(s), \ldots, h^{n-1}(s)R^ih^n(s)$. If $h^2(s) \in \Phi_V$, then it denotes the set of worlds accessible from t; if $h^2(s) \in \Phi_C$, then $t = g(h^2(s))$,[5] in any case, through reflexivity $t \subseteq g(h^2(s))$, so we take t as a representative of the set denoted by $h^2(s)$, which implies $tR^ih^3(s)$. We repeat the same argument until we arrive at $tR^ih^n(s)$: if $h^n(s) \in \Phi_C$, then $t = g(h^2(s))$ and so they denote the same world; if $h^n(s) \in \Phi_V$, then it denotes the set of worlds accessible from t; but t belongs to such a set, therefore, in all cases $g(t) \cap g(s) \neq \emptyset$.

---

[4]Cases ii), iii), and iv) are not found in KEM proofs, but they are useful both for dealing with cases in the inductive step and for case v).

[5]Due to the rules of KEM and the definition of the unifications, this case is possible only if $h^2(s)$ has been obtained by a previous unification, and so they do denote the same world.

For the inductive step we have $\min\{\ell(t), \ell(s)\} = n > 1$. Let us assume inductively that the lemma is valid up to n; if $\ell(t) = \ell(s)$ we shall write t and s as $(h(t), b(t))$ and $(h(s), b(s))$, respectively. If $[t|s]\sigma^D$, by the definition of σ^D we get $[b(t)|b(s)]\sigma^D$, for which the lemma holds; let w_j be one of the worlds shared by $b(t)$ and $b(s)$, whence $w_j R^i h(t)$ and $w_j R^i h(s)$. We have now only to analyze what kind of labels are $h(t)$ and $h(s)$, which falls under the cases i), ii), and iii). Cases i) and ii) are the same as the inductive base. We have thus to examine case iii). Both $h(t)$ and $h(s)$ denotes the set of worlds accessible from w_j, but such a set is not empty because of the seriality of R.

If $\ell(t) \neq \ell(s)$, we shall assume that $\ell(t) < \ell(s)$ (the case $\ell(s) < \ell(t)$ is dealt with in the same way). If $[t|s]\sigma^T$ and $h(t) \in \Phi_C$ then $[t|s^{\ell(t)}(s)]\sigma^D$, therefore, combining the proofs of the previous case and case v) of the inductive base we obtain the desired result. If $h(t) \in \Phi_V$, then for all $s^n, n \leq \ell(t)$, $[h(t)|h(s)]\sigma = [h(t)|s^n]\sigma$ which means $g(t) \cap g(s^n(s)) \neq \emptyset$, and in particular $g(t) \cap g(s^{\ell(t)}(s)) \neq \emptyset$.

If $[t|s]\sigma^4$ then $h(t) \in \Phi_V$ and $[b(t)|s^{\ell(t)-1}(s)]\sigma^D$, for which the inductive hypothesis holds; let w_j be such a shared world. $h(t)$ denotes all the worlds accessible from w_j, but, due to transitivity, the world(s) denoted by $h(s)$ belong(s) to $h(t)$ and so $g(t) \cap g(s) \neq \emptyset$.

This ends the proof of Lemma 24. ■

We return to the proof of the main lemma. If σ_{L_i} consists of a single step of $\sigma^{A_1^i \cdots A_n^i}$, then $[t|s]\sigma_{L_i} = [t|s]\sigma^{A_1^i \cdots A_n^i}$; by Lemma 24 we obtain $g(t) \cap g(s) \neq \emptyset$.

Let us assume, inductively, that the lemma holds up to n. If σ_{L_i} consists of $n + 1$ $\sigma^{A_1^i \cdots A_n^i}$-unifications, $[t|s]\sigma_{L_i} = [c^t(t)|c^s(s)]\sigma^{A_1^i \cdots A_n^i}$ where $[s^t(t)|s^s(s)]\sigma_{L_i}$, which contains n applications of $\sigma^{A_1^i \cdots A_n^i}$, and the lemma holds for it. We can now repeat the argument of Lemma 19 with respect to $[c^t(t)|c^s(s)]\sigma^{A_1^i \cdots A_n^i}$, proving thus that $g(t) \cap g(s) \neq \emptyset$. We have thus proved that the property holds also for σ_{L_i}. $\sigma_{L_I^\delta}$ results from the fibred combination of σ_{L_i}-unifications of i-pure labels. According to the interpretation of the fibred function given in 7.7 we can combine recursively the same argument as before.

This ends the proof of Lemma 23. ■

Lemma 25 *For any $t, s \in \Im$, if $f(SA : t)$, $[t|s]\sigma_{L_I^\delta}$ then $f(SA : [t|s]\sigma_{L_I^\delta})$.*

Proof Let us suppose, by contradiction, that the lemma does not hold, so the proof trivially follows from Lemma 23, and the definition of f. ■

The α-rules and *PB* are obviously sound rules in $\mathbf{m}$ in so far as they work locally. For the β-rules and *PNC*: by the hypothesis $[t|s]\sigma_{L_I^\delta}$, then, by Lemmas 23 and 25, there exists a world in $g([t|s]\sigma_{L_I^\delta})$, let us say w_j, where both β,

β_n $(n = 1, 2)$ and β_{3-n} hold. This implies that β_n and β_n^C hold at w_j, which is a contradiction. The same argument can be applied in the case of *PNC*.

For the ν_i-rules. Let us suppose $\nu_i = \mathsf{T}\Box_i A$ the case of $\nu_i = \mathsf{F}\Diamond_i A$ is identical. Let us suppose it does not hold, then for all $w_i \in g(t)$, $h(w_i, \Box_i A) = 1$ and for some $w_m \in g((s,t))$, $h(A, w_m) = 0$, where (s,t) is i-preferred. If t is i-preferred, then $h^i(w_i, \Box_i A) = 1$ implies $\forall w_j(w_i R^i w_j \rightarrow h^i(w_j, A) = 1)$, By the definitions of g and r we know $w_i R^i w_m$, obtaining thus a contradiction. If t is not i-preferred, then $h^j(w_i, \Box_i A) = 1$ iff $h^i(\mathbf{F}_i(w_i), \Box_i A) = 1$. The label (s,t) is not i-pure, but (s, w_0), where $\mathbf{F}_i(w_i) = g(w_0)$, is. At this point we can repeat the same argument as before with w_0 instead of w_i. The proof for the π_i-rule is similar.

This ends the proof of Theorem 22. ■

5 FINAL REMARKS

In the last few years we have witnessed a luxuriant growth of multidimensional systems in every field of logic, although only few seeds have been sown in the garden of proof theory [4, 5, 7, 8].

In this work we have been mainly concerned with combinations of normal modal logics. It is easy to see that dovetailing of normal modal logics correspond to fusion [19]. However fibring is more general, and we have no constraints on the components. They may be as well combined logics. It's worth noting that, as far propositional normal modal logics are concerned, the system presented here satisfies all the conditions required for combined tableau calculi reported in [5]; in particular the condition stating "the labels that are part of tableau formulae represent words in models and do not contain other information". However KEM-based tableau calculi can be provided for different systems having Kripke-style semantics, e.g., quantified modal logics, non normal modal logics, conditional logics; but such systems require more information in the labels (see [1, 2]). The unification scheme we have proposed is not sensitive of such additional burden, and can be applied also for their combination, in so far as it is closely related to the fibring function.

We have not studied multi-modal logics with interaction axioms specifying how modalities relate each other. But in [3, 16, 17, 18] it has been shown how this formalism can be extended to handle various systems of multi-modal logic with interaction axioms. The key idea consists of providing either new substitutions and unifications or constraints on the already existing ones, both reflecting semantic properties of such axioms.

Let us consider the axiom $\Box_1 A \rightarrow \Box_2 A$; whose corresponding semantic condition is $R_2 \subseteq R_1$. This is achieved in KEM by defining a substitution as follows: $\sigma_1 : \Phi_V^1 \mapsto \Im^1 \cup \Im^2$. The above interaction axiom belongs to the family of inclusion axioms of the form $\langle a\rangle[b]A \to [c]\langle d\rangle A$, where a, b, c, and d are strings of modalities obtained from composition and union (i.e., $\Box_1 \cdot \Box_2 A = \Box_1 \Box_2 A$ and $\Box_1 \cup \Box_2 A = \Box_1 A \wedge \Box_2 A$). Such axioms characterize models satisfying a, b, c, d-incestuality, i.e., $\rho(a)^{-1} \cdot \rho(b) \subseteq \rho(c) \cdot \rho(d)^{-1}$, where ρ is map from string of modalities to accessibility relation, see [6]. In more complex cases we may define an unification schema matching the appropriate a, b, c, d-incestuality condition, e.g., the first label ends with a string of world symbols corresponding to a, b while the second corresponds to c, d; then we can apply (fibre) more unifications inside the strings of labels. For some interaction axioms (with a different form) special inference rules may be required (see [17]). Similarly non-normal modal logics can be dealt with by weakening the substitutions, e.g., $\sigma : \Phi_V \mapsto \Im^*$ where $\Im^* \subseteq \Im$. At the light of what we have already said the study of the proof theory of multi-modal logics reduces to the study of the components.

Acknowledgments

Thanks to the anonymous referees for their valuable comments on an earlier version of this paper. This research was partially supported by the Australian Research Council under Large Grant No. A49803544.

Dov M. Gabbay
Department of Computer Science
King's College London, UK
dg@dcs.kcl.ac.uk

Guido Governatori
School of Computing and Information Technology
Griffith University
Brisbane, Australia
G.Governatori@cit.gu.edu.au

References

[1] Alberto Artosi, Paola Benassi, Guido Governatori, and Antonino Rotolo. Shakespearian modal logic: a labelled treatment of modal identity. In M. Kracht, M. de Rijke, H. Wansing, and M. Zakharyaschev (eds), *Advances in Modal Logic*, CSLI Publications, Stanford, 1998, 1–20.

[2] Alberto Artosi and Guido Governatori. A tableaux methodology for deontic conditional logic. In *ΔEON'98, 4th International Workshop on Deontic Logic in Computer Science*, CIRFID, Bologna, 1998: 65–84.

[3] Alberto Artosi, Guido Governatori, and Giovanni Sartor. Towards a computational treatment of deontic defeasibility. In Mark Brown and José Carmo (eds.), *Deontic Logic Agency and Normative Systems*, Springer-Verlag, Berlin, 1996: 27–46.

[4] Matteo Baldoni, Laura Giordano and Alberto Martelli. A tableau calculus for multimodal logics andl some (un)decidability results. In H. de Swart (ed.), *International Conference on Theorem Proving with Analytic Tableaux and Related Methods*, Springer-Verlag, Berlin, 1998: 44–59.

[5] Bernhard Beckert and Dov M. Gabbay. Fibring semantic tableaux. In H. de Swart (ed.), *International Conference on Theorem Proving with Analytic Tableaux and Related Methods*, Springer-Verlag, Berlin, 1998: 77–92.

[6] Laurent Catach. Normal multimodal logics. In *Proc. Nat. Conf. on AI (AAAI'88)*, 491–495, 1988.

[7] Marcello D'Agostino and Dov M. Gabbay. Fibred tableaux for multi-implication logics. In P. Miglioli, U. Moscato, D. Mundici, and M. Ornaghi, (eds.), *Theorem Proving with Analytic Tableaux and Related Methods*, Springer-Verlag, Berlin, 1996: 16–35.

[8] Marcello D'Agostino, Dov M. Gabbay, and Alessandra Russo. Grafting modalities onto substructural implication systems. *Studia Logica* 59: 65–102, 1997.

[9] Marcello D'Agostino and Marco Mondadori. The taming of the cut. *Journal of Logic and Computation*, 4: 285–319, 1994.

[10] Melvin Fitting. *Proof Methods for Modal and Intuitionistic Logics*. Reidel, Dordrecht, 1983.

[11] Dov M. Gabbay. Fibred semantics and the weaving of logic I: Modal and intuitionistic logics. *Journal of Symbolic Logic*, 61: 1057–1120, 1996.

[12] Dov M. Gabbay. *Labelled Deductive System*. Oxford University Press, 1996.

[13] Dov M. Gabbay. An overview of fibred semantics and the combination of logic. In F. Baader and K. Schulz, (eds.), *Frontiers of Combining Systems*, Kluwer, Dordrecht, 1996: 1–55.

[14] Dov M. Gabbay. *Fibring Logic*. Oxford University Press, 1998.

[15] Olivier Gasquet. Optimization of deduction for multimodal logics. In Masuch, Marx and Plós (eds.), *Applied Logic: What, How and Why?*. Kluwer, Dordrecht, 1993.

[16] Guido Governatori. Labelled tableaux for multi-modal logics. In P. Baumgartner, R. Hähnle and J. Posegga (eds.) *Theorem Proving with Analytic Tableaux and Related Methods*, Springer-Verlag, Berlin, 1995: 79–94.

[17] Guido Governatori. Labelling ideality and subideality. In D. Gabbay, H. J. Ohlbach (eds.), *Practical Reasoning*, Springer-Verlag, Berlin, 1996: 291–304.

[18] Guido Governatori. Un modello formale per il ragionamento giuridico. PhD. Thesis, University of Bologna, 1997.

[19] Marcus Kracht and Frank Wolter. Properties of independently axiomatizable bimodal logics. *Journal of Symbolic Logic*, 56:1485–1991, 1991.

[20] Hans Jürgen Ohlbach. Semantics-based translation methods for modal logics. *Journal of Logic and Computation*, 1:691–746, 1990.

[21] Hans Jürgen Ohlbach. Optimized translation of multi modal logic into predicate Logic. A. Voronkov (ed.) *Proc. of Logic Programming and Automated Reasoning (LPAR)*, Springer-Verlag, Berlin, 253–264, 1993.

LABELLED DEDUCTION FOR THE GUARDED FRAGMENT

Maarten Marx, Szabolcs Mikulás and Stefan Schlobach

Abstract We present the tableau calculus LC_2-TAB which is sound and complete with respect to local square modal logic. The system is a labelled deduction calculus in the spirit of those for modal **S5**. The novelty here is that the calculus works in two interacting dimensions. This 2-dimensional modal logic allows one to simulate different other modal logics, like K, KT, KTB or multi-K in quite an elegant way. The calculus is also strong enough to decide an interesting PSPACE complete sub-fragment of the guarded fragment, which is generally conceived of as the true modal fragment of first order logic. A PROLOG implementation of this calculus is available through the WWW.

Introduction

In [15], a modalised version of first order logic with n variables is presented, called MLR_n (for modal logic of n–ary relations). This modal language can be given different semantics, corresponding to different fragments of the language of first order logic. Of particular interest is the relativized semantics which gives rise to a decidable system. As we will see here, this semantics corresponds to the guarded fragment of first order logic (cf. [2, 3, 17]). This decidable fragment was introduced to account for the fact that modal logics are so robustly decidable.

Here we will focus on the case with only two variables. First –as is well known– two variables are sufficient to translate (multi–) modal logics into first order logic [8]. Secondly, the language MLR_n for $n = 2$ has a PSPACE complete satisfaction problem, while for $n > 2$ this problem becomes EXPTIME–complete [12]. But of greater importance for our purpose here is that for $n = 2$, the language has a special finite model property which gives rise to a straightforward tableaux calculus based on labelled deduction ideas. For $n > 2$, the finite model property also holds [1, 10], but a simple model creating technique (like a naive tableaux system) does not work (see Section 3).

We designed and implemented a sound and complete labelled tableau calculus for LC_2, which can be used via the WWW from URL

D. Basin et al. (eds.), Labelled Deduction, 193–214.
© 2000 *Kluwer Academic Publishers. Printed in the Netherlands.*

`http://www.dcs.kcl.ac.uk/cgi-bin/LC2/lc2.html.`

We will show how the system works by looking at an instructive example proof. We evaluate the strength of the logic by showing how it can be used as a decision procedure for a range of other modal logics, including description logics.

The paper is organized as follows. The next section introduces two–dimensional modal logic and shows its relation with first order logic and other modal systems. Section 2 shows from a complexity point of view why our choice is correct if we want to use the system presented here to do proofs for PSPACE complete modal logics. Section 3 presents an example indicating the difficulties extending our system to higher dimensions. Section 4 presents the tableaux calculus. The final section deals with its implementation in PROLOG.

1 LOCAL CUBIC MODAL LOGIC OF TWO DIMENSIONS

1.1 LANGUAGE AND SEMANTICS

Throughout this paper we will focus on the 2-dimensional case for which we introduce a special notation, due to Segerberg. For the logic MLR_n, we introduce n diamonds $\Diamond_i$, for every function $\sigma : n \longrightarrow n$ a unary modality $\otimes_\sigma$ and for all $i, j < n$ a modal constant id_{ij}. The reader interested in the more general ideas is referred to [15] for details about the logic MLR_n.

Definition 1 *Given a countable set of propositional variables $p_0, p_1, \ldots$, the set of MLR_2-formulas is defined as the smallest set such that*

1. *the propositional variables $p_0, p_1, \ldots$ and the modal constant id are MLR_2-formulas (called atomic formulas),*
2. *if φ and ψ are MLR_2-formulas then also $\neg\varphi, \varphi \vee \psi, \diamondbar\varphi, \diamondminus\varphi$ and $\otimes\varphi$ are MLR_2-formulas.*

We abbreviate $\boxbar\varphi \equiv \neg\diamondbar\neg\varphi$, $\boxminus\varphi \equiv \neg\diamondminus\neg\varphi$ and $\varphi \wedge \psi \equiv \neg(\neg\psi \vee \neg\psi)$ as usual. Two new operators are defined by $\ominus\,\varphi \equiv \diamondminus(id \wedge \varphi)$ and $\obar\,\varphi \equiv \diamondbar(id \wedge \varphi)$.

Semantically, boxes and diamonds of MLR_2 will be interpreted as classical S5 modalities in two dimensions, with two interacting accessibility relations. We are interested in two classes of models for this language: the *squares* and the *local squares*. For U a set, we use ${}^2U$ to denote the product $U \times U$.

Definition 2 *Let U be a set. The set $W = U \times U$ is called a* square. *A set $W \subseteq {}^2U$ is called a* local square *if $\langle a, b\rangle \in W$ implies that ${}^2\{a, b\} \subseteq W$.*

Definition 3 *A model is a pair $\mathcal{M} = (W, v)$, with $W \subseteq {}^2U$ for an arbitrary set U and v a valuation, i.e. a function mapping every propositional variable to a subset of W. Such a model is called a (local) square if W is a (local) square. Truth of formulas at a pair $\langle a, b\rangle \in W$ in such models is defined as follows:*

$$
\begin{array}{lcl}
\mathcal{M}, \langle a, b\rangle \models p & \Leftrightarrow & \langle a, b\rangle \in v(p) \\
\mathcal{M}, \langle a, b\rangle \models id & \Leftrightarrow & a = b \\
\mathcal{M}, \langle a, b\rangle \models \neg\varphi & \Leftrightarrow & \mathcal{M}, \langle a, b\rangle \not\models \varphi \\
\mathcal{M}, \langle a, b\rangle \models \varphi \vee \psi & \Leftrightarrow & \mathcal{M}, \langle a, b\rangle \models \varphi \text{ or } \mathcal{M}, \langle a, b\rangle \models \psi \\
\mathcal{M}, \langle a, b\rangle \models \otimes\varphi & \Leftrightarrow & \mathcal{M}, \langle b, a\rangle \models \varphi \\
\mathcal{M}, \langle a, b\rangle \models \Phi\varphi & \Leftrightarrow & \text{there is a } \langle a, c\rangle \in W, \\
 & & \text{such that } \mathcal{M}, \langle a, c\rangle \models \varphi \\
\mathcal{M}, \langle a, b\rangle \models \text{⟠}\varphi & \Leftrightarrow & \text{there is a } \langle c, b\rangle \in W, \\
 & & \text{such that } \mathcal{M}, \langle c, b\rangle \models \varphi.
\end{array}
$$

A formula φ is square valid (notation $\models_{SQ} \varphi$) if φ is true in all pairs in all square models. A formula φ is locally square valid (notation $\models_{LC_2} \varphi$) if φ is true in all pairs in all locally square models.

The modal logic MLR_2 interpreted on the class of local squares will be denoted by LC_2. A graphical intuition for the $\otimes$ modality is given in figure 8.1, where $\otimes\varphi$ is true in $\langle a, b\rangle$ if and only if φ is true in $\langle b, a\rangle$.

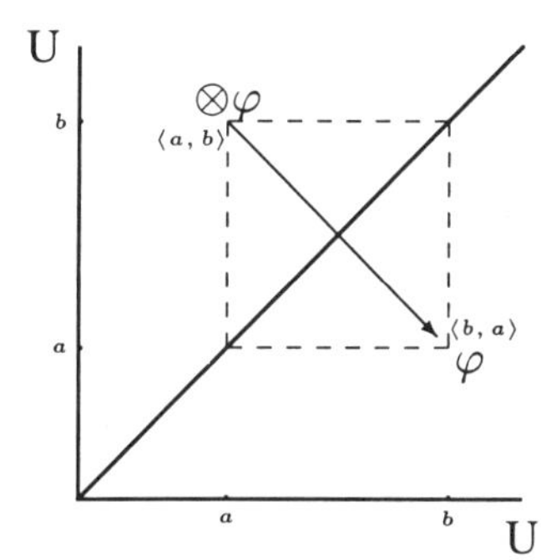

Figure 8.1 The semantics of a converse-formula $\otimes\varphi$

The semantics of the defined operators is an immediate result of definition 3.

$$
\begin{array}{lcl}
\mathcal{M}, \langle a, b\rangle \models \text{⦶}\,\varphi & \Leftrightarrow & \mathcal{M}, \langle a, a\rangle \models \varphi \\
\mathcal{M}, \langle a, b\rangle \models \text{◫}\varphi & \Leftrightarrow & \text{for all pairs } \langle a, x\rangle \in W \\
 & & \text{holds that } \mathcal{M}, \langle a, x\rangle \models \varphi.
\end{array}
$$

The semantics of $\wedge$, $\ominus$ and $\boxminus$ is defined in a similar way.

1.2 RELATION WITH FIRST ORDER LOGIC

The language MLR_2 interpreted on squares is just as expressive as L_2 — first order logic with two variables and equality and at most binary relation symbols (no constant and function symbols). In fact, the language MLR_n was precisely designed to be the modal counterpart of first order logic. For the precise translations back and forth, we refer to [15]. Here we just give the most interesting clauses in the two–dimensional case. For every binary relation symbol R, we introduce a propositional variable r.

$$(Rv_0v_1)^t = r \qquad (Rv_1v_0)^t = \otimes r \quad (Rv_0v_0)^t = \obar\, r \quad (Rv_1v_1)^t = \ominus\, r$$

$$(v_0 = v_1)^t = id \quad (\exists v_0\varphi)^t = \diamondminus\varphi^t \quad (\exists v_1\varphi)^t = \diamondbar\varphi^t.$$

The translation in the other direction is just the converse of $(\cdot)^t$, though we need to take care of $\otimes$. One might wonder how we translate unary predicates like Pv_0. There are several standard ways of handling these, all preserving satisfiability. One way is to transform the first order formula φ into one with only binary predicates by uniformly replacing occurrences of Pv_0 by $P'v_0v_0$ and of Pv_1 by $P'v_1v_1$. For details we refer to e.g., [6].

When we interpret MLR_2 on local squares we lose the one–to–one correspondence with L_2. This is immediate since the L_2 sentence $\exists v_0\exists v_1 Rv_0v_1 \leftrightarrow \exists v_1\exists v_0 Rv_0v_1$ is valid, while its translation $\diamondminus\diamondbar r \leftrightarrow \diamondbar\diamondminus r$ is not valid on local squares (a counter example is provided using our tableaux system in section 4.2).

So the question becomes whether there exists a natural fragment of first order logic which has a similar correspondence to MLR_2 interpreted on local squares as L_2 has to MLR_2 interpreted on squares. We find the answer by looking at the so called guarded fragments introduced in [2, 3, 17]. The guarded fragment is defined as follows:

Definition 4 *The* atomic formulas *are of the usual form:*

1. $v_1 = v_2$*, for variables or constants* v_1, v_2.
2. $Pv_1 \cdots v_n$*, for n-ary* P*, and variables or constants* $v_1, \ldots, v_n$.

Guarded formulas *are defined by induction as follows.*

1. *An atomic formula is a guarded formula.*
2. *If* φ, ψ *are guarded formulas, then* $\varphi \wedge \psi$ *and* $\neg\varphi$ *are guarded formulas.*
3. *If* $\bar{v}$ *is a finite non-empty sequence of variables,* ψ *is a guarded formula, and* G *is an atomic formula such that* $free(\psi) \subseteq free(G)$*, then* $\exists\bar{v}(G \wedge \psi)$ *is a guarded formula.*

The atomic formula G which appears in the above clause (3) is called the *guard*. The guarded fragment restricted to n variables and at most n-ary relation symbols is denoted by GF_n.

Obviously, we have to restrict the guarded fragment to two variables and at most binary relation symbols for our corresponding language. As we will show in Section 2, the full guarded fragment with two variables is still too strong. We define the weak guarded fragment with two variables (notation WGF_2) as the subset of GF_2–formulas in which

1. no constants occur,

2. no generalized quantifier of the form $\exists v_0 v_1$ occur, and

3. every guard is of the form Rv_0v_1 or Rv_1v_0 for an arbitrary relation symbol R.

Theorem 5 *There exist linear translations from WGF_2 to MLR_2 and vice–versa which preserve satisfiability when MLR_2 is interpreted on local squares.*

The important point of this theorem for our purposes here is that any decision algorithm we design for LC_2 will be a decision procedure for WGF_2 as well. As will be clear from the given translation, the computational costs involved are minimal.

Proof We first provide the translation from WGF_2 into MLR_2. We assume that all relation symbols occurring in WGF_2 formulas are binary (as noted above, we can replace the unary predicates by binary ones without loss of satisfiability). The translation is just the one given above. We claim that for every WGF_2 formula φ, φ is satisfiable iff φ^t is satisfiable on a local square. From left to right is immediate: If (D, I) satisfies φ, then the (square) model $(D \times D, v)$ where $v(r) = I(R)$ satisfies φ^t. For the other direction, let $\mathcal{M} = (W, v)$ satisfy φ at (a_0, a_1). Let $W \subseteq U \times U$. We claim that the first order model $\mathcal{M}'$ with domain U and the interpretation I defined by $v(r) = I(R)$ satisfies φ at the assignment which sends v_i to a_i. The only interesting case in the inductive proof is the one for the quantifier. Our induction hypothesis is that for all $(c_0, c_1) \in W, \mathcal{M}, (c_0, c_1) \models \varphi^t$ iff $\mathcal{M}' \models \varphi[v_i/c_i]$.

So let $\mathcal{M}' \models \exists v_0(Rv_0v_1 \wedge \varphi)[v_i/c_i]$. Then there exists a $c_2 \in U$ such that $\mathcal{M}' \models (Rv_0v_1 \wedge \varphi)[v_0/c_2, v_1/c_1]$. Thus $(c_2, c_1) \in I(R)$. Then $(c_2, c_1) \in v(r)$, so $(c_2, c_1) \in W$. But then the induction hypothesis applies and we obtain our desired result. The other direction is proved similarly.

The translation in the other direction is given as follows. For ease of exposition we assume that $\otimes$ is only applied to propositional variables (this restriction can easily be lifted).

$$\begin{aligned}
(p)^t &= Pv_0v_1 \\
(\otimes p)^t &= Pv_1v_0 \\
(id)^t &= (v_0 = v_1) \\
(\Phi\varphi)^t &= \exists v_1(Wv_0v_1 \wedge Wv_1v_0 \wedge Wv_1v_1 \wedge Wv_0v_0 \wedge \varphi^t) \\
(\ominus\varphi)^t &= \exists v_0(Wv_0v_1 \wedge Wv_1v_0 \wedge Wv_1v_1 \wedge Wv_0v_0 \wedge \varphi^t),
\end{aligned}$$

and $(\cdot)^t$ commutes with the booleans. Here W is a fixed new predicate symbol.

Clearly the translation goes to WGF_2. We claim that for every MLR_2 formula φ, φ is locally square satisfiable iff $\varphi^t \wedge Wv_0v_1 \wedge Wv_1v_0 \wedge Wv_1v_1 \wedge Wv_0v_0$ is satisfiable. The direction from left to right is obvious: interpret the new predicate symbol W as the domain of the locally square model. The other direction is easy too: take the local square model whose universe is the smallest local square containing the interpretation of the predicate W. ∎

1.3 SIMULATING OTHER MODAL LOGICS IN LC_2

One can use LC_2 to simulate other logics, among them the modal logics (multi-) K, KT, KB and KTB, and their tensed versions. Special properties of the accessibility relation $\mathcal{R}$ can be encoded into the translation of the diamond operator. The advantage is that we have a single proof–system which works for a wide range of logics. Using the proof–system for LC_2 is without increase in (theoretical) complexity since all other logics are PSPACE hard themselves.

Modal logic S5. Let $(\cdot)^*$ be the following translation from modal formulas to LC_2 formulas:

$$p^* = p, \ (\cdot)^* \text{ commutes with the booleans, } (\Diamond)^* = \Phi\varphi^*.$$

One easily shows that for all φ, φ is **S5** satisfiable iff φ^* is satisfiable on a local square.

Modal logics K, KT, KB and KTB and their tensed versions. Now we give the translations for other modal logics. Note how we just mimic the standard translation for modal logic **K**, but without the cumbersome details of variables.

Definition 6 *Let $(\cdot)^{t_K}$ be the following translation from K to LC_2:*

$$\begin{aligned}
(p)^{t_K} &= p && \textit{for propositional variables } p \\
(\neg\varphi)^{t_K} &= \neg(\varphi)^{t_K} \\
(\varphi \vee \psi)^{t_K} &= (\varphi)^{t_K} \vee (\psi)^{t_K} \\
(\Diamond\varphi)^{t_K} &= \Phi(r \wedge \ominus\varphi^{t_K})
\end{aligned}$$

where r is a new propositional variable, not occurring in the set of the propositional variables of K.

In order to simulate normal modal logics KT, KB or KTB where the accessibility relations have properties like reflexivity, symmetry or both reflexivity and symmetry, one only slightly has to change the translation of the diamond operator from definition 6. We use $\Diamond^{\smile}$ to denote the backwards looking (tense) modality having the same accessibility relation as $\Diamond$.

Definition 7 *Let $(\cdot)^{t_{K_t}}, (\cdot)^{t_{KB}}, (\cdot)^{t_{KT}}$ and $(\cdot)^{t_{KTB}}$ be modifications of the translation from K to LC_2, corresponding exactly to $(\cdot)^K$ for all but the diamond operator and with the following translations for K_t-, KT-, KB- and KTB-diamonds $\Diamond$:*

$$\begin{array}{lcl}
(\Diamond^{\smile}\varphi)^{t_{K_t}} & = & \Phi(\otimes r \wedge \ominus\, \varphi^{t_{K_t}}) \\
(\Diamond\varphi)^{t_{KT}} & = & \Phi((r \vee id) \wedge \ominus\, \varphi^{t_{KT}}) \\
(\Diamond\varphi)^{t_{KB}} & = & \Phi(r \wedge \otimes r \wedge \ominus\, \varphi^{t_{KB}}) \\
(\Diamond\varphi)^{t_{KTB}} & = & \Phi(\ominus\, \varphi^{t_{KTB}}).
\end{array}$$

The brackets $(\cdot)$ will be omitted whenever no confusion is likely to ensue. The translations preserve satisfiability as will be shown in the following theorem.

Theorem 8 *For all modal formulas φ holds: $\models_{K^*} \varphi \Leftrightarrow \models_{LC_2} id \rightarrow \varphi^{t_{K^*}}$, where K^* is any of the logics K, K_t, KT, KB or KTB.*

The idea of the proof is to construct a locally square model $\mathcal{M} = (W, v)$ satisfying $id \wedge \varphi$ from a given K^*-model $\mathcal{M}^{K^*} = (W_{K^*}, \mathcal{R}, v_{K^*})$ for φ and vice versa. The most interesting part of the proof of theorem 8 is the proof for the translation of modal formulas into LC_2, because of the explicit use of the local square property. The construction of the models in the other cases is similar.

Proof (KTB) ($\Rightarrow$) Given a locally square model $\mathcal{M}$, we construct a new KTB-model $\mathcal{M}^{KTB} = (W_{KTB}, \mathcal{R}, v_{KTB})$ where the universe $W_{KTB} = \{u \mid \langle u, u\rangle \in W\}$ is the set of all points on the diagonal in W, and with the following valuation $v_{KTB}(p) = \{u \mid \langle u, u\rangle \in v(p)\}$. The new accessibility relation $\mathcal{R}$ is defined by $\mathcal{R} = W$, which forces $\mathcal{R}$ to be reflexive and symmetric. It remains to show, that for this model and for all modal formulas φ and pairs $\langle a, a\rangle \in W_{LC_2}$: $\mathcal{M}, \langle a, a\rangle \models_{LC_2} \varphi^{t_{KTB}} \Leftrightarrow \mathcal{M}^{KTB}, a \models_{KTB} \varphi$, which is done by induction over the structure of the formula φ.

($\Leftarrow$) Given a KTB-model $\mathcal{M}^{KTB}$, we construct $\mathcal{M}' = (W', v')$, where the universe $W' = \mathcal{R}$, which is by the reflexivity and symmetry of $\mathcal{R}$ a local square. The valuation $v'(p) = \{\langle a, a\rangle \mid a \in v_{KTB}(p)\}$ maps all propositional variables p to the set of diagonal points of W' corresponding to the elements mapped by $v_{KTB}(p)$. For this model, we prove for all $a \in W_{KTB}$ and all modal formulas φ, that $\mathcal{M}^{KTB}, a \models_{KTB} \varphi \Leftrightarrow \mathcal{M}', \langle a, a\rangle \models \varphi^{t_{KTB}}$ by induction on the structure of the formula φ. ■

Description logics. If one uses different new propositional variables r_i it is possible to simulate multi-modal logic K^{multi} within the same translational framework. For each different diamond $\Diamond_i$ the translation function $(\cdot)^{t_{multiK}}$ is defined by:

$$(\Diamond_i \varphi)^{t_{multiK}} = \diamondminus(r_i \wedge \ominus \varphi^{t_{multiK}})$$

the intuition of which is shown in Figure 8.2 for two different diamond formulas $\Diamond_1\varphi$ and $\Diamond_2\psi$, that are multi K–satisfiable at a world $a \in W_K$ and the translated LC_2-formulas $(\Diamond_1\varphi)^{t_{multiK}}$ and $(\Diamond_2\psi)^{t_{multiK}}$ are LC_2-satisfiable in $\langle a, a\rangle \in W$. This shows that LC_2 is also well suited to perform proofs in the basic description logic $\mathcal{ALC}$. Obviously, even satisfiability proofs in the logic $\mathcal{ALCR}$ can be mimicked in LC_2. Conjunctions of roles are translated as follows:

$$(\exists(R_1 \sqcap R_2).C)^t = \diamondbar(r_1 \wedge r_2 \wedge \ominus C^t).$$

Note that LC_2 contains expressive power not available in $\mathcal{ALCR}$. For instance the concept *self–employed* is definable in LC_2 but not in $\mathcal{ALCR}$. The following first order definition (8.1) is translated to the LC_2 definition (8.2). Obviously both express the same concept.

$$\text{self–employed}(x, x) \equiv \text{employed–by}(x, x) \wedge \forall y(\text{employed–by}(x, y) \rightarrow x = y) \quad (8.1)$$

$$\text{self–employed} \equiv \obar\, \text{employed–by} \wedge \boxbar(\text{employed–by} \rightarrow id). \quad (8.2)$$

2 COMPLEXITY

[12] shows that the satisfiability problem of MLR_2 interpreted on local squares is PSPACE complete. As a corollary of the above translation we obtain that the WGF_2 fragment is in PSPACE as well. In fact it is PSPACE complete since the range of the standard translation of modal logic **K** is still inside that fragment. We restricted GF_2 in three different ways. We now show that each of them is necessary for this complexity bound. In other words, relaxing any one of these restrictions leads to an EXPTIME hard satisfaction problem. We shall reduce the EXPTIME-complete *global* **K**-satisfaction problem for unimodal languages. The global **K**-satisfaction problem for unimodal languages is this:

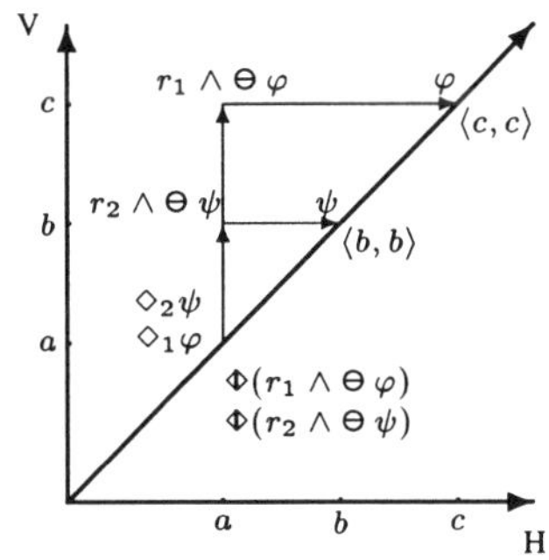

Figure 8.2 Simulation of multi-K with two diamonds $\diamond_1$ and $\diamond_2$ by LC_2.

given a formula φ in the unimodal language, does there exist a Kripke model $\mathcal{M}$ such that $\mathcal{M} \models \varphi$ (in other words, where φ is true in *all* worlds)? The EXPTIME-completeness of this problem is an easy consequence of (the proof of) the EXPTIME-completeness of modal logic **K** expanded with the universal modality in [16].

Adding constants. We just need to add one single constant c to obtain an EXPTIME hard satisfaction problem. For this we modify the standard translation as follows:

$$ST_x(\diamond\varphi) = \exists y(Rxy \wedge Rcy \wedge ST_y(\varphi)).$$

It is not hard to show that φ is globally **K** satisfiable iff $\exists y Rcy \wedge \forall y(Rcy \rightarrow ST_y(\varphi))$ is satisfiable. The idea of the proof is to add a new world to the Kripke model, name it c and connect all old worlds to c by R. This use of a world named by a constant as a "spy-point" originates with [5]. For a complete proof, see [4].

Allowing arbitrary guards. Let φ be a formula and W a fresh unary predicate (that is not corresponding to any proposition letter in φ). Modify the $\diamond$–clause in the standard translation as follows:

$$ST_x(\diamond\varphi) = \exists y(Rxy \wedge Wy \wedge ST_y(\varphi)).$$

Then φ is globally **K** satisfiable iff $\exists x Wx \wedge \forall x(Wx \rightarrow ST_x(\varphi))$ is satisfiable, as is easy to show.

Allowing generalized quantifiers. In this case use the following translation:

$$ST_x(\diamond\varphi) = \exists y(Rxy \wedge Wyy \wedge ST_y(\varphi)).$$

Then φ is globally **K** satisfiable iff $\exists xy Wxy \wedge \forall xy(Wxy \rightarrow x = y) \wedge \forall xy(Wxy \rightarrow ST_x(\varphi))$ is satisfiable. The proof is an obvious modification of the previous one.

3 DIMENSIONS HIGHER THAN TWO

If we consider MLR_n for n larger than 2, we obtain a modal analogue of first order logic restricted to n variables and n-ary relation symbols. The MLR_n language consists of n diamonds $\Diamond_i$, identity constants id_{ij} $(i, j < n)$, and substitution and permutation operators. The obvious extensions of the translations and modal classes apply, cf. [15]. The semantic analogue of the local squares is the class of local cubes, satisfying

$$(s_0, s_1, \ldots, s_{n-1}) \in W \Rightarrow {}^n\{s_0, s_1, \ldots, s_{n-1}\} \subseteq W,$$

i.e., the domains of the models are closed under all permutations and substitutions. MLR_n for n at least 3 has an undecidable satisfaction problem when interpreted on the class of cubic models with universes of the form ${}^nU$. This follows simply from similar results in first order logic (see [11] for a strong result). When we interpret MLR_n on local cubes, the satisfaction problem becomes EXPTIME complete [12]. In fact, an expansion of MLR_n can be used to decide the guarded fragment, as is shown in [12].

Not only does the complexity increase when we go from two to higher dimensions, it is also much harder to prove the finite model property (fmp). In two dimensions, the fmp is an immediate result of the termination of the tableaux system presented below. The obtained model has several nice "tree–like" properties comparable to the models one obtains by a standard tableaux system for modal **K**. For higher dimensions this tree–like property is lost. Stated simply, the situation is comparable with **S4** where one also has to re–use worlds to obtain finite models. In fact the proof for fmp requires quite heavy machinery from finite model theory [1, 10]. This fact has its repercussions when one wants to design a terminating tableaux system for MLR_n or the guarded fragment. The following example indicates that a naive tableaux system which builds a model using fresh constants all the time (thereby making as little connections as possible) cannot be guaranteed to terminate.

Example. We give an example for 3 dimensions; straightforward modifications in the argument below yield the same result for any higher dimension.

We define the formula φ as

$$id \wedge p\,;(p\,;\otimes q) \wedge \neg(q\,;\neg(p\,;\otimes q))$$

where $x\,;y$ is the abbreviation for $\Diamond_2(\Diamond_1(x \wedge id_{12}) \wedge \Diamond_0(y \wedge id_{02}))$ and id stands for $id_{01} \wedge id_{12}$. $\otimes$ represents the permutation operator which swaps the first two coordinates of a triple.

Now we will construct a model $\mathfrak{M} = (W, v)$ for φ by a step-by-step procedure that uses fresh points at every step when we need witnesses for $\Diamond_i\psi$ formulas.

Our goal is to define $\mathfrak{M} = (W, v)$ such that $s \models \varphi$ (with some $s \in W$). Clearly, we can assume that $W \subseteq {}^3\omega$, and we will denote the elements of W by natural numbers. We start with a partial model $s \in {}^3\omega$, and require that φ must hold at s. To achieve this, we have to extend our partial model.

First note that $s_0 = s_1 = s_2$ by $s \models id$. We denote s_0 by 0. By $(0,0,0) \models p\,;(p\,;\otimes q)$, we need a fresh point 1 such that $(0,1,1) \models p$ and $(1,0,1) \models p\,;\otimes q$. Now $(1,0,1) \models p\,;\otimes q$ implies that there is another fresh point 2 such that $(1,2,2) \models p$ and $(2,0,2) \models \otimes q$. Thus $(0,2,2) \models q$. Recall that $(0,0,0) \models \neg(q\,;\neg(p\,;\otimes q))$. Hence $(2,0,2) \models p\,;\otimes q$. So there exists a new point 3 such that $(2,3,3) \models p$ and $(0,3,3) \models q$.

Now let us assume that we are in the kth step of the step-by-step construction and that we already defined a partial model $\mathfrak{M}_k = (W_k, v_k)$ such that W_k is the local cube generated by the set $\{(0,i,i), (i,i+1,i+1), (i,0,i+1) : i \leq k\}$ and $v_k(p) = \{(i,i+1,i+1) : i \leq k\}$ and $v_k(q) = \{(0,i+2,i+2) : i < k\}$.

Observe that $(0,k+1,k+1) \models q$. Since we want $(0,0,0) \models \neg(q\,;\neg(p\,;\otimes q))$, we see that $(k+1,0,k+1) \models p\,;\otimes q$ must hold. Using the definition of $;$ and our strategy that we take a fresh point whenever we need a witness for a $\Diamond_i$-formula, we have to extend W_k by $k+2$ such that $(k+1,k+2,k+2) \models p$ and $(k+2,0,k+2) \models \otimes q$. Thus we arrive at a partial model $\mathfrak{M}_{k+1} = (W_{k+1}, v_{k+1})$ such that W_{k+1} is the local cube generated by the set $\{(0,i,i), (i,i+1,i+1), (i,0,i+1) : i \leq k+1\}$ and $v_k(p) = \{(i,i+1,i+1) : i \leq k+1\}$ and $v_k(q) = \{(0,i+2,i+2) : i < k+1\}$.

Continuing the same way, we arrive at the structure $\mathfrak{M} = (W, v)$, where W is the local cube generated by $\{(0,i,i), (i,i+1,i+1), (i,0,i+1) : i < \omega\}$ and $v(p) = \{(i,i+1,i+1) : i < \omega\}$ and $v(q) = \{(0,i+2,i+2) : i < \omega\}$. It should be clear from the above argument that s satisfies φ in $\mathfrak{M}$.

We note that it is rather easy to find a finite model satisfying φ. Indeed, the model with the singleton universe $\{(0,0,0)\}$ and the valuation $v(p) = v(q) = \{(0,0,0)\}$ satisfies φ.

Our point with the above example was to show that the straightforward step-by-step (or tableau) construction does not yield finite models, and that it is not clear how to modify the model-construction so that it yields a finite model for *every* formula.

In the case of two dimensions, matters are considerably simpler. In the next section we present a simple tableaux system which does terminate.

4 LC_2-TAB, A DECISION PROCEDURE FOR LC_2

The tableau calculus LC_2-TAB follows roughly the semantics given in definition 3. We will use a labelled tableau calculus in which each formula is labelled by a pair (the actual "world") and a set of pairs (the actual universe).

In order to show for example that a branch of a tableau containing the formula $\otimes\varphi$ labelled with $\langle a, b\rangle, V$ closes, it is enough to show, that a branch containing φ, labelled with $\langle b, a\rangle, V$ closes. The tableau rule for converse formulas thus only adds $\varphi, \langle b, a\rangle, V$ to the actual branch and ticks $\otimes\varphi, \langle a, b\rangle, V$ (meaning that no more rule is to be applied on this formula). This closely corresponds to the `converse rule` in our PROLOG implementation:

```
prove(TickedFormulas,Dom,[lf(⊗φ,X,Y)|Formulas],Res) :-
   prove([(⊗φ,X,Y)|TickedFormulas],Dom,[(φ,Y,X)|Formulas],Res).
```

The predicate `prove([],[(0,0)],[lf(¬⊖φ,0,0)],Result)` succeeds with `Result = theorem` if and only if contradictions in all possible branches are detected.

4.1 THE TABLEAUX SYSTEM LC_2-TAB

The idea of LC_2-TAB is to start with the negated formula, theoremhood of which has to be proven, and to label it with an arbitrary point (which we will call $\langle 0, 0\rangle$)[1] on the diagonal and the smallest universe containing this point. This universe $V = {}^2\{0\}$ is a local square. The proof consists in explicitly constructing a tableau with an open branch corresponding to a counter model for the original formula. This means that the formula φ is a theorem if and only if there is a closed tableau for $\ominus\neg\varphi$, labelled with $\langle 0, 0\rangle$.

To be able to construct a counter model and to keep track of all existing pairs in a branch, we also label the formulas in the calculus with a set of pairs V, to construct the universe for a possible model for φ. New pairs are created if and only if formulas of the form $\ominus\varphi$ or $\obar\varphi$ occur, where a witness pair $\langle c, b\rangle$ for the labelled formula $\ominus\varphi, \langle a, b\rangle, V$ has to be given, such that φ holds in $\langle c, b\rangle$ where $\langle c, b\rangle \notin V$.

If a contradiction occurs, one of the closure rules put $\bot$ in the branch. ($\checkmark$) means, that a formula (or a set of formulas) is marked *after* application of the rule and cannot be used any more. We follow Fitting [7] in his uniform notation for conjunctions and disjunctions. Fitting groups all propositional formulas of the form $X \circ Y$ and $\neg(X \circ Y)$ into two categories, those who act conjunctively, which are called α formulas and those who act disjunctively and are called β formulas. He then defines components $\alpha_1, \alpha_2, \beta_1$ and β_2 such that $\alpha \equiv \alpha_1 \wedge \alpha_2$ and $\beta \equiv \beta_1 \vee \beta_2$. The advantage of this approach is, that you only need two tableau rules for all propositional connectives: $\vee, \wedge, \rightarrow, \leftrightarrow$ and their negations.

[1] We use the natural numbers as base for the universe which is constructed in the step-by-step approach of the calculus LC_2-TAB, because they reflect the geometrical nature of the construction in a very natural way. Due to the structure of our rules, we do only need one variable over $\mathbb{N}$, which we will call x.

All LC_2-TAB rules are so called extension rules, which are displayed as:

$$
\begin{array}{c|c|c}
\multicolumn{3}{c}{\xi_1} \\
\multicolumn{3}{c}{\vdots} \\
\multicolumn{3}{c}{\xi_k} \\
\hline
\zeta_{11} & \cdots & \zeta_{n1} \\
\vdots & \vdots & \vdots \\
\zeta_{1m_1} & & \zeta_{nm_n}
\end{array}
\qquad \text{or} \qquad
\begin{array}{c|c}
\multicolumn{2}{c}{B} \\
\multicolumn{2}{c}{\xi} \\
\hline
B_1 & B_2 \\
 & \\
\zeta_1 & \zeta_2
\end{array}
$$

The left-hand side means that if there are labelled formulas $\xi_1, \ldots, \xi_k$ in the branch, then the branch will be split into n sub-branches and all labelled formulas $\zeta_{i1}, \ldots \zeta_{im_i}$ will be added to the i–th branch. n and the m_i may be 1. Since we identify the branch B with the set of labelled formulas occurring in B, the meaning of a rule of the right-hand type is, that a new set B_1 (B_2) of labelled formulas is added to the first (second) branch. These B_1 and B_2 are usually defined from B in one of the conditions of the rules. Most of the rules come with additional restrictions for their application.

To avoid cumbersome notation we use the following convention for application of the simple rules. A rule of the form $B, \varphi / B, \varphi', \varphi(\checkmark)$ will be abbreviated by $\varphi(\checkmark)/\varphi'$. The only rules which are not of this form are the diamond rules and the label rules.

Definition 9 *A formula φ is an LC_2-**theorem** (denoted by $\vdash \varphi$), if and only if there is tableau proof for it, where a **tableau proof** for a formula φ in the calculus LC_2-TAB is a closed tableau starting with the labelled formula $\ominus(\neg\varphi), \langle 0, 0\rangle, {}^2\{0\}$, using the tableau rules stated below. Below $\mathbb{N}$ denotes the natural numbers and x a variable.*

- Propositional rules.

$(\neg)$	(α)	(β)
$\dfrac{\neg\neg Z, \langle a, b\rangle, V(\checkmark)}{Z, \langle a, b\rangle, V}$	$\dfrac{\alpha, \langle a, b\rangle, V(\checkmark)}{\begin{array}{c}\alpha_1, \langle a, b\rangle, V \\ \alpha_2, \langle a, b\rangle, V\end{array}}$	$\dfrac{\beta, \langle a, b\rangle, V(\checkmark)}{\beta_1, \langle a, b\rangle, V \mid \beta_2, \langle a, b\rangle, V}$
where $a, b \in \mathbb{N} \cup \{x\}$	$a, b \in \mathbb{N} \cup \{x\}$	$a, b \in \mathbb{N}$

- Closing rules

(id_a)	(id_b)	(con)
$\dfrac{\neg id, \langle a, a\rangle, V(\checkmark)}{\bot}$	$\dfrac{id, \langle a, b\rangle, V(\checkmark)}{\bot}$	$\dfrac{\begin{array}{c}\neg\varphi, \langle a, b\rangle, V(\checkmark) \\ \varphi, \langle a, b\rangle, V(\checkmark)\end{array}}{\bot}$
where $a \in \mathbb{N}$	$a, b \in \mathbb{N}$ and $a \neq b$	$a, b \in \mathbb{N}$

- Transition rules

$(\otimes)\ \dfrac{\otimes\varphi, \langle a,b\rangle, V(\checkmark)}{\varphi, \langle b,a\rangle, V}$ where $a, b \in \{x\} \cup \mathbb{N}$	$(\neg\otimes)\ \dfrac{\neg\otimes\varphi, \langle a,b\rangle, V(\checkmark)}{\neg\varphi, \langle b,a\rangle, V}$ where $a, b \in \{x\} \cup \mathbb{N}$
$(\obar)\ \dfrac{\obar\,\varphi, \langle a,b\rangle, V(\checkmark)}{\varphi, \langle a,a\rangle, V}$ where $b \in \{x\} \cup \mathbb{N}, a \in \mathbb{N}$	$(\neg\obar)\ \dfrac{\neg\obar\,\varphi, \langle a,b\rangle, V(\checkmark)}{\neg\varphi, \langle a,a\rangle, V}$ where $b \in \{x\} \cup \mathbb{N}, a \in \mathbb{N}$
$(\Diamond\!\!\!\mid)\ \dfrac{B(\checkmark) \quad \Diamond\!\!\!\mid\varphi, \langle a,b\rangle, V(\checkmark)}{B,\ \varphi, \langle a,a\rangle, V \ \mid\ B',\ \varphi, \langle a,c\rangle, V'}$	$(\boxbar)\ \dfrac{\boxbar\varphi, \langle a,b\rangle, V(\checkmark)}{\varphi, \langle a,x\rangle, V}$ where $a \in \mathbb{N},\ b \in \mathbb{N} \cup \{x\}$

The $(\Diamond\!\!\!\mid)$-rule can only be applied under the conditions that:

- B is the set of all labelled formulas occurring in the branch,
- $a \in \mathbb{N},\ b \in \mathbb{N} \cup \{x\}$, c is a **new** constant, i.e. $\langle c, c\rangle \notin V$,
- $V' = V \cup \{\langle c,a\rangle, \langle a,c\rangle, \langle c,c\rangle\}$,
- there is no pair $\langle a,d\rangle \in V$, such that $\varphi, \langle a,d\rangle, V \in B$,
- $B' = \{\psi, \langle x,y\rangle, V' \mid \psi, \langle x,y\rangle, V \in B\} \cup \{\Diamond\!\!\!\mid\varphi, \langle a,b\rangle, V'(\checkmark)\}$.

By definition of $\boxbar$ (and $\boxminus$), the $(\boxbar)$-rule (and $(\boxminus)$-rule) correspond to the negation of the diamond rules $(\Diamond\!\!\!\mid)$-rule (and $(\Diamond\!\!\!-)$-rule). There are corresponding rules for the horizontal $\Diamond\!\!\!-$, $\boxminus$ and $\ominus$ modalities.

- Label rules (lr_1) and (lr_2)

$(lr_1)\ \dfrac{B \quad \varphi, \langle x,b\rangle, V}{B \quad \varphi, \langle a,b\rangle, V}$	$(lr_2)\ \dfrac{B \quad \varphi, \langle a,x\rangle, V}{B \quad \varphi, \langle a,b\rangle, V}$

where $a, b \in \mathbb{N},\ \langle a,b\rangle \in V$ and $\varphi, \langle a,b\rangle, V \notin B$.

The rules (lr_1) and (lr_2) are the only rules in the tableau, where the original formulas $\varphi, \langle x,b\rangle, V$ and $\varphi, \langle a,x\rangle, V$ cannot be ticked off, i.e. they can be applied as long as there are pairs in V, that have not been used yet.

It is worth mentioning, that the variable x in a branch is independent of all other occurrences in any other branch. This is by definition of the branching rules $(\beta), (\ominus)$ and (Φ). The definition of the diamond-rules blows the tableau up in a way, that all labelled formulas of a branch are copied into every new branch and no labelled formulas are deleted. But in practice (that is, in the PROLOG implementation), the formulas of the old branch are replaced by the updated labelled formulas. Even more, since the universes of all newly created labelled formulas are all the same, the domain of all these formulas is represented as only one global set of pairs.

4.2 A DOCUMENTED SAMPLE PROOF

We will construct (the open branch of) a proof in LC_2-TAB to illustrate the use of the rules. We prove, that the formula $\Phi\ominus\varphi \rightarrow \ominus\Phi\varphi$ is not a theorem of LC_2 by showing, that there is an open branch for the tableau, starting with the negation of the formula. For reasons of readability, we only do the proof for a point on the diagonal at $\langle 0, 0\rangle$, which is sufficient to find a counter model.

We present a complete calculation until we reach a diamond formula, which forces us to branch and update the universe of the formulas in the right-hand branch. The numbers and letters in parenthesis are not part of the calculus but are needed to make it easier to talk about the example. $(\checkmark_{(i)})$ means, that a formula is ticked after application of a rule creating the new formula (i). In addition, we give a brief explanation, why we created a new formula, on the right hand side of the tableau.

(1)	$\neg(\Phi\ominus\varphi \rightarrow \ominus\Phi\varphi), \langle 0,0\rangle, {}^2\{0\}(\checkmark_{(2)(3)})$		Negated input
(2)	$\Phi\ominus\varphi, \langle 0,0\rangle, {}^2\{0\}(\checkmark_{(5a)(5b)})$		(α) on (1)
(3)	$\neg\ominus\Phi\varphi, \langle 0,0\rangle, {}^2\{0\}(\checkmark_{(4)})$		(α) on (1)
(4)	$\neg\Phi\varphi, \langle x,0\rangle, {}^2\{0\}(\checkmark_{(5a)(5b)})$		$(\boxminus)$ on (3)
	$(5a)\quad \ominus\varphi, \langle 0,0\rangle, {}^2\{0\}$	$(5b)\quad \ominus\varphi, \langle 0,1\rangle, {}^2\{0,1\}$	(Φ) on (2)
	$(1a)$	$(1b)$	copies of (1)
	⋮	⋮	

In the left branch (a), exact copies of the labelled formulas (1) to (4) are made. In the right-hand branch (b), the universes of (1b) to (4b) have been updated according to the definition of the (Φ)-rule. The new important formulas of the branches are (1a) to (5a) and (1b) to (5b). We only proceed with the right-hand branch (b).[2] After branching on (5b) by applying the Φ–rule, the left branch (c) closes with a contradiction on $\langle 1, 1\rangle$. But the second branch can be fully expanded, until no further rule can be applied.

[2]Both sub-branches of (a) close by contradiction on pairs of the row $\langle x, 0\rangle$.

$(1d)$	$\neg(\Phi\ominus\varphi \to \ominus\Phi\varphi), \langle 0,0\rangle, {}^2\{0,1\} \cup {}^2\{1,2\}(\checkmark)$	copy of $(1b)$
$\vdots$	$\vdots$	$\vdots$
$(6d)$	$\varphi, \langle 2,1\rangle, {}^2\{0,1\} \cup {}^2\{1,2\}$	
$(7d)$	$\neg\Phi\varphi, \langle 0,0\rangle, {}^2\{0,1\} \cup {}^2\{1,2\}(\checkmark_{(8d)})$	(lr_1) on $(4d)$
$(8d)$	$\neg\varphi, \langle 0,x\rangle, {}^2\{0,1\} \cup {}^2\{1,2\}$	$(\boxbar)$ on $(7d)$
$(9d)$	$\neg\varphi, \langle 0,0\rangle, {}^2\{0,1\} \cup {}^2\{1,2\}$	(lr_2) on $(8d)$
$(10d)$	$\neg\varphi, \langle 0,1\rangle, {}^2\{0,1\} \cup {}^2\{1,2\}$	(lr_2) on $(8d)$
$(11d)$	$\neg\Phi\varphi, \langle 1,0\rangle, {}^2\{0,1\} \cup {}^2\{1,2\}(\checkmark_{(12d)})$	(lr_1) on $(4d)$
$(12d)$	$\neg\varphi, \langle 1,x\rangle, {}^2\{0,1\} \cup {}^2\{1,2\}$	$(\boxbar)$ on $(11d)$
$(13d)$	$\neg\varphi, \langle 1,0\rangle, {}^2\{0,1\} \cup {}^2\{1,2\}$	(lr_2) on $(12d)$
$(14d)$	$\neg\varphi, \langle 1,1\rangle, {}^2\{0,1\} \cup {}^2\{1,2\}$	(lr_2) on $(12d)$

At this stage there are no more rules applicable, so the branch is finished. This proves that there is a model, that satisfies $\neg(\Phi\ominus\varphi \to \ominus\Phi\varphi)$ such that $\Phi\ominus\varphi \to \ominus\Phi\varphi$ is not a theorem of LC_2. The model $\mathcal{M} = ({}^2\{0,1\} \cup {}^2\{1,2\}, v)$, where $v(\varphi) = \{\langle 2,1\rangle\}$, is the counter model, which has been constructed in the proof and which is shown in figure 8.3. As is clear from the picture we can create this counterexample precisely because we are not forced to include the pair $(2,0)$ in the universe (as we would in the square case). The reader interested in implementations might want to compare the proof and the counter example given here with the one generated by our PROLOG program.

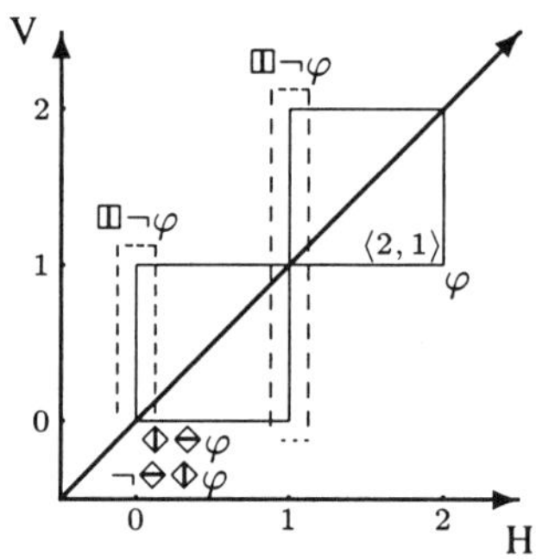

Figure 8.3 A counter model

4.3 SOUNDNESS, COMPLETENESS AND TERMINATION

LC_2-TAB is sound and complete with respect to the logic LC_2 and terminates. The idea behind the soundness proof is that a formula φ is not valid if and only if $\neg\ominus\varphi$ is satisfiable at an arbitrary pair $\langle 0,0\rangle$ on the diagonal. It is easily verified that all rules preserve satisfiability. For completeness, we construct a model for a formula from an open branch of the tableau, where we use the set V of pairs, which we construct during application of the rules. The completeness proof is a straightforward induction on the formulas occurring in the labels. Full proofs are available in [14]. Termination guarantees that the procedure builds a model for every satisfiable formula. It is the only property which is a bit harder to show.

Termination. We now show that, for any formula ξ, the tableau construction terminates (either with a closed or with an open tableau), and that an effective bound on the size of the counter model can be given.

First we note that the only danger for non-termination is that we create an infinite counter model because we are forced to apply the diamond rules again and again (as a repercussion of the box and label rules).

Second, note that, by condition of the diamond-rules, whenever we extend the current universe V we take some $\langle a, a\rangle \in V$ and some b such that $\langle b, b\rangle \notin V$ and form the new universe:

$$V' = V \cup {}^2\{a, b\}.$$

Thus, at any node of the tableau, the current universe is a local square with a tree-like form in the following sense. There is a reflexive tree $T = (I, \leq_T)$ for some finite subset I of the natural numbers such that the root is 0 and

$$V = \bigcup_{i \in I} \{{}^2\{i, j\} : i \leq_T j\}.$$

We define the *distance* $d(i)$ of a node i from the root 0 in T in the standard way: $d(0) = 0$ and $d(i) = d(j) + 1$ for $j \leq_T i$. Let $\langle x, y\rangle \in V$. We define

$$\delta(x, y) = \begin{cases} 2 \cdot \min\{d(x), d(y)\} + 1 & \text{if } x \neq y \\ 2 \cdot d(x) & \text{if } x = y. \end{cases}$$

That is, the *distance* of $\langle x, y\rangle$ from the root $\langle 0, 0\rangle$ is the number of horizontal and vertical steps in the path leading to $\langle x, y\rangle$ from $\langle 0, 0\rangle$. Using this definition we will show that labelled formulas become simpler and simpler during the tableau construction.

Let ξ be a fixed formula. Below by a tableau we mean a tableau starting with $\ominus\neg\xi, \langle 0, 0\rangle, {}^2\{0\}$. For any formula φ, let $m(\varphi)$ denote the modal depth of φ. Let $c(m)$ be the number of diamond subformulas of $\ominus\xi$ of modal depth at most m. Note that any formula occurring in the tableau is a (perhaps negated) subformula of $\xi \wedge id$.

Lemma 10 *Let $\varphi, \langle c, d\rangle, V'$ be a ground labelled formula such that it occurs in a tableau. Then*

$$m(\varphi) + \delta(c, d) \leq m(\ominus\xi).$$

Proof The key observation is that if we have a tableau proof for $\varphi, \langle c, d\rangle, V'$ and $\langle c, d\rangle \neq \langle 0, 0\rangle$, then we can find a tableau proof for $\varphi, \langle c, d\rangle, V'$ such that there is a ground labelled formula $\psi, \langle a, b\rangle, V$ satisfying

- $\psi, \langle a,b\rangle, V$ precedes $\varphi, \langle c,d\rangle, V'$ in the branch where $\varphi, \langle c,d\rangle, V'$ occurs and
- $m(\psi) + \delta(a,b) \geq m(\varphi) + \delta(c,d)$.

We show the lemma by induction on the complexity of the proof for $\varphi, \langle c,d\rangle, V'$. The base case is when $\langle c,d\rangle = \langle 0,0\rangle$ is obvious.

The induction hypothesis says that the lemma holds for any $\psi, \langle a,b\rangle, V$ such that $\psi, \langle a,b\rangle, V$ precedes $\varphi, \langle c,d\rangle, V'$ in a tableau.

First assume that $\varphi, \langle c,d\rangle, V'$ is the result of an application of a rule on a ground formula $\psi, \langle a,b\rangle, V$. It is easy to check that every such rule has the property that $m(\psi) + \delta(a,b) \geq m(\varphi) + \delta(c,d)$. Then, by the induction hypothesis, we have $m(\varphi) + \delta(c,d) \leq m(\diamondminus\xi)$.

Next assume that $\varphi, \langle c,d\rangle, V'$ is the result of applying a non-label rule ρ to a non-ground formula $\psi, \langle a,b\rangle, V$. Then consider the following modified tableau proof for $\varphi, \langle c,d\rangle, V'$: first apply the label rule to $\psi, \langle a,b\rangle, V$ resulting in a ground formula $\psi, \langle a',b'\rangle, V$ and then apply ρ to $\psi, \langle a',b'\rangle, V$ obtaining $\varphi, \langle c,d\rangle, V'$. By the induction hypothesis, $m(\psi)+\delta(a',b') \leq m(\diamondminus\xi)$, whence the lemma follows as in the previous paragraph.

Finally assume that $\varphi, \langle c,d\rangle, V'$ is the result of applying a label rule to a non-ground formula $\psi, \langle a,b\rangle, V$. Say, we applied the (lr_2)-rule to $\psi, \langle a,x\rangle, V$ resulting in $\psi, \langle a,b\rangle, V$. The reason why $\psi, \langle a,x\rangle, V$ occurs in the tableau is that, in a previous step, a box rule has been applied: for instance, the $(\boxbar)$-rule has been applied to the ground formula $\boxbar\psi, \langle a,b'\rangle, V$. By the induction hypothesis, $m(\boxbar\psi) + \delta(a,b') \leq m(\diamondminus\xi)$. Since $\delta(a,b) \leq \delta(a,b') + 1$ and $m(\psi) = m(\boxbar\psi) - 1$, the lemma easily follows. The other cases, when after the box rule some propositional or converse rules have been applied before the label rule, can be treated similarly. ■

By the lemma above, every point $\langle c,d\rangle$ has at most $c(m(\diamondminus\xi) - \delta(c,d))$ many horizontal or vertical successors which are created to witness the diamond formulas labelled with $\langle c,d\rangle$. Indeed, if $\diamondminus\psi, \langle c,d\rangle, V$ occurs in the tableau and we once created a witness $\psi, \langle a,d\rangle, V'$ by an application of the $(\diamondminus)$-rule, then we do not create more witnesses of the form $\psi, \langle b,d\rangle, V''$. In fact, the fourth condition of the diamond rule forbids creating unnecessary witnesses. Hence we can find a bound on the size of the universe of any counter model created: the size of the tree with depth $m(\diamondminus\xi)$ and branching factor $c(m(\diamondminus\xi) - k)$ on level k provides an upper bound for $\{a : \langle a,a\rangle \in V\}$ (V the universe of the created model).

The square case. The tree–property with *polynomially* bounded depth and branching shown in the termination proof provides for simple inference algorithms. This tree property is lost when we consider square models as is clear from the lower bound proofs in [13]. Here we briefly sketch how the polynomial depth of the trees is lost when we consider square models. First we realise that on square models the formula $\boxbar\boxminus\varphi$ ensures that φ is true *everywhere* in the model. In other words $\boxbar\boxminus$ works as a universal modality. So we can interpret —using the reductions from Section 1.3— modal logic **K** expanded with the universal modality in LC_2 interpreted on square models. In [9] (Proposition 6.18) a formula in that language is given which, when satisfied, forces an *exponential* deep R–path. The intuition behind the formula is easily explained. We take n propositional variables to count in binary from 0 to 2^n. We ensure that at the point of evaluation all of them are false, thus coding the number 0, and make sure that always (here the universal modality is needed) when we make a transition we increase the counter by one. The formula itself is only quadratic in n.

5 IMPLEMENTATION

The system `lc2` is a tableau theorem prover for the logic LC_2. The implementation method applied in `lc2` is a simple subgoaling method. This means that we prove satisfiability of a (labelled) formula φ by proving satisfiability for one or more formulas that have been created according to the rules of LC_2-TAB. The core of `lc2` is the Prolog program `lc2.pl` using a depth-first search method to look for an open branch in the tableau, which is constructed by application of all possible rules on the set of actual formulas. If a contradiction is found in the branch, the search stops on it and is continued on a new branch. If a branch is fully evaluated, i.e. that no rule can be applied, and no contradiction has been detected, the search stops and a counter model has been found. With this method no backtracking is necessary.

The easiest way to use `lc2` is to fill out the form on the WWW page of LC_2-TAB. The contents of the form are sent to a central server, which activates the corresponding interface program and then `lc2` itself. The answer is immediately sent back to the user. The user can ask for satisfiability of LC_2 formulas but also for satisfiability of formulas in any of the modal logics discussed in Section 1.3 and in the weak guarded fragment with two variables. In these cases they are automatically translated. In case the formula is satisfiable, the system provides a model in the respective class of frames.

The URL of the document is

`http://www.dcs.kcl.ac.uk/cgi-bin/LC2/lc2.html.`

More information and examples can be found following the links on that page.

Further research

The most pressing research question is whether a simple tableaux system for MLR_n interpreted on local cubes can be designed for any n. As indicated here it is not possible to simply extend the system LC_2-TAB if termination is required. It seems feasible to us to turn the mosaic deletion algorithm from [12] into a tableaux system, but we do not dare guess about its user–friendliness or even its behavior when implemented.

Acknowledgments

This research has been supported by UK EPSRC grant No.GR/K54946 and GR/L82441. Maarten Marx has also been supported by the Spinoza project "Logic in Action", Institute for Logic, Language and Computation, Amsterdam.

Maarten Marx
Department of Artificial Intelligence
Faculty of Sciences
Free University
Amsterdam, The Netherlands
marx@cs.vu.nl

Szabolcs Mikulás
Department of Computer Science
King's College
Strand
London WC2R 2LS
United Kingdom
szabolcs@dcs.kcl.ac.uk

Stefan Schlobach
Department of Computer Science
King's College
Strand
London WC2R 2LS
United Kingdom
schlobac@dcs.kcl.ac.uk

References

[1] H. Andréka, I. Hodkinson, and I. Németi. Finite algebras of relations are representable on finite sets. *Journal of Symbolic Logic*, 64(1):243–267, 1999.

[2] H. Andréka, J. van Benthem, and I. Németi. Back and forth between modal logic and classical logic. *Bulletin of the Interest Group in Pure and Applied Logics*, 3(3):685–720, 1995.

[3] H. Andréka, J. van Benthem, and I. Németi. Modal languages and bounded fragments of predicate logic. *Journal of Philosophical Logic*, 27(3):217–274, 1998.

[4] C. Areces, P. Blackburn, and M. Marx. Hybrid logics. Characterization, interpolation and complexity. Technical report, Institute for Logic, Language and Computation, University of Amsterdam, 1999.

[5] P. Blackburn and J. Seligman. Hybrid languages. *Journal of Logic, Language and Information*, 4(3):251–272, 1995.

[6] E. Börger, E. Grädel, and Y. Gurevich. *The Classical Decision Problem*. Springer Verlag, 1997.

[7] M. Fitting. *First Order Logic and Automated Theorem Proving*. Springer Verlag, 1990.

[8] D.M. Gabbay. Expressive functional completeness in tense logic. In U. Mönnich, editor, *Aspects of Philosophical Logic*, pages 91–117. D. Reidel Publishing Company, 1981.

[9] J.Y. Halpern and Y.O. Moses. A guide to completeness and complexity for modal logics of knowledge and belief. *Artificial Intelligence*, 54:319–379, 1992.

[10] R. Hirsch, I. Hodkinson, M. Marx, Sz Mikulás, and M. Reynolds. Mosaics and step–by–step. Remarks on "A modal logic of relations". In E. Orlowska, editor, *Logic at Work. Essays Dedicated to the Memory of Elena Rasiowa*, Studies in Fuzziness and Soft Computing, pages 158–167. Springer Verlag, Berlin/Heidelberg, 1999.

[11] R. Maddux. The equational theory of CA_3 is undecidable. *Journal of Symbolic Logic*, 45(2):311–316, 1980.

[12] M. Marx. Complexity of modal logics of relations. Technical Report ML–97–02, Institute for Logic, Language and Computation, University of Amsterdam, May 1997.

[13] M. Marx. Complexity of products of modal logics. *Journal of Logic and Computation*, 9(2):221–238, 1999. See http://www.oup.co.uk/logcom/hdb/Volume_09/Issue_02/.

[14] M. Marx, S. Mikulás, and S. Schlobach. A tableau calculus for a local cubic modal logic and it's implementation. Technical report, King's College, Department of Computer Science, London., 1999.

[15] M. Marx and Y. Venema. *Multi-dimensional Modal Logic*. Applied Logic Series. Kluwer Academic Publishers, 1997.

[16] E. Spaan. *Complexity of modal logics*. PhD thesis, University of Amsterdam, Institute for Logic, Language and Computation, 1993.

[17] J. van Benthem. *Exploring Logical Dynamics*. Studies in Logic, Language and Information. CSLI Publications, Stanford, 1996.

SEMANTICS FOR TEMPORAL ANNOTATED CONSTRAINT LOGIC PROGRAMMING

Alessandra Raffaetà and Thom Frühwirth

Abstract We investigate the semantics of a considerable subset of Temporal Annotated Constraint Logic Programming (TACLP), a class of languages that allows us to reason about qualitative and quantitative, definite and indefinite temporal information using time points and time periods as labels for atoms.

After illustrating the power of TACLP with some non-trivial examples, TACLP is given two different kinds of semantics, an operational one based on meta-logic (top-down semantics) and, for the first time, a fixpoint one based on an immediate consequence operator (bottom-up semantics). We prove the top-down semantics to be sound and complete with respect to the bottom-up semantics.

1 INTRODUCTION

Temporal reasoning is at the heart of human activity and not surprisingly it has raised a lot of interest in computer science, be it in the form of temporal logics [39, 21, 17], temporal programming languages [32, 34, 18, 1, 9, 11, 15, 40] or temporal databases [36, 38, 5, 7, 10, 20, 33]. No matter if one programs with temporal information or stores data with temporal information, in most cases the formal underpinnings will be logic, and often be variants or extensions of first order logic.

In a logical formulation and formalization of temporal information and reasoning it is quite natural to think of formulae that are labelled with temporal information and about proof procedures that take into account these labels. In other words, we separate the temporal from the non-temporal aspects. What D. Gabbay enthusiastically says in the introduction of his book on labelled deductive systems [19], that allow to label a general class of logics, "This sounds very simple but it ... makes a serious difference, like the difference between using one hand only or allowing for the coordinated use of two hands", can apply to the special case of temporal labels as well.

In our case, the logic and the labels are familiar structures: First-order logic (FOL) and lattices. The labels are called annotations, and the overall

D. Basin et al. (eds.), Labelled Deduction, 215–243.
© 2000 *Kluwer Academic Publishers. Printed in the Netherlands.*

class of logics is called annotated logics [26]. Based on this framework and on constraint logic programming concepts [24, 25, 29, 30, 16], the family of temporal annotated constraint logic programming (TACLP) languages has been developed in [14, 13, 15, 35, 28].

The pieces of temporal information are given by *temporal annotations* which say at what time(s) the formula to which they are applied is valid. The annotations of TACLP make time explicit but avoid the proliferation of temporal variables and quantifiers of the first order approach. In this way, TACLP supports qualitative and quantitative (metric) temporal reasoning involving both time points and time periods (time intervals) and their duration. Moreover, it allows us to represent definite, indefinite and periodic temporal information. In Figure 9.1, an example from [14] illustrates the expressiveness and conceptual simplicity of TACLP. As shown in [14] by asking the query $murder(x, y)$ we obtain two solutions (i.e., two suspects) of the workshop murder mystery: $x = Lepov, y = Lepov$ and $x = Maringer, y = Lepov$. The first one means that Prof. Lepov could have committed suicide. This unexpected solution is found because Prof. Lepov does not have an alibi for the time of his death. The second answer means that Dr. Maringer could be the murderer, because his alibi does not hold: The copying would have taken 30 minutes, so it cannot have happened during a talk of 25 minutes.

In [15] TACLP is presented as an instance of annotated constraint logic (ACL) for reasoning about time. ACL is a generalization of generalized annotated programs [26, 27], and extends first-order languages with a distinguished class of predicates, called *constraints*, and a distinguished class of terms, called *annotations*, used to label formulae. Moreover ACL provides inference rules for annotated formulae and a constraint theory for handling annotations. One advantage of a language in the ACL framework is that its clausal fragment can be efficiently implemented: Given a logic in this framework, there is a systematic way to make a clausal fragment executable as a constraint logic program. Both an interpreter and a compiler can be generated and implemented in standard constraint logic programming languages.

Constraint logic programming (CLP) [24, 25, 29, 30, 16] is an extension of logic programming, where in addition to predicates, which are defined by clauses and reasoned about by resolution (a form of Modus Ponens), there is a distinguished class of predicates called *constraints*. Their meaning is defined by a *constraint theory* whose reasoning capability is implemented by some efficient algorithm in the so-called *constraint solver*. In this way, efficient special-purpose algorithms can be integrated in a sound way into logic programming.

Overview of the paper. In this paper, the TACLP language is given two different kinds of semantics, an operational one based on meta-logic (top-down

There is a workshop at the Plaza hotel.

(1) In the afternoon session, after the coffee break (3:00pm - 3:25pm), there were four more talks, 25 minutes each - *time periods*. Dr. Maringer gave the 3rd talk. The last talk was to be given by Prof. Lepov.

$\mathit{coffeebreak}\ \mathtt{th}\,[3{:}00, 3{:}25].$
$\mathit{talk}(1, Hunon, AlgebraicSemantics...)\ \mathtt{th}\,[3{:}25, 3{:}50].$
$\mathit{talk}(2, ..., ...)\ \mathtt{th}\,[3{:}50, 4{:}15].$
$\mathit{talk}(3, Maringer, ...)\ \mathtt{th}\,[4{:}15, 4{:}40].$
$\mathit{talk}(4, Lepov, P = NP)\ \mathtt{th}\,[4{:}40, 5{:}05].$

(2) Prof. Lepov was found dead in his hotel room at 5:35pm - *time point*.

$\mathit{founddead}(Lepov)\ \mathtt{at}\ 5{:}35.$

(3) The doctor said he was dead for one to one and a half hours - *duration and indefinite information.*

$\mathit{murdered}(p)\ \mathtt{in}\,[t_1 - 1{:}30, t_2 - 1{:}00] \leftarrow \mathit{founddead}(p)\ \mathtt{in}\,[t_1, t_2]$

There are two suspects, Dr. Kosta and Dr. Maringer. They have alibis.

(4) Dr. Kosta took the last shuttle to the airport possible to reach the 5:10pm plane - *time point.*

$\mathit{plane}(Kosta)\ \mathtt{at}\ 5{:}10.$

(5) The shuttle from the hotel leaves every half hour between noon and 11pm - *recurrent (periodic) data.*

$\mathit{shuttle}\ \mathtt{at}\ 0{:}00.$
$\mathit{shuttle}\ \mathtt{at}\ t + 30 \leftarrow 0{:}00 \le t, t \le 11{:}00, \mathit{shuttle}\ \mathtt{at}\ t$

(6) It takes at least 50 minutes to get to the airport - *duration and indefinite information.*

$\mathit{onshuttle}(p)\mathtt{th}\,[t_1, t_2] \leftarrow t_2 = t_1 + 50, \mathit{shuttle}\,\mathtt{at}\ t_1, \mathit{plane}(p)\mathtt{in}\,[t_2, t_2 + 50]$

(7) During the 2nd talk Dr. Maringer realized that he had forgotten to copy his 30 slides - *related time periods.*

So he picked up the slides from his hotel room and copied them. It takes 5 minutes to get to the room, another 5 minutes to get to the copy room from there, and 5 more minutes to get back to the lecture hall - *durations.*

A copy takes half a minute - *repeated durations.*

$$\mathit{copying}(Maringer)\ \mathtt{th}\,[t_1, t_2] \leftarrow t_2 = t_1 + 5 + 5 + 30 * 0.5 + 5, \quad \mathit{talk}(2, p, w)\ \mathtt{th}\,[t_1, t_2]$$

(8) *Who murdered Prof. Lepov?* The murderer is a person who is involved in the case and does not have an alibi during the time of murder.

$\mathit{murder}(p_1, p_2) \leftarrow \mathit{murdered}(p_2)\mathtt{in}\,[t_1, t_2], \mathit{invol}(p_1), \neg(\mathit{alibi}(p_1)\mathtt{th}\,[t_1, t_2])$
$\mathit{invol}(Kosta). \qquad \mathit{invol}(Lepov). \qquad \mathit{invol}(Maringer).$
$\mathit{alibi}(p)\ \mathtt{th}\,[t_1, t_2] \leftarrow (\mathit{onshuttle}(p) \vee \mathit{copying}(p) \vee \mathit{talk}(n, p, w))\mathtt{th}\,[t_1, t_2]$

Figure 9.1 The Workshop Murder Mystery

semantics) using a meta-interpreter and a fixpoint one obtained by extending the definition of the immediate consequence operator of *CLP* to deal with annotated atoms (bottom-up semantics). The meta-interpreter is proved to be sound and complete with respect to the bottom-up semantics. While top-down semantics are known from the previous work on TACLP [14, 13, 15], this paper presents for the first time a bottom-up semantics and, consequently, for the first time soundness and completeness results for TACLP.

The paper is organized as follows. Section 2 introduces the TACLP framework. Section 3 defines the two semantics for TACLP and proves soundness and completeness. Section 4 shortly presents related work and Section 5 concludes the paper.

2 TEMPORAL ANNOTATED CONSTRAINT LOGIC PROGRAMMING

This subsection briefly reviews TACLP. TACLP is a constraint logic programming language where formulae can be annotated with temporal labels and where temporal constraints express relations between these labels. In TACLP, the choice of the temporal ontology is free. In this paper, we consider the subset of TACLP, where time points are totally ordered, and sets of time points are convex and non-empty. Moreover only atomic formulae can be annotated and clauses are free of negation. For a more detailed treatment of TACLP and for the general theory of ACL we refer the reader to [15].

Time can be discrete or dense. Time points are totally ordered by the relation $\leq$. We call the set of time points D and we suppose that a set of operations (such as the binary operations $+, -$) to manage such points are associated with it. We assume that the time-line is left-bounded by the number 0 and open to the future, with the symbol ∞ used to denote a time point that is later than any other. A *time period* is an interval $[r, s]$ with $0 \leq r \leq s \leq \infty, r \in D,\ s \in D$ that represents the convex, non-empty set of time points $\{t \mid r \leq t \leq s\}$[1]. Thus the interval $[0, \infty]$ denotes the whole time line.

An *annotated formula* is of the form $A\,\alpha$ where A is an atomic formula and α an annotation. In TACLP, there are three kinds of annotations based on time points and on time periods. Let t be a time point and let $I = [r, s]$ be a time period.

(`at`) The annotated formula $A\,\texttt{at}\,t$ means that A holds at time point t.

(`th`) The annotated formula $A\,\texttt{th}\,I$ means that A holds *throughout*, i.e., at *every* time point in the time period I. The definition of a `th`-annotated

[1] The results in this paper naturally extend to time lines that are bounded or unbounded in other ways and to time periods that are open on one or both sides.

formula in terms of at is:

$$A\,\mathtt{th}\,I \;\Leftrightarrow\; \forall t\,(t \in I \rightarrow A\,\mathtt{at}\,t).$$

(in) The annotated formula $A\,\mathtt{in}\,I$ means that A holds at *some* time point(s) - but we do not know exactly when - in the time period I. The definition of an in-annotated formula in terms of at is:

$$A\,\mathtt{in}\,I \;\Leftrightarrow\; \exists t\,(t \in I \wedge A\,\mathtt{at}\,t).$$

The in temporal annotation accounts for indefinite temporal information.

The set of annotations is endowed with a partial order relation $\sqsubseteq$ which turns it into a lattice. Given two annotations α and β, the intuition is that $\alpha \sqsubseteq \beta$ if α is "less informative" than β in the sense that for all formulae A, $A\,\beta \Rightarrow A\,\alpha$. More precisely, being an instance of ACL, in addition to Modus Ponens, TACLP has two further inference rules: The rule ($\sqsubseteq$) and the rule ($\sqcup$).

$$\frac{A\,\alpha \qquad \gamma \sqsubseteq \alpha}{A\,\gamma}\; \textit{rule}\ (\sqsubseteq) \qquad\qquad \frac{A\,\alpha \qquad A\,\beta \qquad \gamma = \alpha \sqcup \beta}{A\,\gamma}\; \textit{rule}\ (\sqcup)$$

The rule ($\sqsubseteq$) states that if a formula holds with some annotation, then it also holds with all annotations that are smaller according to the lattice ordering. The rule ($\sqcup$) says that if a formula holds with some annotation and the same formula holds with another annotation then it holds with the least upper bound of the annotations.

Now, we define the *constraint theory for temporal annotations*. A constraint theory is a non-empty, consistent first order theory that axiomatizes the meaning of the constraints. First of all, our constraint theory includes an axiomatization of the total order relation $\leq$ on time points D. Then it contains the following axioms defining the partial order on temporal annotations.

$$\begin{array}{llcl}
(\mathtt{at}\,\mathtt{th}) & \mathtt{at}\,t = \mathtt{th}\,[t,t] & & \\
(\mathtt{at}\,\mathtt{in}) & \mathtt{at}\,t = \mathtt{in}\,[t,t] & & \\
(\mathtt{th}\ \sqsubseteq) & \mathtt{th}\,[s_1,s_2] \sqsubseteq \mathtt{th}\,[r_1,r_2] & \Leftrightarrow & r_1 \leq s_1,\ s_1 \leq s_2,\ s_2 \leq r_2 \\
(\mathtt{in}\ \sqsubseteq) & \mathtt{in}\,[r_1,r_2] \sqsubseteq \mathtt{in}\,[s_1,s_2] & \Leftrightarrow & r_1 \leq s_1,\ s_1 \leq s_2,\ s_2 \leq r_2
\end{array}$$

The first two axioms state that $\mathtt{th}\,I$ and $\mathtt{in}\,I$ are equivalent to $\mathtt{at}\,t$ when the time period I consists of a single time point t.[2] Next, if a formula holds at every element of a time period, then it holds at every element in all sub-periods of that period ((th $\sqsubseteq$) axiom). On the other hand, if a formula holds at some

[2]Especially in dense time, one may disallow singleton periods and drop the two axioms. This restriction has no effects on the results of the paper.

points of a time period then it holds at some points in all periods that include this period ((in $\sqsubseteq$) axiom).

To summarize the partial order relation on annotations, the axioms can be arranged in the following chain, assuming $r_1 \leq s_1,\ s_1 \leq s_2,\ s_2 \leq r_2$:

$$\begin{aligned} \mathtt{in}\,[r_1, r_2] \sqsubseteq \mathtt{in}\,[s_1, s_2] \sqsubseteq \mathtt{in}\,[s_1, s_1] = \mathtt{at}\, s_1 = \\ = \mathtt{th}\,[s_1, s_1] \sqsubseteq \mathtt{th}\,[s_1, s_2] \sqsubseteq \mathtt{th}\,[r_1, r_2] \end{aligned}$$

Now we axiomatize the least upper bound $\sqcup$ of temporal annotations over time points and time periods. As explained in [15], the least upper bound exists but sometimes may be "too large". In fact, rule ($\sqcup$) is correct only if the lattice order ensures that $A\,\alpha \wedge A\,\beta \wedge (\gamma = \alpha \sqcup \beta) \implies A\,\gamma$ whereas, in general, this is not true in our case. For instance, according to the lattice, $\mathtt{th}\,[1,2] \sqcup \mathtt{th}\,[4,5] = \mathtt{th}\,[1,5]$, but according to the definition of th-annotated formulae in terms of at , the conjunction $A\,\mathtt{th}\,[1,2] \wedge A\,\mathtt{th}\,[4,5]$ does not imply $A\,\mathtt{th}\,[1,5]$, since it does not express that $A\,\mathtt{at}\,3$ holds. From a theoretical point of view, this problem can be overcome by enriching the lattice of annotations with expressions involving $\sqcup$. In practice, it suffices to consider the least upper bound for time periods that produce another different meaningful time period. Concretely, we can restrict ourselves to th annotations with overlapping time periods that do not include one another:

$$(\mathtt{th}\,\sqcup) \qquad \mathtt{th}\,[s_1, s_2] \sqcup \mathtt{th}\,[r_1, r_2] = \mathtt{th}\,[s_1, r_2] \Leftrightarrow s_1 < r_1, r_1 \leq s_2, s_2 < r_2$$

We can now define the clausal fragment of TACLP that can be used as an efficient temporal programming language. A *TACLP program* is a finite set of TACLP clauses. A *TACLP clause* is a TACLP formula of the form:

$$A\,\alpha \leftarrow C_1, \ldots, C_n, B_1\,\alpha_1, \ldots, B_m\,\alpha_m \quad (n, m \geq 0)$$

where A is an atom (not a constraint), α and α_i are (optional) temporal annotations, the C_j's are the constraints and the B_i's are atomic formulae. Constraints C_j cannot be annotated. Following the usual syntax of logic programming, commas "," denote conjunctions. The conclusion of the implication is called the *head* of the clause and the premise the *body* of the clause. Variables in a clause are implicitly assumed to be universally quantified at the outermost scope.

We conclude the introduction of TACLP with some examples. In the following for programs, we adopt the convention of denoting variables by identifiers starting with a lower-case letter and constant symbols by identifiers starting with an upper-case letter.

Example 1 *In a company, there are managers and a secretary who has to manage their meetings. A manager is busy if he is in a meeting or if he is out.*

$$busy(p)\,\texttt{th}\,[t_1,t_2] \leftarrow in\text{-}meeting(p)\,\texttt{th}\,[t_1,t_2]$$
$$busy(p)\,\texttt{th}\,[t_1,t_2] \leftarrow out\text{-}of\text{-}office(p)\,\texttt{th}\,[t_1,t_2]$$

Suppose the schedule for today to be the following: Smith and Jones have a meeting at 9am and at 9:30am respectively, each lasting one hour. In the afternoon Smith goes out for lunch at 2pm and comes back at 3pm:

$$in\text{-}meeting(Smith)\,\texttt{th}\,[9am, 10am].$$
$$in\text{-}meeting(Jones)\,\texttt{th}\,[9{:}30am, 10{:}30am].$$
$$out\text{-}of\text{-}office(Smith)\,\texttt{th}\,[2pm, 3pm].$$

If the secretary wants to know whether Smith is busy between 9:30am and 10:30am she can ask for $busy(Smith)\,\texttt{in}\,[9{:}30am, 10{:}30am]$. *Since Smith is in a meeting from 9am till 10am, one can indeed derive that Smith is busy. Notice that this query exploits indefinite information: since Smith is busy at least in one instant of the period* $[9{:}30am, 10{:}30am]$, *the secretary cannot schedule an appointment for him for that period.*

On the other hand, $busy(Smith)\,\texttt{th}\,[9{:}30am, 10{:}30am]$ *does not hold, because Smith is not busy between 10am and 10:30am. Moreover, also* $busy(Smith)\,\texttt{in}\,[10{:}30am, 1{:}30pm]$ *does not hold, because Smith is not busy in that time period at all.*

The query $(busy(Smith)\,\texttt{th}\,[t_1,t_2],\ busy(Jones)\,\texttt{th}\,[t_1,t_2])$ *reveals that both managers are busy throughout the time period* $[9{:}30am, 10am]$, *because this is the largest interval that is included in the time periods where both managers are busy.*

Now assume that we define

$$busy\,\texttt{th}\,[t_1,t_2] \leftarrow busy(p)\,\texttt{th}\,[t_1,t_2]$$

Then $busy$ *holds when either manager is busy, namely for the time periods* $[9am, 10{:}30am]$ *(which is the least upper bound of the time periods for the two overlapping meetings of Smith and Jones) and* $[2pm, 3pm]$.

In [35] TACLP is successfully applied to a system for calculating the liquid flow in a network of water tanks from some events specifying when the taps were switched on and off. The following example involving continuous change is also presented.

Example 2 *We model information about the growth of trees.*

1. *Tree 1 sprouts at time* 3.5 *(the middle of year 3).*

   $$sprouts(\mathit{Tree1})\,\texttt{at}\,3.5.$$

2. *Tree 1 is an oak tree.*

$$tree\_type(Tree1, Oak).$$

3. *The growth rate of oak trees is 3 meters per year.*

$$growth\_rate(Oak, 3).$$

4. *If a tree is of a type that has a given growth rate r, and the tree sprouts at time s then at time t it has a height h, where $h = (t - s) \times r$.*

$$\begin{aligned} &height(tree, h) \;\texttt{at}\; t \leftarrow \\ &\quad h = (t - s) \times r, \\ &\quad tree\_type(tree, type), \\ &\quad growth\_rate(type, r), \\ &\quad sprouts(tree) \;\texttt{at}\; s \end{aligned}$$

5. *If a tree has height h m at time t, where $h \geq 6.75$, then it is mature.*

$$mature(tree) \;\texttt{th}\; [t, \infty] \leftarrow h \geq 6.75, height(tree, h) \;\texttt{at}\; t$$

In the last clause, the maturity of the tree at an instant is implied by a constraint on the height of the tree at that instant. Height is the continuously changing quantity. The query

$$mature(Tree1) \;\texttt{th}\; [6, 7]$$

can be proved. This means that Tree1 is mature throughout the time period which begins at year 6 *and ends at year* 7.

The query

$$mature(Tree1) \;\texttt{th}\; [t_1, t_2]$$

yields $t_1 \geq 5.75, t_2 = \infty$.

3 SEMANTICS OF TACLP

In this section we define the operational (top-down) semantics of the language TACLP by presenting a meta-interpreter for it. Then we provide TACLP with a fixpoint (bottom-up) semantics, based on the definition of an immediate consequence operator, and we prove that the meta-interpreter is sound and complete with respect to the bottom-up semantics.

In the definition of the semantics, without loss of generality, we assume all atoms to be annotated with `th` or `in` labels. In fact `at` t annotations can be replaced with `th` $[t, t]$ by exploiting the (`at` `th`) axiom. Moreover, each atom which is not annotated in the object level program is intended to be true throughout the whole temporal domain, and thus can be labelled with `th` $[0, \infty]$. Constraints stay unchanged.

3.1 OPERATIONAL SEMANTICS VIA META-INTERPRETER

The *vanilla* meta-interpreter presented in [37] is the simplest application of meta-programming in logic. A general formulation of the vanilla meta-interpreter can be given by means of the *demo* predicate used to represent provability. Namely, $demo(g)$ means that the formula g is provable in the object program.

$$\begin{aligned} & demo(Empty). \\ demo((b_1, b_2)) & \leftarrow \; demo(b_1),\; demo(b_2) \\ demo(a) & \leftarrow \; clause(a, b),\; demo(b) \end{aligned}$$

The unit clause states that the empty goal, represented by the constant symbol $Empty$, is always solved. The second clause deals with conjunctive goals. It states that a conjunction (B_1, B_2) is solved if B_1 is solved and B_2 is solved. Finally, the third clause deals with the case of atomic goal reduction. To solve an atomic goal A, a clause from the program is chosen whose head unifies with A and the body of the clause is recursively solved. An object level program P is represented at the meta-level by a set of axioms of the kind $clause(A, B)$, one for each object level clause $A \leftarrow B$ in P.

The extended meta-interpreter for our subset of TACLP is defined by the following clauses:

$$demo(Empty). \tag{9.1}$$

$$demo((b_1, b_2)) \leftarrow demo(b_1), demo(b_2) \tag{9.2}$$

$$\begin{aligned} demo(a \,\mathtt{th}\, [t_1, t_2]) \leftarrow & \; s_1 \leq t_1, t_2 \leq s_2, t_1 \leq t_2, \\ & clause(a \,\mathtt{th}\, [s_1, s_2], b), demo(b) \end{aligned} \tag{9.3}$$

$$\begin{aligned} demo(a \,\mathtt{th}\, [t_1, t_2]) \leftarrow & \; s_1 \leq t_1, t_1 < s_2, s_2 < t_2, \\ clause(a \,\mathtt{th}\, [s_1, s_2], b), & \; demo(b), demo(a \,\mathtt{th}\, [s_2, t_2]) \end{aligned} \tag{9.4}$$

$$\begin{aligned} demo(a \,\mathtt{in}\, [t_1, t_2]) \leftarrow & \; t_1 \leq s_2, s_1 \leq t_2, t_1 \leq t_2, \\ & clause(a \,\mathtt{th}\, [s_1, s_2], b), demo(b) \end{aligned} \tag{9.5}$$

$$\begin{aligned} demo(a \,\mathtt{in}\, [t_1, t_2]) \leftarrow & \; t_1 \leq s_1, s_2 \leq t_2, \\ & clause(a \,\mathtt{in}\, [s_1, s_2], b), demo(b) \end{aligned} \tag{9.6}$$

$$demo(c) \leftarrow constraint(c), c \tag{9.7}$$

A clause $A\,\alpha \leftarrow B$ of a TACLP program P is represented at the meta-level by

$$clause(A\,\alpha, B) \leftarrow t_1 \leq t_2. \tag{9.8}$$

where $\alpha = \mathtt{th}\,[t_1, t_2]$ or $\alpha = \mathtt{in}\,[t_1, t_2]$.

This meta-interpreter can be written in any CLP language that provides a suitable constraint solver for temporal annotations (see Section 2 for the constraint theory). Hence the first difference with the vanilla meta-interpreter is that our meta-interpreter handles constraints, which can either occur explicitly in its clauses, e.g. $s_1 \leq t_1, t_1 \leq t_2, t_2 \leq s_2$ in clause (9.3), or can come from the resolution steps. The latter kind of constraints is managed by clause (9.7) which passes each constraint C to be solved directly to the constraint solver.

The second difference is that our meta-interpreter implements not only Modus Ponens but also the rule ($\sqsubseteq$) and the rule ($\sqcup$). This is the reason why the third *demo* clause of the vanilla meta-interpreter is now split into four clauses. Clauses (9.3), (9.5) and (9.6) implement the inference rule ($\sqsubseteq$): The atomic goal to be solved is required to be labelled with an annotation which is smaller than the one labelling the head of the clause used in the resolution step. For instance, clause (9.3) states that given a clause $A\,\mathtt{th}\,[s_1, s_2] \leftarrow B$ whose body B is solvable, we can derive the atom A annotated with any $\mathtt{th}\,[t_1, t_2]$ such that $\mathtt{th}\,[t_1, t_2] \sqsubseteq \mathtt{th}\,[s_1, s_2]$, i.e., according to axiom ($\mathtt{th}\ \sqsubseteq$), $[t_1, t_2] \subseteq [s_1, s_2]$, as expressed by the constraint $s_1 \leq t_1, t_2 \leq s_2, t_1 \leq t_2$. Clauses (9.5) and (9.6) are built in an analogous way by exploiting axioms ($\mathtt{in}\,\mathtt{th}\ \sqsubseteq$) and ($\mathtt{in}\ \sqsubseteq$), respectively.

Rule ($\sqcup$) is implemented by clause (9.4). According to the discussion in Section 2, it is applicable only to th annotations with overlapping time periods which do not include one another. More precisely, clause (9.4) states that if we can find a clause $A\,\mathtt{th}\,[s_1, s_2] \leftarrow B$ such that the body B is solvable, and if moreover the atom A can be proved *throughout* the time period $[s_2, t_2]$ (i.e., $demo(A\,\mathtt{th}\,[s_2, t_2])$ is solvable) then we can derive the atom A labelled with any annotation $\mathtt{th}\,[t_1, t_2] \sqsubseteq \mathtt{th}\,[s_1, t_2]$. The constraints on temporal variables ensure that the time period $[t_1, t_2]$ is a *new* time period different from $[s_1, s_2]$ and $[s_2, t_2]$ and their subintervals.

Finally, in the meta-level representation of object clauses, clause (9.8), we have to add the constraint $t_1 \leq t_2$ to ensure that the head of the object clause has a well-formed, namely non-empty, annotation.

Example 3 *Consider a library database containing information about loans. Mary first borrowed the book Hamlet from May 12, 1995 to June 12, 1995 and then on June 12, 1995 she extended her loan:*

$$borrow(Mary, Hamlet)\,\mathtt{th}\,[May\ 12\ 1995, Jun\ 12\ 1995].$$
$$borrow(Mary, Hamlet)\,\mathtt{th}\,[Jun\ 12\ 1995, Aug\ 1\ 1995].$$

The period of time in which Mary borrowed Hamlet can be obtained by the query

$$demo(borrow(Mary, Hamlet)\ \mathtt{th}\ [t_1, t_2]).$$

By using clause (9.4), we can derive the interval $[May\ 12\ 1995, Aug\ 1\ 1995]$ *(more precisely, the constraints* $May\ 12\ 1995 \leq t_1$, $t_1 < Jun\ 12\ 1995$, $Jun\ 12\ 1995 < t_2$, $t_2 \leq Aug\ 1\ 1995$ *are derived) that otherwise would be never generated. In fact, by applying clause (9.3) alone, we can prove only that Mary borrowed Hamlet in the intervals* $[May\ 12\ 1995, Jun\ 12\ 1995]$ *and* $[Jun\ 12\ 1995, Aug\ 1\ 1995]$ *separately.*

3.2 FIXPOINT SEMANTICS

There are several ways of defining a bottom-up semantics of TACLP, related to the different possible choices of the semantic domain where the immediate consequence operator is defined. The simpler solution consists in using the powerset $\wp(\mathcal{A}\text{-}base^3 \times Ann)$ with set-theoretic inclusion, disregarding the partial order structure of the set of annotations Ann. Alternative solutions (as shown for generalized annotated programs in [26]) may consider a more abstract domain, which is obtained by endowing $\mathcal{A}$-*base* $\times Ann$ with the product order (induced by the discrete order on $\mathcal{A}$-*base* and the order on Ann) and then by taking as elements of power domain only those subsets of annotated atoms which satisfy some closure properties with respect to such an order. For instance, one can require "downward-closedness", which amounts to including subsumption in the $\mathcal{T}_P$ operator. Another possible property is "limit-closedness", namely the presence of the least upper bound of all directed sets which, from a computational point of view, amounts to consider computations which possibly require more than ω steps. For space limitations, we treat here the first, simpler solution.

The intended interpretation of constraints is defined by fixing a structure $\mathcal{A}$. In our case $\mathcal{A}$ surely contains a structure $\mathcal{D}$ (with domain D) in which we interpret the temporal constants and functions. However, since the constraints which can occur in a TACLP program are not only on temporal data, in general the structure $\mathcal{A}$ will be multi-sorted.

Let $Dom_{\mathcal{A}}$ be the domain of the structure $\mathcal{A}$. An $\mathcal{A}$-*valuation* is a (multi-sorted) mapping from variables to $Dom_{\mathcal{A}}$, and its natural extension maps terms to $Dom_{\mathcal{A}}$ and formulae to formulae whose predicates have arguments ranging over $Dom_{\mathcal{A}}$. An $\mathcal{A}$-*ground instance* A' of an atom A (resp. of a constraint or of a clause) is obtained by applying an $\mathcal{A}$-valuation to the atom (resp. to the

[3] The formal definition of $\mathcal{A}$-*base* is given later. Briefly, it is the natural generalization of the notion of Herbrand Base in constraint logic programming.

constraint or to the clause), thus producing a construct of the form $p(a_1, \ldots, a_n)$ with $a_1, \ldots, a_n$ elements from $Dom_{\mathcal{A}}$. We denote by $ground_{\mathcal{A}}(P)$ the set of $\mathcal{A}$-ground instances of clauses from a program P.

We first define the standard fixpoint operator of constraint logic programming and then extend it to deal with TACLP. An *$\mathcal{A}$-interpretation* for a $CLP(\mathcal{A})$ program P is a subset of the *$\mathcal{A}$-base* of P, written $\mathcal{A}\text{-}base_P$, which is the set

$$\left\{ p(a_1, \ldots, a_n) \mid \begin{array}{l} p \text{ is a } n\text{-ary user-defined predicate in } P \\ \text{and each } a_i \text{ is an element of } Dom_{\mathcal{A}} \end{array} \right\}$$

Then the standard immediate consequence operator [29] for a $CLP(\mathcal{A})$ program P is a function $T_P^{\mathcal{A}} : \wp(\mathcal{A}\text{-}base_P) \to \wp(\mathcal{A}\text{-}base_P)$ defined as follows:

$$T_P^{\mathcal{A}}(I) = \left\{ A \mid \begin{array}{l} A \leftarrow C_1, \ldots, C_k, B_1, \ldots, B_n, \in ground_{\mathcal{A}}(P), \\ \{B_1, \ldots, B_n\} \subseteq I,\ \mathcal{A} \models C_1, \ldots, C_k \end{array} \right\}$$

The operator $T_P^{\mathcal{A}}$ is continuous [29], and therefore it has a least fixpoint which can be computed as the least upper bound of the chain $\{(T_P^{\mathcal{A}})^i\}_{i \geq 0}$ of the iterated applications of $T_P^{\mathcal{A}}$ starting from the empty set.[4] The least fixpoint is denoted by $(T_P^{\mathcal{A}})^\omega$.

To generalize the above operator to deal with temporal annotations we consider a kind of extended interpretations, basically consisting of sets of annotated elements of *$\mathcal{A}$-base*. Formally we define the set of (semantical) annotations

$$Ann = \{\mathtt{th}\,[t_1, t_2], \mathtt{in}\,[t_1, t_2] \mid t_1 \in D, t_2 \in D, \mathcal{D} \models t_1 \leq t_2\}$$

Then given a TACLP program P, the lattice of interpretations is defined as $(\wp(\mathcal{A}\text{-}base_P \times Ann), \subseteq)$ where $\wp$ is the powerset operator and $\subseteq$ is the usual relation of set-theoretic inclusion.

Definition 4 *Let P be a TACLP program, the function $\mathcal{T}_P^{\mathcal{A}} : \wp(\mathcal{A}\text{-}base_P \times Ann) \to \wp(\mathcal{A}\text{-}base_P \times Ann)$ is defined as follows.*

$$\mathcal{T}_P^{\mathcal{A}}(I) =$$
$$\left\{ (A, \alpha) \mid \begin{array}{l} (\alpha = \mathtt{th}\,[s_1, s_2] \ \vee\ \alpha = \mathtt{in}\,[s_1, s_2]), \\ A\ \alpha \leftarrow C_1, \ldots, C_k, B_1\alpha_1, \ldots, B_n\alpha_n \in ground_{\mathcal{A}}(P), \\ \{(B_1, \beta_1), \ldots, (B_n, \beta_n)\} \subseteq I, \\ \mathcal{A} \models C_1, \ldots, C_k, \alpha_1 \sqsubseteq \beta_1, \ldots, \alpha_n \sqsubseteq \beta_n, s_1 \leq s_2 \end{array} \right\}$$
$$\cup$$
$$\left\{ (A, \alpha) \mid \begin{array}{l} \alpha = \mathtt{th}\,[s_1, r_2], \\ A\mathtt{th}\,[s_1, s_2] \leftarrow C_1, \ldots, C_k, B_1\alpha_1, \ldots, B_n\alpha_n \in ground_{\mathcal{A}}(P), \\ \{(B_1, \beta_1), \ldots, (B_n, \beta_n)\} \subseteq I,\ (A, \mathtt{th}\,[r_1, r_2]) \in I, \\ \mathcal{A} \models C_1, \ldots, C_k, \alpha_1 \sqsubseteq \beta_1, \ldots, \alpha_n \sqsubseteq \beta_n, \\ \qquad s_1 < r_1, r_1 \leq s_2, s_2 < r_2 \end{array} \right\}$$

[4]Formally, for a function $T : \wp(S) \to \wp(S)$ we define $T^0 = \emptyset$ and $T^{i+1} = T(T^i)$.

This definition properly extends the standard definition of the immediate consequence operator. In fact, in a sense, it captures not only the Modus Ponens rule, as the standard operator does, but also rule ($\sqcup$) (second set in the above definition). In addition, rule ($\sqsubseteq$) is used to prove that an annotated atom holds in an interpretation: To derive the head $A\,\alpha$ of a clause it is not necessary to find in the interpretation exactly the atoms $B_1\,\alpha_1, \ldots, B_n\,\alpha_n$ occurring in the body of the clause, but it suffices to find atoms $B_i\,\beta_i$ which implies $B_i\,\alpha_i$, i.e., such that each β_i is an annotation stronger than α_i ($\mathcal{A} \models \alpha_i \sqsubseteq \beta_i$). Finally, notice that $\mathcal{T}_P^{\mathcal{A}}(I)$ is not downward closed, namely, it is not true that if $(A, \alpha) \in \mathcal{T}_P^{\mathcal{A}}(I)$ then for all (A, γ) such that $\mathcal{A} \models \gamma \sqsubseteq \alpha$, we have $(A, \gamma) \in \mathcal{T}_P^{\mathcal{A}}(I)$. However such a closure is done at the end of the computation of the fixpoint of $\mathcal{T}_P^{\mathcal{A}}$. In this way the rule ($\sqsubseteq$) is completely captured.

An important property of the $\mathcal{T}_P^{\mathcal{A}}$ operator, which is at the core of the definition of the fixpoint semantics, is continuity over the lattice of interpretations.

Theorem 5 (Continuity) *For any TACLP program P the function $\mathcal{T}_P^{\mathcal{A}}$ is continuous (on $(\wp(\mathcal{A}\text{-}base \times Ann), \subseteq)$).*

Proof The result is a direct consequence of the definition of $\mathcal{T}_P^{\mathcal{A}}$ and of the partial order $\subseteq$ on the interpretations. For more details see the Appendix. ∎

The bottom-up semantics for a program P is defined as the downward closure of the least fixpoint of $\mathcal{T}_P^{\mathcal{A}}$ which, by Theorem 5, is the least upper bound of the chain $\{(\mathcal{T}_P^{\mathcal{A}})^i\}_{i \geq 0}$.

Definition 6 *Let P be a TACLP program. Then the fixpoint semantics of P is defined as*

$$\mathcal{F}^{\mathcal{A}}(P) = \{(A, \alpha) \mid (A, \beta) \in (\mathcal{T}_P^{\mathcal{A}})^{\omega},\ \mathcal{A} \models \alpha \sqsubseteq \beta\}$$

where $(\mathcal{T}_P^{\mathcal{A}})^{\omega} = \bigcup_{i \geq 0} (\mathcal{T}_P^{\mathcal{A}})^i$.

3.3 SOUNDNESS AND COMPLETENESS

The semantics of meta-logic is a quite debated issue (see e.g [23, 31, 3]). In the spirit of [8, 31], we define the semantics of the extended vanilla meta-interpreter by relating the semantics of an object program to the semantics of the corresponding vanilla meta-program (i.e., including the meta-level representation of the object program). When stating the correspondence between the object program and the meta-program we consider only formulae of interest, i.e., elements of $\mathcal{A}$-*base* annotated with labels from Ann. We show that given a TACLP program P (object program) for any $A \in \mathcal{A}\text{-}base_P$ and any $\alpha \in Ann$, $demo(A\,\alpha)$ is provable at the meta-level if and only if (A, α) is provable in the object program. Formally, we are going to prove that

$$demo(A\,\alpha) \in (T_{\mathcal{V}}^{\mathcal{A}_m})^{\omega} \iff (A, \alpha) \in \mathcal{F}^{\mathcal{A}}(P) \tag{9.9}$$

where $\mathcal{V}$ is the meta-program containing the meta-level representation of the object program P according to (9.8) and the clauses (9.1)–(9.7), and $T_{\mathcal{V}}^{\mathcal{A}_m}$ is the standard immediate consequence operator of CLP. It is worth noting that $\mathcal{V}$ is a $CLP(\mathcal{A}_m)$ program where the structure $\mathcal{A}_m$ in which we interpret the constraints is composed by two structures: The Herbrand structure associated with $\mathcal{V}$ and the structure $\mathcal{A}$ where the constraints of P are interpreted. Therefore if C is an $\mathcal{A}$-ground instance of a constraint then $\mathcal{A}_m \models C \Leftrightarrow \mathcal{A} \models C$.

Since the given meta-logical definition axiomatizes a top-down operational semantics for TACLP programs, the proof of the statement (9.9) corresponds to showing the equivalence of computing top-down and bottom-up.

In the following for simplicity we drop the reference to $\mathcal{A}$ and $\mathcal{A}_m$ in the name of the immediate consequence operators. Furthermore P denotes a TACLP program, $\mathcal{A}$ the structure where the constraints of P are interpreted, A, B elements of $\mathcal{A}$-$base_P$, α, β, γ elements of Ann and C an $\mathcal{A}$-ground instance of a constraint. All symbols may have subscripts. The proof of the next two subsections are just sketched in the main text. The details can be found in the Appendix.

Soundness In order to show the soundness of the meta-interpreter (restricted to the atoms of interest), we first prove the following lemma stating that if a conjunctive goal is provable at the meta-level then also its atomic conjuncts are provable at the meta-level.

Lemma 7 *Let P be a program and let $\mathcal{V}$ be the corresponding meta-interpreter. For any $B_1\,\alpha_1, \ldots, B_n\,\alpha_n$ with $B_i \in \mathcal{A}$-$base_P$ and $\alpha_i \in Ann$ and for any $C_1, \ldots, C_k$, with C_i an $\mathcal{A}$-ground instance of a constraint, we have:*

$$\textit{For all } h \quad demo((C_1, \ldots, C_k, B_1\,\alpha_1, \ldots, B_n\,\alpha_n)) \in T_{\mathcal{V}}^h$$
$$\Longrightarrow \{demo(B_1\,\alpha_1), \ldots, demo(B_n\,\alpha_n)\} \subseteq T_{\mathcal{V}}^h \;\wedge\; \mathcal{A} \models C_1, \ldots, C_k$$

Proof The result easily follows from the definition of $T_{\mathcal{V}}$ and from clauses (9.2) and (9.7) of the meta-interpreter. ∎

The soundness of the meta-interpreter is an easy corollary of the following theorem stating that if $demo(A\,\alpha)$ is provable at the meta-level then (A, γ) is a consequence of the program P for some γ such that $A\,\gamma \Rightarrow A\,\alpha$, i.e., the annotation α is less or equal to γ.

Theorem 8 *Let P be a program and let $\mathcal{V}$ be the corresponding meta-program. For any $A \in \mathcal{A}$-$base_P$ and $\alpha \in Ann$, the following statement holds:*

$$demo(A\,\alpha) \in T_{\mathcal{V}}^{\omega} \quad \Longrightarrow \quad \exists \gamma \in Ann : (A, \gamma) \in \mathcal{T}_P^{\omega} \;\wedge\; \mathcal{A} \models \alpha \sqsubseteq \gamma.$$

Proof We first show that for all h

$$demo(A\,\alpha) \in T_{\mathcal{V}}^h \Longrightarrow \exists \gamma \in Ann : \; (A, \gamma) \in \mathcal{T}_P^{\omega} \;\wedge\; \mathcal{A} \models \alpha \sqsubseteq \gamma. \quad (9.10)$$

The proof is by induction on h.
(Base case). Trivial since $T_{\mathcal{V}}^0 = \emptyset$.
(Inductive case). Assume that

$$demo(A\,\alpha) \in T_{\mathcal{V}}^h \qquad \Longrightarrow \qquad \exists \gamma \in Ann : (A, \gamma) \in \mathcal{T}_P^\omega \wedge \mathcal{A} \models \alpha \sqsubseteq \gamma.$$

Then:

$$\begin{array}{l} demo(A\,\alpha) \in T_{\mathcal{V}}^{h+1} \\ \Longleftrightarrow \quad \{\text{definition of } T_{\mathcal{V}}^i\} \\ demo(A\,\alpha) \in T_{\mathcal{V}}(T_{\mathcal{V}}^h) \end{array}$$

We have four cases corresponding to clauses (9.3), (9.4), (9.5) and (9.6) of the meta-interpreter. We only show the first two cases since the others can be treated as clause (9.3).

(clause (9.3)) $\{\alpha = \mathtt{th}\,[t_1, t_2]$, definition of $T_{\mathcal{V}}$ and clause (9.3)$\}$

$$\begin{array}{l}
\{clause(A\,\mathtt{th}\,[s_1, s_2], G), demo(G)\} \subseteq T_{\mathcal{V}}^h \wedge \\
\mathcal{A} \models s_1 \leq t_1, t_2 \leq s_2, t_1 \leq t_2 \\
\Longrightarrow \{\text{meta-level representation of clauses of } P, \text{ given by clause (9.8)}\} \\
\exists (B_1\,\alpha_1), \ldots, (B_n\,\alpha_n), C_1, \ldots, C_k : \\
A\,\mathtt{th}\,[s_1, s_2] \leftarrow C_1, \ldots, C_k, B_1\,\alpha_1, \ldots, B_n\,\alpha_n \in ground_{\mathcal{A}}(P) \wedge \\
demo((C_1, \ldots, C_k, B_1\,\alpha_1, \ldots, B_n\,\alpha_n)) \in T_{\mathcal{V}}^h \wedge \\
\mathcal{A} \models s_1 \leq t_1, t_2 \leq s_2, t_1 \leq t_2 \\
\Longrightarrow \{\text{Lemma 7}\} \\
A\,\mathtt{th}\,[s_1, s_2] \leftarrow C_1, \ldots, C_k, B_1\,\alpha_1, \ldots, B_n\,\alpha_n \in ground_{\mathcal{A}}(P) \wedge \\
\{demo(B_1\,\alpha_1), \ldots, demo(B_n\,\alpha_n)\} \subseteq T_{\mathcal{V}}^h \wedge \\
\mathcal{A} \models C_1, \ldots, C_k \wedge \mathcal{A} \models s_1 \leq t_1, t_2 \leq s_2, t_1 \leq t_2 \\
\Longrightarrow \{\text{inductive hypothesis}\} \\
\exists \beta_1, \ldots, \beta_n : \\
A\,\mathtt{th}\,[s_1, s_2] \leftarrow C_1, \ldots, C_k, B_1\,\alpha_1, \ldots, B_n\,\alpha_n \in ground_{\mathcal{A}}(P) \wedge \\
\{(B_1, \beta_1), \ldots, (B_n, \beta_n)\} \subseteq \mathcal{T}_P^\omega \wedge \mathcal{A} \models \alpha_1 \sqsubseteq \beta_1, \ldots, \alpha_n \sqsubseteq \beta_n \wedge \\
\mathcal{A} \models C_1, \ldots, C_k \wedge \mathcal{A} \models s_1 \leq t_1, t_2 \leq s_2, t_1 \leq t_2 \\
\Longrightarrow \{\text{definition of } \mathcal{T}_P \text{ since } \mathcal{A} \models s_1 \leq s_2\} \\
(A, \mathtt{th}\,[s_1, s_2]) \in \mathcal{T}_P(\mathcal{T}_P^\omega) \wedge \mathcal{A} \models s_1 \leq t_1, t_2 \leq s_2, t_1 \leq t_2 \\
\Longrightarrow \{\mathcal{T}_P^\omega \text{ is a fixpoint of } \mathcal{T}_P \text{ and } \mathcal{A} \models s_1 \leq t_1, t_2 \leq s_2, t_1 \leq t_2\} \\
(A, \mathtt{th}\,[s_1, s_2]) \in \mathcal{T}_P^\omega \wedge \mathcal{A} \models \mathtt{th}\,[t_1, t_2] \sqsubseteq \mathtt{th}\,[s_1, s_2]
\end{array}$$

(clause (9.4)) $\{\alpha = \mathtt{th}\,[t_1, t_2]$, definition of $T_{\mathcal{V}}$ and clause (9.4)$\}$

$$\begin{array}{l}
\{clause(A\,\mathtt{th}\,[s_1, s_2], G), demo(G), demo(A\,\mathtt{th}\,[s_2, t_2])\} \subseteq T_{\mathcal{V}}^h \wedge \\
\mathcal{A} \models s_1 \leq t_1, t_1 < s_2, s_2 < t_2 \\
\Longrightarrow \{\text{meta-level representation of clauses of } P, \text{ given by clause (9.8)}\} \\
\exists (B_1\,\alpha_1), \ldots, (B_n\,\alpha_n), C_1, \ldots, C_k : \\
A\,\mathtt{th}\,[s_1, s_2] \leftarrow C_1, \ldots, C_k, B_1\,\alpha_1, \ldots, B_n\,\alpha_n \in ground_{\mathcal{A}}(P) \wedge
\end{array}$$

$$
\begin{array}{l}
\quad demo((C_1,\ldots,C_k,B_1\,\alpha_1,\ldots,B_n\,\alpha_n)) \in T^h_{\mathcal{V}} \wedge \\
\quad demo(A\,\mathtt{th}\,[s_2,t_2]) \in T^h_{\mathcal{V}} \;\wedge\; \mathcal{A} \models s_1 \le t_1, t_1 < s_2, s_2 < t_2 \\
\Longrightarrow \{\text{Lemma 7}\} \\
\quad A\,\mathtt{th}\,[s_1,s_2] \leftarrow C_1,\ldots,C_k,B_1\,\alpha_1,\ldots,B_n\,\alpha_n \in ground_{\mathcal{A}}(P) \wedge \\
\quad \{demo(B_1\,\alpha_1),\ldots,demo(B_n\,\alpha_n)\} \subseteq T^h_{\mathcal{V}} \;\wedge\; \mathcal{A} \models C_1,\ldots,C_k \wedge \\
\quad demo(A\,\mathtt{th}\,[s_2,t_2]) \in T^h_{\mathcal{V}} \;\wedge\; \mathcal{A} \models s_1 \le t_1, t_1 < s_2, s_2 < t_2 \\
\Longrightarrow \{\text{inductive hypothesis}\} \\
\quad \exists \beta, \beta_1,\ldots,\beta_n : \\
\quad A\,\mathtt{th}\,[s_1,s_2] \leftarrow C_1,\ldots,C_k,B_1\,\alpha_1,\ldots,B_n\,\alpha_n \in ground_{\mathcal{A}}(P) \wedge \\
\quad \{(B_1,\beta_1),\ldots,(B_n,\beta_n),(A,\beta)\} \subseteq \mathcal{T}^\omega_P \wedge \\
\quad \mathcal{A} \models \alpha_1 \sqsubseteq \beta_1,\ldots,\alpha_n \sqsubseteq \beta_n, \mathtt{th}\,[s_2,t_2] \sqsubseteq \beta \;\wedge\; \mathcal{A} \models C_1,\ldots,C_k \wedge \\
\quad \mathcal{A} \models s_1 \le t_1, t_1 < s_2, s_2 < t_2.
\end{array}
$$

Since $\mathcal{A} \models \mathtt{th}\,[s_2,t_2] \sqsubseteq \beta$ then $\beta = \mathtt{th}\,[w_1,w_2]$ with $\mathcal{A} \models w_1 \le s_2, t_2 \le w_2$.

According to the relation between w_1 and s_1 we can distinguish two subcases

(a) $\mathcal{A} \models w_1 \le s_1$. Then $\mathcal{A} \models w_1 \le t_1, t_1 < t_2, t_2 \le w_2$ which allows us to conclude
$(A, \mathtt{th}\,[w_1,w_2]) \in \mathcal{T}^\omega_P \;\wedge\; \mathcal{A} \models \mathtt{th}\,[t_1,t_2] \sqsubseteq \mathtt{th}\,[w_1,w_2]$

(b) $\mathcal{A} \models s_1 < w_1$. Then $\mathcal{A} \models s_1 < w_1, w_1 \le s_2, s_2 < t_2, t_2 \le w_2$ then by definition of $\mathcal{T}_P$

$$
\begin{array}{l}
\quad (A, \mathtt{th}\,[s_1,w_2]) \in \mathcal{T}_P(\mathcal{T}^\omega_P) \wedge \mathcal{A} \models s_1 \le t_1, t_1 < s_2, s_2 < t_2, t_2 \le w_2 \\
\Longrightarrow \quad \{\mathcal{T}^\omega_P \text{ is a fixpoint of } \mathcal{T}_P \text{ and } \mathcal{A} \models s_1 \le t_1, t_1 < t_2, t_2 \le w_2\} \\
\quad (A, \mathtt{th}\,[s_1,w_2]) \in \mathcal{T}^\omega_P \;\wedge\; \mathcal{A} \models \mathtt{th}\,[t_1,t_2] \sqsubseteq \mathtt{th}\,[s_1,w_2]
\end{array}
$$

We are now able to conclude the proof.

$$
\begin{array}{l}
\quad demo(A\,\alpha) \in T^\omega_{\mathcal{V}} \\
\Longrightarrow \{T^\omega_{\mathcal{V}} = \bigcup_{i \ge 0} T^i_{\mathcal{V}}\} \\
\quad \exists h : \; demo(A\,\alpha) \in T^h_{\mathcal{V}} \\
\Longrightarrow \{\text{Statement (9.10)}\} \\
\quad \exists \beta : (A,\beta) \in \mathcal{T}^\omega_P \;\wedge\; \mathcal{A} \models \alpha \sqsubseteq \beta
\end{array}
$$

■

Theorem 9 (Soundness) *Let P be a program and let $\mathcal{V}$ be the corresponding meta-program. For any $A \in \mathcal{A}$-base$_P$ and $\alpha \in Ann$, the following statement holds:*

$$demo(A\,\alpha) \in T^\omega_{\mathcal{V}} \qquad \Longrightarrow \qquad (A,\alpha) \in \mathcal{F}^{\mathcal{A}}(P)$$

Proof By Theorem 8 there exists $\gamma \in Ann$ such that $(A,\gamma) \in \mathcal{T}^\omega_P$ and $\mathcal{A} \models \alpha \sqsubseteq \gamma$. Hence by definition of $\mathcal{F}^{\mathcal{A}}$, we have $(A,\alpha) \in \mathcal{F}^{\mathcal{A}}(P)$. ■

Completeness In order to show the completeness of the meta-interpreter we first prove a lemma stating that if an annotated atom $A\,\alpha$ is provable at the meta-level then all A labelled with a weaker annotation ($\gamma \sqsubseteq \alpha$) are provable at the meta-level, too.

Lemma 10 *Let P be a program and let $\mathcal{V}$ be the corresponding meta-program. For any $A \in \mathcal{A}$-base$_P$ and $\alpha \in Ann$, the following statement holds:*

$$demo(A\,\alpha) \in T_{\mathcal{V}}^{\omega} \quad \Longrightarrow \quad \{demo(A\,\gamma) \mid \gamma \in Ann, \mathcal{A} \models \gamma \sqsubseteq \alpha\} \subseteq T_{\mathcal{V}}^{\omega}$$

Proof The proof is carried out by induction on the number of iteration of $T_{\mathcal{V}}$ and by considering the different cases corresponding to clauses (9.3), (9.4), (9.5) and (9.6) of the meta-interpreter. ■

Now we can prove the completeness of the meta-interpreter with respect to the bottom-up semantics.

Theorem 11 (Completeness) *Let P be a program and let $\mathcal{V}$ be the corresponding meta-program. For any $A \in \mathcal{A}$-base$_P$ and $\alpha \in Ann$, the following statement holds:*

$$(A, \alpha) \in \mathcal{F}^{\mathcal{A}}(P) \quad \Longrightarrow \quad demo(A\,\alpha) \in T_{\mathcal{V}}^{\omega}.$$

Proof We first show that for all h

$$(A, \alpha) \in \mathcal{T}_P^h \quad \Longrightarrow \quad demo(A\,\alpha) \in T_{\mathcal{V}}^{\omega}. \tag{9.11}$$

The proof is by induction on h.

(Base case). Trivial since $\mathcal{T}_P^0 = \emptyset$.

(Inductive case). Assume that

$$(A, \alpha) \in \mathcal{T}_P^h \quad \Longrightarrow \quad demo(A\,\alpha) \in T_{\mathcal{V}}^{\omega}$$

Then:

$$\begin{aligned} & (A, \alpha) \in \mathcal{T}_P^{h+1} \\ \Longleftrightarrow\ & \{\text{definition of } \mathcal{T}_P^i\} \\ & (A, \alpha) \in \mathcal{T}_P(\mathcal{T}_P^h) \end{aligned}$$

We distinguish two cases according to the definition of $\mathcal{T}_P$.

(1) $\exists (B_1, \beta_1), \ldots, (B_n, \beta_n), C_1, \ldots, C_k, \alpha_1, \ldots, \alpha_n :$
$(\alpha = \mathtt{th}[s_1, s_2] \;\vee \alpha = \mathtt{in}[s_1, s_2]) \;\wedge$
$A\,\alpha \leftarrow C_1, \ldots, C_k, B_1\,\alpha_1, \ldots, B_n\,\alpha_n \in ground_{\mathcal{A}}(P) \;\wedge$
$\{(B_1, \beta_1), \ldots, (B_n, \beta_n)\} \subseteq \mathcal{T}_P^h \;\wedge$
$\mathcal{A} \models C_1, \ldots, C_k, \alpha_1 \sqsubseteq \beta_1, \ldots, \alpha_n \sqsubseteq \beta_n, s_1 \leq s_2$

$\implies$ {definition of meta-level representation}
$\exists (B_1,\beta_1),\ldots,(B_n,\beta_n),C_1,\ldots,C_k,\alpha_1,\ldots,\alpha_n :$
$(\alpha = \mathtt{th}\,[s_1,s_2] \vee \alpha = \mathtt{in}\,[s_1,s_2]) \wedge$
$\mathit{clause}(A\,\alpha,(C_1,\ldots,C_k,$
$B_1\,\alpha_1,\ldots,B_n\,\alpha_n)) \leftarrow s_1 \leq s_2 \in \mathit{ground}_{\mathcal{A}_m}(\mathcal{V}) \wedge$
$\{(B_1,\beta_1),\ldots,(B_n,\beta_n)\} \subseteq \mathcal{T}_P^h \wedge$
$\mathcal{A} \models C_1,\ldots,C_k,\alpha_1 \sqsubseteq \beta_1,\ldots,\alpha_n \sqsubseteq \beta_n, s_1 \leq s_2$
$\implies$ {definition and monotonicity of $T_{\mathcal{V}}$ and $\mathcal{A} \models s_1 \leq s_2$}
$\exists (B_1,\beta_1),\ldots,(B_n,\beta_n),C_1,\ldots,C_k,\alpha_1,\ldots,\alpha_n :$
$\mathit{clause}(A\,\alpha,(C_1,\ldots,C_k,B_1\,\alpha_1,\ldots,B_n\,\alpha_n)) \in T_{\mathcal{V}}^{\omega} \wedge$
$\{(B_1,\beta_1),\ldots,(B_n,\beta_n)\} \subseteq \mathcal{T}_P^h \wedge$
$\mathcal{A} \models C_1,\ldots,C_k,\alpha_1 \sqsubseteq \beta_1,\ldots,\alpha_n \sqsubseteq \beta_n$
$\implies$ {inductive hypothesis}
$\mathit{clause}(A\,\alpha,(C_1,\ldots,C_k,B_1\,\alpha_1,\ldots,B_n\,\alpha_n)) \in T_{\mathcal{V}}^{\omega} \wedge$
$\{\mathit{demo}(B_1\,\beta_1),\ldots,\mathit{demo}(B_n\,\beta_n)\} \subseteq T_{\mathcal{V}}^{\omega} \wedge$
$\mathcal{A} \models C_1,\ldots,C_k,\alpha_1 \sqsubseteq \beta_1,\ldots,\alpha_n \sqsubseteq \beta_n$
$\implies$ {Lemma 10 and $\mathcal{A} \models \alpha_1 \sqsubseteq \beta_1,\ldots,\alpha_n \sqsubseteq \beta_n$}
$\mathit{clause}(A\,\alpha,(C_1,\ldots,C_k,B_1\,\alpha_1,\ldots,B_n\,\alpha_n)) \in T_{\mathcal{V}}^{\omega} \wedge$
$\{\mathit{demo}(B_1\,\alpha_1),\ldots,\mathit{demo}(B_n\,\alpha_n)\} \subseteq T_{\mathcal{V}}^{\omega} \;\wedge\; \mathcal{A} \models C_1,\ldots,C_k$
$\implies$ {definition of $T_{\mathcal{V}}$, clause (9.7) used k times and $T_{\mathcal{V}}^{\omega}$ is a fixpoint of $T_{\mathcal{V}}$}
$\mathit{clause}(A\,\alpha,(C_1,\ldots,C_k,B_1\,\alpha_1,\ldots,B_n\,\alpha_n)) \in T_{\mathcal{V}}^{\omega} \wedge$
$\{\mathit{demo}(B_1\,\alpha_1),\ldots,\mathit{demo}(B_n\,\alpha_n),\mathit{demo}(C_1),\ldots,\mathit{demo}(C_k)\} \subseteq T_{\mathcal{V}}^{\omega}$
$\implies$ {definition of $T_{\mathcal{V}}$, clause (9.2) used $n+k-1$ times and $T_{\mathcal{V}}^{\omega}$ is a fixpoint of $T_{\mathcal{V}}$}
$\mathit{clause}(A\,\alpha,(C_1,\ldots,C_k,B_1\,\alpha_1,\ldots,B_n\,\alpha_n)) \in T_{\mathcal{V}}^{\omega} \wedge$
$\mathit{demo}((C_1,\ldots,C_k,B_1\,\alpha_1,\ldots,B_n\,\alpha_n)) \in T_{\mathcal{V}}^{\omega}$

Now by exploiting the fact that $T_{\mathcal{V}}^{\omega}$ is a fixpoint of $T_{\mathcal{V}}$ we can conclude $\mathit{demo}(A\,\alpha) \in T_{\mathcal{V}}^{\omega}$ by using respectively clause (9.3) if α is a `th` annotation and clause (9.6) if α is an `in` annotation.

(2) $\exists (B_1,\beta_1),\ldots,(B_n,\beta_n),C_1,\ldots,C_k,\alpha_1,\ldots,\alpha_n,\mathtt{th}\,[s_1,r_2],s_2,r_1 :$
$\alpha = \mathtt{th}\,[s_1,r_2] \wedge$
$A\,\mathtt{th}\,[s_1,s_2] \leftarrow C_1,\ldots,C_k,B_1\,\alpha_1,\ldots,B_n\,\alpha_n \in \mathit{ground}_{\mathcal{A}}(P) \wedge$
$\{(B_1,\beta_1),\ldots,(B_n,\beta_n),(A,\mathtt{th}\,[r_1,r_2])\} \subseteq \mathcal{T}_P^h \wedge$
$\mathcal{A} \models C_1,\ldots,C_k,\alpha_1 \sqsubseteq \beta_1,\ldots,\alpha_n \sqsubseteq \beta_n, s_1 < r_1, r_1 \leq s_2, s_2 < r_2$
Following the same steps of (1) we obtain
$\mathit{clause}(A\,\mathtt{th}\,[s_1,s_2],(C_1,\ldots,C_k,B_1\,\alpha_1,\ldots,B_n\,\alpha_n)) \in T_{\mathcal{V}}^{\omega} \wedge$
$\mathit{demo}((C_1,\ldots,C_k,B_1\,\alpha_1,\ldots,B_n\,\alpha_n)) \in T_{\mathcal{V}}^{\omega} \wedge$
$\mathit{demo}(A\,\mathtt{th}\,[r_1,r_2]) \in T_{\mathcal{V}}^{\omega} \;\wedge\; \mathcal{A} \models s_1 < r_1, r_1 \leq s_2, s_2 < r_2)$

By Lemma 10, $demo(A\,\mathtt{th}\,[s_2, r_2]) \in T_{\mathcal{V}}^{\omega}$ since $\mathcal{A} \models \mathtt{th}\,[s_2, r_2] \sqsubseteq \mathtt{th}\,[r_1, r_2]$. Now $\mathcal{A} \models s_1 \leq s_1, s_1 < s_2, s_2 < r_2$ then by using clause (9.4) we conclude $demo(A\,\mathtt{th}\,[s_1, r_2]) \in T_{\mathcal{V}}^{\omega}$.

We are now able to conclude the proof of completeness.

$$
\begin{array}{l}
\quad (A, \alpha) \in \mathcal{F}^{\omega}(P) \\
\Longrightarrow \{\text{definition of } \mathcal{F}^{\omega}(P)\} \\
\quad \exists \gamma \in Ann:\ (A, \gamma) \in \mathcal{T}_P^{\omega}\ \wedge\ \mathcal{A} \models \alpha \sqsubseteq \gamma \\
\Longrightarrow \{\mathcal{T}_P^{\omega} = \bigcup_{i \geq 0} \mathcal{T}_P^{i}\} \\
\quad \exists h:\ (A, \gamma) \in \mathcal{T}_P^{h}\ \wedge\ \mathcal{A} \models \alpha \sqsubseteq \gamma \\
\Longrightarrow \{\text{statement } (9.11)\} \\
\quad demo(A\,\gamma) \in T_{\mathcal{V}}^{\omega}\ \wedge\ \mathcal{A} \models \alpha \sqsubseteq \gamma \\
\Longrightarrow \{\text{Lemma } 10\} \\
\quad demo(A\,\alpha) \in T_{\mathcal{V}}^{\omega}
\end{array}
$$

■

4 RELATED WORK

In [26], Templog [1] and an interval based temporal logic are translated into generalized annotated logic programs. The annotations used there correspond to the th annotations of TACLP. To implement the annotated logic language, the paper proposed to use "reductants", additional clauses which are derived from existing clauses to express all possible least upper bounds. The problem was that a finite program may generate infinitely many such reductants. Then, "ca-resolution" for annotated logic programs was proposed [27]. The idea is to compute dynamically and incrementally the least upper bounds by collecting partial answers. Operationally this is similar to the meta-interpreter presented here which relies on recursion to collect the partial answers. However, in [27] the intermediate stages of the computation are not sound with respect to the standard *CLP* semantics.

Moreover, in [26] two fixpoint semantics, defined in terms of two different operators, are presented for generalized annotated programs (GAP). The first operator, called T_P, is based on interpretations which associate to each element of the Herbrand Base of the program P a *set* of annotations which is an ideal, i.e., a set downward closed and closed with respect to *finite* least upper bounds. The computed ideal is the least one containing the annotations α of annotated atoms $A\,\alpha$ which are heads of (instances of) clauses whose body holds in the interpretation. The other operator, R_P, is based on interpretations which associate to each atom of the Herbrand Base a *single* annotation which is the least upper bound of the set of annotations computed as in the previous case. Our fixpoint operator for TACLP works similarly to the T_P operator: At each step we close with respect to (representable) finite least upper bounds, and, although we perform the downward closure only at the end of the computation, this does not reduce the set of derivable consequences. The main difference resides in the language: TACLP is an extension of CLP, which focuses on temporal aspects, taking from GAP the basic ideas for the handling of annotations, whereas GAP is a general language with negation and arbitrary annotations but without constraints.

Our temporal annotations correspond to some of the predicates proposed by Galton in [22], which is a critical examination of Allen's classical work on a theory of action and time [2]. Galton provides for both time points and time periods in dense linear time. Assuming that the intervals I are not singletons, Galton's predicate *holds-in(A,I)* can be mapped into TACLP's $A\ \mathtt{in}\ I$, *holds-on(A,I)* into $A\ \mathtt{th}\ I$, and *holds-at(A,t)* into $A\ \mathtt{at}\ t$, where A is an atomic formula.

From this mapping it becomes clear that TACLP can be seen as reified FOL where annotated formulae, for example $born(John)\mathtt{at}\ t$, correspond to binary meta-relations between predicates and temporal information, for example $\mathtt{at}(born(John), t)$. But also, TACLP can be regarded as a modal logic, where the annotations are seen as parameterized modal operators, e.g. $born(John)\ (\mathtt{at}\ t)$.

In [9], a powerful temporal logic named MTL (tense logic extended by parameterized temporal operators) is translated into first order constraint logic. The resulting language subsumes Templog, as does TACLP. The parameterized temporal operators of MTL correspond to the temporal annotations of TACLP. The constraint theory of MTL is rather complex as it involves quantified variables and implication, whose treatment goes beyond standard CLP implementations. On the other hand, TACLP inherits an efficient standard constraint-based implementation of annotations from the ACL framework.

The ACL framework shares ideas with the work in [4]. There, a proof system for a large class of (propositional and) quantified modal logics is formalized. Soundness, completeness, and normalization can be proved uniformly for every logic in the class. The class is modular with respect to properties of the accessibility relation in the Kripke frame and leads to a simple implementation of a modal logic theorem prover in standard logical frameworks. Obviously, this approach is motivated by the same ideas that motivated ACL, though on a much wider and thus more abstract class of languages. What [4] calls *base logic*, corresponds to FOL in ACL, what is called *relational theory*, corresponds to the constraint theory of ACL. The relational theory defines the accessibility relation in the Kripke frame by (Horn) rules, while the constraint theory can be any FOL theory, as long as there are efficient algorithms to implement it. Already in [12] it has been argued that the accessibility relation can be regarded as constraint and the associated axioms as constraint theory.

In [6], translations between *signed logic* and FOL are investigated. Annotated logic [26] can be embedded in signed logic. Both formalisms work with labels whose structure forms a lattice. The two rules of regular unit resolution for signed logic correspond exactly to the inference rules of constraint annotated logics. In [6], the lattice structure is encoded by giving FOL clauses for each pair of lattice elements (i.e., labels). This is only feasible for finite lattices. In ACL, the use of a constraint theory also allows for infinite lattices using a small number of axioms.

5 CONCLUSIONS

We investigated the semantics of a considerable subset of the language TACLP that allows us to reason about qualitative and quantitative, definite and indefinite temporal information using time points and time periods. Its expressive power has been illustrated with some non-trivial examples.

We defined the operational (top-down) semantics of TACLP by presenting a meta-interpreter for it. Then we provided TACLP for the first time with a fixpoint (bottom-up) semantics, based on the definition of an immediate consequence operator. We proved that the meta-interpreter is sound and complete with respect to the bottom-up semantics. As future work it would be interesting to investigate operators similar to the function R_P defined in [26], and adapt our approach to the bottom-up semantics to the general framework of annotated constraint logic.

Moreover in this paper, we considered the subset of TACLP where time points are totally ordered, sets of time points are convex and non-empty, and only atomic formulae can be annotated. Furthermore clauses are free of negation. In general, in TACLP arbitrary formulae can be annotated. In some cases, as shown in [15], the annotations can be pushed inside disjunctions, conjunctions and negation. This means that the omission of negation is the main restriction of the current work. Consequently, we want to investigate next if and how the proofs relating the operational and fixpoint semantics can be adapted to deal with negation.

Finally, there are many applications that need the ability of storing and manipulating geometric and temporal data, such as geographic information systems (GIS), geometric modeling systems (CAD), and temporal databases. In such applications space and time are often closely interconnected: Much information which is referenced to space is also referenced to time. Therefore another interesting direction for future research is the representation and handling of spatio-temporal data using annotations.

Acknowledgments

We thank Paolo Baldan and Roberta Gori for their useful comments and suggestions and the anonymous referees for their constructive remarks. This work has been partially supported by Esprit Working Group 28115 - DeduGIS.

Alessandra Raffaetà
Dipartimento di Informatica
Università di Pisa
Corso Italia, 40
I-56125 Pisa, Italy
raffaeta@di.unipi.it

Thom Frühwirth
Institut für Informatik
Ludwig-Maximilians-Universität (LMU)
Oettingenstrasse 67
D-80538 Munich, Germany
fruehwir@informatik.uni-muenchen.de

References

[1] M. Abadi and Z. Manna. Temporal logic programming. *Journal of Symbolic Computation*, 8:277–295, 1989.

[2] J.F. Allen. Towards a general theory of action and time. *Artificial Intelligence*, 23:123–154, 1984.

[3] K. Apt and F. Turini, editors. *Meta-logics and Logic Programming*. MIT Press, London, 1995.

[4] D. Basin, S. Matthews, and L. Viganò. Labelled Modal Logics: Quantifiers. *Journal of Logic, Language, and Information*, 7(3), 1998.

[5] M. Baudinet, J. Chomicki, and P. Wolper. Temporal Deductive Databases. In *[38]*, pages 294–320.

[6] B. Beckert, R. Hähnle, and F. Manyá. The SAT Problem of Signed CNF Formulas. This volume.

[7] M. Böhlen and R. Marti. On the Completeness of Temporal Database Query Languages. In *Temporal Logic: Proceedings of the First International Conference, ICTL'94*, volume 827 of *Lecture Notes in Artificial Intelligence*, pages 283–300, 1994.

[8] A. Brogi and F. Turini. Meta-Logic for Program Composition: Semantics Issues. In *[3]*, pages 83–109.

[9] C. Brzoska. Temporal Logic Programming with Metric and Past Operators. In *[11]*, pages 21–39.

[10] J. Chomicki. Temporal Query Languages: A Survey. In *Temporal Logic: Proceedings of the First International Conference, ICTL'94*, volume 827 of *Lecture Notes in Artificial Intelligence*, pages 506–534. Springer Verlag, 1994.

[11] M. Fisher and R. Owens, editors. *Executable Modal and Temporal Logics*, volume 897 of *Lecture Notes in Artificial Intelligence*. Springer Verlag, 1995.

[12] A. M. Frisch and R. B. Scherl. A General framework for Modal Deduction. In *Proceedings of the second KR '91*, pages 196–207. Morgan Kaufmann, 1991.

[13] T. Frühwirth. Temporal Logic and Annotated Constraint Logic Programming. In *[11]*, pages 58–68.

[14] T. Frühwirth. Annotated Constraint Logic Programming Applied to Temporal Reasoning. In *Programming Language Implementation and Logic Programming (PLILP)*, volume 844 of *Lecture Notes in Computer Science*, pages 230–243. Springer Verlag, 1994.

[15] T. Frühwirth. Temporal Annotated Constraint Logic Programming. *Journal of Symbolic Computation*, 22:555–583, 1996.

[16] T. Frühwirth and S. Abdennadher. *Constraint-Programmierung: Grundlagen und Anwendungen.* Springer, Berlin, 1997.

[17] D. M. Gabbay, I. Hodkinson, and M. Reynolds. *Temporal Logic.* Clarendon Press, Oxford, 1994.

[18] D. M. Gabbay. Modal and temporal logic programming. In *[21]*, pages 197–237.

[19] D. M. Gabbay. *Labelled deductive systems: volume 1*, volume 33 of *Oxford logic guides.* Clarendon Press, Oxford, 1996.

[20] D. M. Gabbay and P. McBrien. Temporal Logic & Historical Databases. In *Proceedings of the Seventeenth International Conference on Very Large Databases*, pages 423–430, September 1991.

[21] A. Galton, editor. *Temporal Logics and Their Applications.* Academic Press, 1987.

[22] A. Galton. A Critical Examination of Allen's Theory of Action and Time. *Artificial Intelligence*, 42:159–188, 1990.

[23] P. Hill and J. W. Lloyd. Analysis of Metaprograms. In H.D. Abramson and M.H. Rogers, editors, *Metaprogramming in Logic Programming*, pages 23–52. MIT Press, 1989.

[24] J. Jaffar and J. L. Lassez. Constraint Logic Programming. In *Proceedings of the Fourteenth Annual ACM Symposium on Principles of Programming Languages*, pages 111–119, 1987.

[25] J. Jaffar and M. J. Maher. Constraint Logic Programming: A Survey. *Journal of Logic Programming*, 19 & 20:503–582, May 1994.

[26] M. Kifer and V. S. Subrahmanian. Theory of Generalized Annotated Logic Programming and its Applications. *Journal of Logic Programming*, 12:335–367, 1992.

[27] S. M. Leach and J. J. Lu. Computing Annotated Logic Programs. In *Proceedings of the eleventh ICLP*, pages 257–271, 1994.

[28] P. Mancarella, A. Raffaetà, and F. Turini. Temporal Annotated Constraint Logic Programming with Multiple Theories. In *Proceedings of the Tenth International Workshop on Database and Expert Systems Applications*, pages 501–508, IEEE Computer Society Press, 1999.

[29] K. Marriott, J. Jaffar, M. J. Maher, and P. J. Stuckey. The Semantics of Constraint Logic Programs. *Journal of Logic Programming*, 37(1-3):1–46, 1998.

[30] K. Marriott and P. J. Stuckey. *Programming with Constraints.* MIT Press, USA, 1998.

[31] B. Martens and D. De Schreye. Why Untyped Nonground Metaprogramming Is Not (Much Of) A Problem. *Journal of Logic Programming*, 22(1):47–99, January 1995.

[32] B. Moszkowski. *Execution Temporal Logic Programs.* Cambridge University Press, 1986.

[33] M. A. Orgun. On temporal deductive databases. *Computational Intelligence*, 12(2):235–259, May 1996.

[34] M. A. Orgun and W. Ma. An Overview of Temporal and Modal Logic Programming. In *Temporal Logic: Proceedings of the First International Conference, ICTL'94*, volume 827 of *Lecture Notes in Artificial Intelligence*, pages 445–479, 1994.

[35] J. Singer. Constraint-Based Temporal Logic Programming. BSc. dissertation, Department of Artificial Intelligence, University of Edinburgh, May 1996.

[36] R. Snodgrass. Temporal Databases. In *Proceedings of the International Conference on GIS - From Space to Territory: Theories and Methods of Spatio-Temporal Reasoning in Geographic Space*, pages 22–64, 1992.

[37] L. Sterling and E. Shapiro. *The Art of Prolog.* MIT Press, 1986.

[38] A. Tansel, J. Clifford, S. Gadia, S. Jajodia, A. Segev, and R. Snodgrass editors. *Temporal Databases: Theory, Design, and Implementation.* Benjamin/Cummings, 1993.

[39] J. F. A. K. van Benthem. *The logic of time: a model-theoretic investigation into the varieties of temporal ontology and temporal discourse*, volume 156 of *Synthese Library*. Reidel, Dordrecht, 1983.

[40] W. W. Wadge. Tense Logic Programming: a Respectable Alternative. In *Proceedings of the 1988 International Symposium on Lucid and Intensional Programming*, pages 26–32, 1988.

Appendix: Full Proofs

Theorem 5 (Continuity) *For any TACLP program P the function $\mathcal{T}_P$ is continuous (on $(\wp(\mathcal{A}\text{-}base_P \times Ann), \subseteq)$).*

Proof Let $\{I_i\}_{i\geq 0}$ be a chain in $(\wp(\mathcal{A}\text{-}base_P \times Ann), \subseteq)$, i.e., $I_0 \subseteq I_1 \subseteq \ldots \subseteq I_i \ldots$. Then we have to prove

$$(A,\alpha) \in \mathcal{T}_P\left(\bigcup_{i\geq 0} I_i\right) \Longleftrightarrow (A,\alpha) \in \bigcup_{i\geq 0} \mathcal{T}_P(I_i).$$

$(A,\alpha) \in \mathcal{T}_P(\bigcup_{i\geq 0} I_i)$
$\Longleftrightarrow$ {definition of $\mathcal{T}_P$}
$((\alpha = \mathtt{th}\,[s_1,s_2] \ \vee \ \alpha = \mathtt{in}\,[s_1,s_2]) \wedge$
$A\,\alpha \leftarrow C_1,\ldots,C_k, B_1\alpha_1,\ldots,B_n\alpha_n \in ground_{\mathcal{A}}(P) \wedge$
$\{(B_1,\beta_1),\ldots,(B_n,\beta_n)\} \subseteq \bigcup_{i\geq 0} I_i \wedge$
$\mathcal{A} \models C_1,\ldots,C_k, \alpha_1 \sqsubseteq \beta_1,\ldots,\alpha_n \sqsubseteq \beta_n, s_1 \leq s_2) \vee$
$(\alpha = \mathtt{th}\,[s_1,r_2] \wedge$
$A\,\mathtt{th}\,[s_1,s_2] \leftarrow C_1,\ldots,C_k, B_1\alpha_1,\ldots,B_n\alpha_n \in ground_{\mathcal{A}}(P) \wedge$
$\{(B_1,\beta_1),\ldots,(B_n,\beta_n)\} \subseteq \bigcup_{i\geq 0} I_i \ \wedge \ (A,\mathtt{th}\,[r_1,r_2]) \in \bigcup_{i\geq 0} I_i \wedge$
$\mathcal{A} \models C_1,\ldots,C_k, \alpha_1 \sqsubseteq \beta_1,\ldots,\alpha_n \sqsubseteq \beta_n, s_1 < r_1, r_1 \leq s_2, s_2 < r_2)$
$\Longleftrightarrow$ {property of set-theoretic union and $\{I_i\}_{i\geq 0}$ is a chain. Notice that for $(\Longrightarrow)$ j can be any element of the set $\{k \mid (B_i,\beta_i) \in I_k, i = 1,\ldots,n\}$ which is clearly not empty}
$((\alpha = \mathtt{th}\,[s_1,s_2] \ \vee \ \mathtt{in}\,[s_1,s_2]) \wedge$
$A\,\alpha \leftarrow C_1,\ldots,C_k, B_1\alpha_1,\ldots,B_n\alpha_n \in ground_{\mathcal{A}}(P) \wedge$
$\{(B_1,\beta_1),\ldots,(B_n,\beta_n)\} \subseteq I_j \wedge$
$\mathcal{A} \models C_1,\ldots,C_k, \alpha_1 \sqsubseteq \beta_1,\ldots,\alpha_n \sqsubseteq \beta_n, s_1 \leq s_2) \vee$
$(\alpha = \mathtt{th}\,[s_1,r_2] \wedge$
$A\,\mathtt{th}\,[s_1,s_2] \leftarrow C_1,\ldots,C_k, B_1\alpha_1,\ldots,B_n\alpha_n \in ground_{\mathcal{A}}(P) \wedge$
$\{(B_1,\beta_1),\ldots,(B_n,\beta_n)\} \subseteq I_j \ \wedge \ (A,\mathtt{th}\,[r_1,r_2]) \in I_j \wedge$
$\mathcal{A} \models C_1,\ldots,C_k, \alpha_1 \sqsubseteq \beta_1,\ldots,\alpha_n \sqsubseteq \beta_n, s_1 < r_1, r_1 \leq s_2, s_2 < r_2)$
$\Longleftrightarrow$ {definition of $\mathcal{T}_P$}
$(A,\alpha) \in \mathcal{T}_P(I_j)$
$\Longleftrightarrow$ {set-theoretic union}
$(A,\alpha) \in \bigcup_{i\geq 0} \mathcal{T}_P(I_i)$

■

Lemma 7 *Let P be a program and let $\mathcal{V}$ be the corresponding meta-interpreter. For any $B_1\,\alpha_1,\ldots,B_n\,\alpha_n$ with $B_i \in \mathcal{A}\text{-}base_P$ and $\alpha_i \in Ann$ and for any $C_1,\ldots,C_k$ with C_i an $\mathcal{A}$-ground instance of a constraint, the following statement holds:*

For all h $\quad demo((C_1,\ldots,C_k, B_1\,\alpha_1,\ldots,B_n\,\alpha_n)) \in T^h_{\mathcal{V}}$

$$\Longrightarrow \{demo(B_1\,\alpha_1),\ldots,demo(B_n\,\alpha_n)\} \subseteq T_{\mathcal{V}}^h \;\wedge\; \mathcal{A} \models C_1,\ldots,C_k$$

Proof Notice that if $n = 1$ and $k = 0$ the proof is immediate and if $n = 0$ and $k = 1$ the claim trivially follows by the definition of $T_{\mathcal{V}}$ and clause (9.7).

For $n + k \geq 2$, the proof is by induction on h.

(Base case). Trivial since $T_{\mathcal{V}}^0 = \emptyset$.

(Inductive case). Assume that

$$\begin{array}{l} demo((C_1,\ldots,C_k,B_1\,\alpha_1,\ldots,B_n\,\alpha_n)) \in T_{\mathcal{V}}^h \\ \quad \Longrightarrow \{demo(B_1\,\alpha_1),\ldots,demo(B_n\,\alpha_n)\} \subseteq T_{\mathcal{V}}^h \;\wedge\; \mathcal{A} \models C_1,\ldots,C_k \end{array}$$

Then:

$$\begin{array}{l}
demo((C_1,\ldots,C_k,B_1\,\alpha_1,\ldots,B_n\,\alpha_n)) \in T_{\mathcal{V}}^{h+1} \\
\Longleftrightarrow \{\text{definition of } T_{\mathcal{V}}^i\} \\
demo((C_1,\ldots,C_k,B_1\,\alpha_1,\ldots,B_n\,\alpha_n)) \in T_{\mathcal{V}}(T_{\mathcal{V}}^h) \\
\Longleftrightarrow \{\text{definition of } T_{\mathcal{V}}, \text{ clause (9.2) and} \\
\qquad (G_1,\ldots,G_{n+k}) = (C_1,\ldots,C_k,B_1\,\alpha_1,\ldots,B_n\,\alpha_n)\} \\
\exists\, 1 \leq r < n+k : \\
\{demo((G_1,\ldots,G_r)), demo((G_{r+1},\ldots,G_{n+k}))\} \subseteq T_{\mathcal{V}}^h \\
\Longrightarrow \{\text{inductive hypothesis or previous remark if } r = 1 \text{ or } r = n+k-1\} \\
\{demo(G_i) \mid 1 \leq i \leq r,\ G_i \text{ is an annotated atom}\} \subseteq T_{\mathcal{V}}^h \;\wedge \\
\mathcal{A} \models G_{m_1},\ldots,G_{m_i} \text{ with } G_{m_1},\ldots,G_{m_i} \text{ the constraints in } \{G_1,\ldots,G_r\} \\
\wedge\, \{demo(G_i) \mid r+1 \leq i \leq n+k,\ G_i \text{ is an annotated atom}\} \subseteq T_{\mathcal{V}}^h \;\wedge \\
\mathcal{A} \models G_{l_1},\ldots,G_{l_j} \text{ with } G_{l_1},\ldots,G_{l_j} \text{ the constraints in } \{G_{r+1},\ldots,G_{n+k}\} \\
\Longrightarrow \{\text{monotonicity of } T_{\mathcal{V}}\} \\
\{demo(B_1\,\alpha_1),\ldots,demo(B_n\,\alpha_n)\} \subseteq T_{\mathcal{V}}^{h+1} \;\wedge\; \mathcal{A} \models C_1,\ldots,C_k
\end{array}$$

■

Lemma 10 *Let P be a program and let $\mathcal{V}$ be the corresponding meta-program. For any $A \in \mathcal{A}\text{-}base_P$ and $\alpha \in Ann$, the following statement holds:*

$$demo(A\,\alpha) \in T_{\mathcal{V}}^\omega \quad \Longrightarrow \quad \{demo(A\,\gamma) \mid \gamma \in Ann, \mathcal{A} \models \gamma \sqsubseteq \alpha\} \subseteq T_{\mathcal{V}}^\omega$$

Proof We first show that for all h

$$demo(A\alpha) \in T_{\mathcal{V}}^h \Longrightarrow \{demo(A\gamma) \mid \gamma \in Ann, \mathcal{A} \models \gamma \sqsubseteq \alpha\} \subseteq T_{\mathcal{V}}^h \quad (9.A.1)$$

The proof is by induction on h.

(Base case). Trivial since $T_{\mathcal{V}}^0 = \emptyset$.

(Inductive case). Assume that

$$demo(A\,\alpha) \in T_{\mathcal{V}}^{h} \quad \Longrightarrow \quad \{demo(A\,\gamma) \mid \gamma \in Ann, \mathcal{A} \models \gamma \sqsubseteq \alpha\} \subseteq T_{\mathcal{V}}^{h}.$$

Then:

$$\begin{array}{ll} & demo(A\,\alpha) \in T_{\mathcal{V}}^{h+1} \\ \Longleftrightarrow & \quad \{\text{definition of } T_{\mathcal{V}}^{i}\} \\ & demo(A\,\alpha) \in T_{\mathcal{V}}(T_{\mathcal{V}}^{h}) \end{array}$$

We have four cases corresponding to clauses (9.3), (9.4), (9.5) and (9.6) of the meta-interpreter.

(clause (9.3)) $\{\alpha = \mathtt{th}\,[t_1, t_2]$, definition of $T_{\mathcal{V}}$ and clause (9.3)$\}$
$\{clause(A\,\mathtt{th}\,[s_1, s_2], B), demo(B)\} \subseteq T_{\mathcal{V}}^{h} \wedge \mathcal{A} \models s_1 \leq t_1, t_2 \leq s_2, t_1 \leq t_2$

The annotations β less than $\mathtt{th}\,[t_1, t_2]$, namely such that $\mathcal{A} \models \beta \sqsubseteq \mathtt{th}\,[t_1, t_2]$, are of two kinds:

- $\beta = \mathtt{th}\,[w_1, w_2]$ with $\mathcal{A} \models t_1 \leq w_1, w_1 \leq w_2, w_2 \leq t_2$. Hence $\mathcal{A} \models s_1 \leq w_1, w_1 \leq w_2, w_2 \leq s_2$ and by using definition $T_{\mathcal{V}}$, clause (9.3) and $\{clause(A\,\mathtt{th}\,[s_1, s_2], B), demo(B)\} \subseteq T_{\mathcal{V}}^{h}$ we can conclude $demo(A\,\mathtt{th}\,[w_1, w_2]) \in T_{\mathcal{V}}(T_{\mathcal{V}}^{h}) = T_{\mathcal{V}}^{h+1}$.
- $\beta = \mathtt{in}\,[w_1, w_2]$ with $\mathcal{A} \models w_1 \leq t_2, t_1 \leq w_2, w_1 \leq w_2$. We reason as in the previous case, but now using clause (9.5) and the fact that $\mathcal{A} \models w_1 \leq t_2, t_2 \leq s_2, s_1 \leq w_2, w_1 \leq w_2$. Hence we can conclude $demo(A\,\mathtt{in}\,[w_1, w_2]) \in T_{\mathcal{V}}^{h+1}$.

(clause (9.4)) $\{\alpha = \mathtt{th}\,[t_1, t_2]$, definition of $T_{\mathcal{V}}$ and clause (9.4)$\}$
$\{clause(A\,\mathtt{th}\,[s_1, s_2], B), demo(B), demo(A\,\mathtt{th}\,[s_2, t_2])\} \subseteq T_{\mathcal{V}}^{h} \wedge$
$\mathcal{A} \models s_1 \leq t_1, t_1 < s_2, s_2 < t_2$

The annotations β less than $\mathtt{th}\,[t_1, t_2]$ are of two kinds:

- $\beta = \mathtt{th}\,[w_1, w_2]$ with $\mathcal{A} \models t_1 \leq w_1, w_1 \leq w_2, w_2 \leq t_2$. We have three cases according to the relation of s_2 with w_1 and w_2:
  - $\mathcal{A} \models w_2 \leq s_2$. Then $\mathcal{A} \models s_1 \leq t_1, t_1 \leq w_1, w_1 \leq w_2, w_2 \leq s_2$, and by using the definition of $T_{\mathcal{V}}$, clause (9.3) and the fact that $clause(A\,\mathtt{th}\,[s_1, s_2], B) \in T_{\mathcal{V}}^{h}$ and $demo(B) \in T_{\mathcal{V}}^{h}$ we conclude that $demo(A\,\mathtt{th}\,[w_1, w_2]) \in T_{\mathcal{V}}^{h+1}$.
  - $\mathcal{A} \models s_2 \leq w_1, s_2 < w_2$. Then $\mathcal{A} \models \mathtt{th}\,[w_1, w_2] \sqsubseteq \mathtt{th}\,[s_2, t_2]$. Moreover, since $demo(A\,\mathtt{th}\,[s_2, t_2]) \in T_{\mathcal{V}}^{h}$ by inductive hypothesis we have that $demo(A\,\mathtt{th}\,[w_1, w_2]) \in T_{\mathcal{V}}^{h}$. By monotonicity of $T_{\mathcal{V}}$ we can conclude.

- $\mathcal{A} \models w_1 < s_2, s_2 < w_2$. Then $\mathcal{A} \models s_1 \leq w_1, w_1 < s_2, s_2 < w_2, w_2 \leq t_2$. Since $demo(A\,\mathtt{th}\,[s_2, t_2]) \in T_{\mathcal{V}}^h$ by inductive hypothesis $demo(A\,\mathtt{th}\,[s_2, w_2]) \in T_{\mathcal{V}}^h$ and by using the definition of $T_{\mathcal{V}}$, clause (9.4) and $\{clause(A\,\mathtt{th}\,[s_1, s_2], B), demo(B)\} \subseteq T_{\mathcal{V}}^h$ we can conclude that $demo(A\,\mathtt{th}\,[w_1, w_2]) \in T_{\mathcal{V}}^{h+1}$.

- $\beta = \mathtt{in}\,[w_1, w_2]$ with $\mathcal{A} \models w_1 \leq t_2, t_1 \leq w_2, w_1 \leq w_2$. We have two cases:

  - $[w_1, w_2] \cap [s_1, s_2] \neq \emptyset$. Hence $\mathcal{A} \models w_1 \leq s_2, s_1 \leq w_2, w_1 \leq w_2$, and by using the definition of $T_{\mathcal{V}}$, clause (9.5) and the fact that $\{clause(A\,\mathtt{th}\,[s_1, s_2], B), demo(B)\} \subseteq T_{\mathcal{V}}^h$ we can conclude that $demo(A\,\mathtt{in}\,[w_1, w_2]) \in T_{\mathcal{V}}^{h+1}$.
  - $[w_1, w_2] \cap [s_2, t_2] \neq \emptyset$ then $\mathcal{A} \models \mathtt{in}\,[w_1, w_2] \sqsubseteq \mathtt{th}\,[s_2, t_2]$. Moreover, since $demo(A\,\mathtt{th}\,[s_2, t_2]) \in T_{\mathcal{V}}^h$ by inductive hypothesis we have that $demo(A\,\mathtt{in}\,[w_1, w_2]) \in T_{\mathcal{V}}^h$. By monotonicity of $T_{\mathcal{V}}$ we can conclude.

(clause (9.5)) $\{\alpha = \mathtt{in}\,[t_1, t_2]$, definition of $T_{\mathcal{V}}$ and clause (9.5)$\}$
$\{clause(A\,\mathtt{th}\,[s_1, s_2], B), demo(B)\} \subseteq T_{\mathcal{V}}^h \wedge \mathcal{A} \models t_1 \leq s_2, s_1 \leq t_2, t_1 \leq t_2$

If $\mathcal{A} \models \gamma \sqsubseteq \mathtt{in}\,[t_1, t_2]$ then $\gamma = \mathtt{in}\,[w_1, w_2]$ with $\mathcal{A} \models w_1 \leq t_1, t_2 \leq w_2$. Hence we have $\mathcal{A} \models w_1 \leq s_2, s_1 \leq w_2, w_1 \leq w_2$ and by using definition $T_{\mathcal{V}}$, clause (9.5) and $\{clause(A\,\mathtt{th}\,[s_1, s_2], B), demo(B)\} \subseteq T_{\mathcal{V}}^h$ we can conclude that $demo(A\,\mathtt{in}\,[w_1, w_2]) \in T_{\mathcal{V}}^{h+1}$.

(clause (9.6))] $\alpha = \mathtt{in}\,[t_1, t_2]$. Reasoning as in the previous case, by using the definition of $T_{\mathcal{V}}$ and clause (9.6), we can conclude.

Now the claim of the theorem immediately follows from the above statement and the definition of $T_{\mathcal{V}}^{\omega}$.

$$
\begin{array}{l}
\quad demo(A\,\alpha) \in T_{\mathcal{V}}^{\omega} \\
\Longrightarrow \{T_{\mathcal{V}}^{\omega} = \bigcup_{i \geq 0} T_{\mathcal{V}}^{i}\} \\
\quad \exists h:\ demo(A\,\alpha) \in T_{\mathcal{V}}^{h} \\
\Longrightarrow \{\text{Statement (9.A.1)}\} \\
\quad \{demo(A\,\gamma) \mid \gamma \in Ann, \mathcal{A} \models \gamma \sqsubseteq \alpha\} \subseteq T_{\mathcal{V}}^{h} \\
\Longrightarrow \{T_{\mathcal{V}}^{\omega} = \bigcup_{i \geq 0} T_{\mathcal{V}}^{i}\} \\
\quad \{demo(A\,\gamma) \mid \gamma \in Ann, \mathcal{A} \models \gamma \sqsubseteq \alpha\} \subseteq T_{\mathcal{V}}^{\omega}
\end{array}
$$

■

THE LOGIC OF REUSABLE PROPOSITIONAL OUTPUT WITH THE FULFILMENT CONSTRAINT

Leendert van der Torre

Abstract This paper shows the equivalence of three ways of expressing a certain strong consistency constraint – called the fulfilment constraint – on proofs of the logic of reusable propositional output: as a global requirement on proofs, as a local requirement on labels of formulas, and by phasing of proof rules. More specifically, we first show that the fulfilment constraint may be expressed either as a requirement on the historical structure of the proof tree or as a requirement on the contents of labels attached to its nodes. Second, we show that labelled proofs may be rewritten into a tightly phased form in which rules are applied in a fixed order. Third, we show that when a proof is in such a phased form, the consistency check on labels becomes redundant.

1 INTRODUCTION

In this paper we consider the logic of reusable propositional output [5] with a strong consistency constraint called the fulfilment constraint. It has the following two characteristic properties.

The identity 'if input a then output a' is *not* a theorem. To avoid confusion with conditional or counterfactual logic we therefore write (a, x) for 'if input a then output x' instead of $a \to x$ or $a > x$. The logic has the usual rules of strengthening of the input, weakening of the output, cumulative transitivity or cut, the conjunction rule for the output and the disjunction rule for the input.

Proofs are constrained in the sense that for each case each derived conditional can be achieved or fulfilled together with the conditionals it is derived from. The restriction to cases – as we explain in detail later – enables complex forms of reasoning by cases.

Consistency constraints usually lead to non-monotonic behavior. However, no close relationship to well-known non-monotonic logics is to be expected, not only because the underlying logic (of reusable output) is different from classical

D. Basin et al. (eds.), Labelled Deduction, 245–266.
© 2000 *Kluwer Academic Publishers. Printed in the Netherlands.*

logic, but also because the form of our fulfilment constraint is different from consistency constraints used in non-monotonic reasoning.

The fulfilment constraint was originally proposed to formalize preferences, desires and goals of a planning agent in a decision-theoretic context [8, 9, 10], though it can also formalize for example obligations and beliefs in normative and epistemic contexts. Desirable properties for this decision-theoretic context are that the logic validates strong proof rules like strengthening of the input and transitivity, and that it – as we again explain in detail later – supports context-sensitive and conflict-tolerant reasoning. As a reminder of this original motivation we refer to the premise set of conditionals as the goal base or G.

This paper shows the equivalence of three ways of expressing the fulfilment constraint on derivations in the logic of reusable propositional output: as a global requirement on derivations, as a local requirement on labels, and by phasing. The second and third way are implemented in a labelled deductive system [2]. Labelled conditionals $(a, x)_L$ can *roughly* be read as 'if a then x, against the proof background of L.' The label L contains information about the context in which the goal is derived.

- In the second way, the label L keeps track of the *conditionals* used to derive $(a, x)_L$. The global consistency constraint on fulfilments of conditionals in proofs is shown to be equivalent to a local consistency constraint on fulfilments of conditionals in this label.
- In the third way, the label L keeps track of the *proof rules* used to derive $(a, x)_L$. Each proof rule is associated with a so-called phase, an integer, and a proof rule may no longer be used once a proof rule from a higher phase has been applied.

The equivalence between the global and local constraint is not surprising, in contrast to their equivalence to phasing. The latter can also be expressed as follows. In the logic of reusable propositional output with the fulfilment constraint, the conditional (a, x) can be derived from the conditional base G if and only if there is a phase-4 conditional (a, x) of G, which is stepwise defined as follows. Ignoring replacements by logical equivalents for input and output, the conditional (a, x) is

a phase-1 conditional of G if $a \wedge x$ is consistent and there is a conditional $(b, x) \in G$ such that a classically implies b, i.e. $a \vdash b$;

a phase-2 conditional of G if there is a set of phase-1 conditionals of G $\{(a, x_1), (a \wedge x_1, x_2), \ldots, (a \wedge x_1 \wedge \ldots \wedge x_{n-1}, x_n)\}$ such that x is classically equivalent with $x_1 \wedge \ldots \wedge x_n$, i.e. $x \dashv\vdash x_1 \wedge \ldots \wedge x_n$;

a phase-4 conditional of G if there is a set of phase-2 conditionals of G $\{(a_1, x_1), \ldots, (a_n, x_n)\}$ such that $a \dashv\vdash a_1 \vee \ldots \vee a_n$ and $x_i \vdash x$ for all $i = 1 \ldots n$.

This paper is organized as follows. We first introduce the logic of reusable propositional output (Section 2) and the fulfilment constraint as a global requirement (Section 3), and we illustrate them by several examples. Then we introduce the labelled deductive system and we show that the fulfilment constraint may be expressed indifferently as a requirement on the historical structure of the proof tree or as a requirement on the contents of labels attached to its nodes (Section 4). Thereafter we show that labelled proofs may be brought into a tightly phased form in which rules are applied in a fixed order, and when a proof is in such a phased form, the consistency check on the labels becomes redundant (Section 5). Finally, we illustrate phasing by reconsidering the examples (Section 6).

2 REUSABLE PROPOSITIONAL OUTPUT

David Makinson (private communication) introduced the concept of 'reusable output' to formalize a black box, in which we may feed propositions, and that produces propositions as output, with every element of output available for re-use as input. For example, classical consequence $\vdash$ may be seen in this way. Every element of the output may be recycled as input, for it satisfies the rule known as cumulative transitivity or cut, that if $a \vdash x$ and $a \wedge x \vdash y$ then $a \vdash y$. This tells us that output x may be used as input alongside a and anything thus obtained is considered to be an output of a itself. Classical consequence, as well as other conditional logics satisfying the cut rule, has a further feature: inputs are themselves outputs, since $a \vdash a$ for any consequence relation. However, in certain contexts this is undesired. For example, in the logic of obligations – called deontic logic – cumulative transitivity – called (cautious) deontic detachment – is desired, but deriving is from ought or vice versa is undesired. In [5] the following logic of reusable propositional output has been introduced.[1]

Definition 1 (The logic of reusable propositional output) *The language consists of conditionals (a, x), with (input) a and (output) x sentences of a propositional base logic $\mathcal{L}$. We say that a conditional (a, x) is consistent if and only if $a \wedge x$ is consistent in $\mathcal{L}$, and a set of conditionals or conditional base is consistent if and only if each of its elements is. The consistent conditional base G implies (a, x) in the logic of reusable propositional output, written as $G \vdash_{\text{RO}} (a, x)$, if and only if (a, x) can be derived from G with replacements by logical equivalents (for input and output) and the following five rules.*

$$\text{SI}: \frac{(a,x)}{(a \wedge b, x)} \qquad \text{CT}: \frac{(a,x),(a \wedge x, y)}{(a,y)} \qquad \text{AND}: \frac{(a,x),(a,y)}{(a, x \wedge y)}$$

[1]In [5] the logic of reusable propositional output also satisfies the axiom $(a, \top)$ to facilitate the semantic presentation. Since we do not consider semantics in this paper, we do not consider this axiom either.

$$\text{WO}: \frac{(a,x)}{(a,x \vee y)} \qquad\qquad \text{OR}: \frac{(a,x),(b,x)}{(a \vee b,x)}$$

We give two examples of typical derivations of the logic. The first derivation illustrates how cumulative transitivity together with strengthening of the input formalizes that conditional rules can be applied one after the other. Derivations go 'as far as possible.'

$$\dfrac{(a,b) \quad \dfrac{(b,c)}{(a \wedge b,c)}\,\text{SI}}{(a,c)}\,\text{CT} \qquad \dfrac{(c,d)}{(a \wedge c,d)}\,\text{SI}$$
$$\dfrac{(a,c) \qquad (a \wedge c,d)}{(a,d)}\,\text{CT}$$

The second derivation (taken from [4]) illustrates how the disjunction rule supports reasoning by cases, where we write $\top$ for any tautology like $p \vee \neg p$.

$$\dfrac{\dfrac{(a,x \wedge y)}{(a,x)}\,\text{WO} \qquad \dfrac{(\neg a,x \wedge \neg y)}{(\neg a,x)}\,\text{WO}}{(\top,x)}\,\text{OR}$$

The basic idea of the additional consistency constraints on proofs of the logic of reusable output is to block derivations from conflicting conditionals, without blocking the latter derivation for reasoning by cases.

3 A GLOBAL CONSTRAINT ON PROOFS

To deal with reasoning by cases, the OR rule is thought of as creating different cases or parallel dependency tracks. Given a node in a derivation tree, we travel upwards the derivation tree generated by the node, splitting in two cases at every application of the OR rule, such that one case contains one of the two premise nodes and the other case contains the other premise node.

Definition 2 (Cases) *Consider a tree-like proof of a conditional (the root) from a set of premises (the leaves). A case or dependency tree of a proof is represented by a set of nodes from the tree. The set of cases of the proof can be computed with the following procedure.*

1. *Initially, the set C contains one element, the set with the root.*
2. *As long as there is a non-leave node in some set of C without one of its parents, then either:*
    - *if the node is not derived by* OR, *then add its parents to the set;*
    - *otherwise create two copies of the set, one for each of its parents; add this parent.*

3. *Each set of nodes in C represents a case of the proof.*

In [5] three different options for additional consistency constraints on derivations have been discussed. They are either based on materializations ($a \rightarrow x$), outputs (x) or fulfilments ($a \wedge x$) of the nodes (a, x) of a case. It is suggested that the materialization constraint is best suited to formalize normative reasoning, due to its relation with violations: (a, x) is violated if its materialization $a \rightarrow x$ is false and thus $a \wedge \neg x$ is true. In this paper we only consider the latter most cautious fulfilment constraint. As argued extensively in [12, 14, 15] we think that this constraint is best suited for decision-theoretic reasoning, because it not only refers to qualitative abstractions of penalties of violations but also of rewards of fulfilments.

Definition 3 (The fulfilment constraint) *A consistent conditional base G implies (a, x) in the logic of reusable propositional output with the fulfilment constraint, written as $G \vdash_{\text{ROF}} (a, x)$, if and only if $G \vdash_{\text{RO}} (a, x)$ and for each case of this proof $\{(b_1, y_1), \ldots, (b_m, y_m)\}$ we have that $b_1 \wedge y_1 \wedge \ldots \wedge b_m \wedge y_m$ is consistent in $\mathcal{L}$.*

Due to the consistency constraint, the inference relation is no longer idempotent. That is, if we write $out(G)$ for $\{(a, x) \mid G \vdash_{\text{RO}} (a, x)\}$ and $outF(G)$ for $\{(a, x) \mid G \vdash_{\text{ROF}} (a, x)\}$, then we have $out(G) = out(out(G))$ because out is a closure operator, but we do not have $outF(G) = outF(outF(G))$. The inference relation $\vdash_{\text{ROF}}$, just like $\vdash_{\text{RO}}$, is monotonic ($outF(G) \subseteq outF(G \cup G')$) and compact ($(a, x) \in outF(G)$ implies $(a, x) \in outF(G')$ for some finite $G' \subseteq G$).

The fulfilment consistency constraint has originally been proposed to formalize preferences, desires and goals of a planning agent [8, 9, 10]. Two important properties for this decision-theoretic context are that the logic with this constraint supports context-sensitive and conflict-tolerant reasoning. First, context-sensitive reasoning – called contrary-to-duty reasoning in a normative context [13] – refers to the situation in which a conditional has been 'violated', i.e. in which its materialization $a \rightarrow x$ is false. We say that (b, y) is a secondary conditional of the primary conditional (a, x) if and only if $b \wedge (a \rightarrow x)$ is classically inconsistent. Due to the consistency check, primary and secondary conditionals cannot be combined in a proof, nor can one be derived from the other. For example, consider a decision-theoretic variant of the so-called Chisholm paradox [1]. A robot may have the goal 'to get me some coffee' $(\top, c)$, 'to inform me that coffee will arrive if it gets me some coffee' (c, i), and 'not to inform me that coffee will arrive if it does not get me coffee' $(\neg c, \neg i)$. The secondary conditional $(\neg c, \neg i)$ refers to the 'no coffee' context in which the primary conditional is violated. The coffee machine may be broken, or – if the robot is autonomous – it may decide that other actions have a higher priority.

The consistency check blocks the second step of the following counterintuitive derivation of the goal 'to inform me that coffee will arrive if it does not get me coffee' in the unrestricted logic of reusable propositional output.

$$\dfrac{\dfrac{\dfrac{(\top,c)\quad(c,i)}{(\top,i)}\ \text{CT}}{(\neg c,i)}\ \text{SI}\qquad(\neg c,\neg i)}{(\neg c,i\wedge\neg i)}\ \text{AND}$$

Second, the logic supports reasoning with conflicting preferences, for example due to conflicting objectives. The consistency check ensures that we have $(\top,x),(\top,\neg x)\not\vdash_{\text{ROF}}(\top,x\wedge\neg x)$ and $(\top,x),(\top,\neg x)\not\vdash_{\text{ROF}}(\top,y)$ if y is not logically implied by x or $\neg x$. In particular, the second derivation step in the following counterintuitive derivation in the unrestricted logic of reusable propositional output is blocked.

$$\dfrac{\dfrac{\dfrac{(\top,x)}{(\top,x\vee y)}\ \text{WO}\qquad(\top,\neg x)}{(\top,y\wedge\neg x)}\ \text{AND}}{(\top,y)}\ \text{WO}$$

The latter derivation also illustrates why the logic has a global consistency constraint instead of a local one at each derivation step. Each conditional in the proof is consistent, but still counterintuitive conclusions can be derived. Finally, notice that the following derivations – so-called forbidden conflict, proof by contradiction and contraposition – would be blocked by the consistency constraint if they were valid in the underlying logic. The first two combine conditionals with contradictory fulfilments and the latter derives a conditional whose input is inconsistent with the fulfilment of the premise.

$$\text{FC}:\ \dfrac{(a,x),(a\wedge b,\neg x)}{(a,\neg b)}\qquad\text{PBC}:\ \dfrac{(a,\bot)}{(\top,\neg a)}\qquad\text{CP}:\ \dfrac{(a,x)}{(\neg x,\neg a)}$$

In the following section we formalize the fulfilment consistency constraint as a local requirement on labels in a labelled deductive system, and we show that the global requirement on proofs and the local requirement on labels are equivalent.

4 A LOCAL CONSTRAINT ON LABELS

The examples in the previous section raise the question whether the concept of reusable output can be combined with context sensitivity in a logic that is – besides monotonic and compact – also idempotent. This combination (without conflict tolerance) has been discussed during three decades of research in deontic logic, following the publication of Chisholm's paradox [1] and several other notorious deontic paradoxes, and more recently it has been discussed (together with conflict tolerance) in qualitative decision theory. Several approaches have been proposed, see e.g. [13] for a survey, but no consensus has been reached. The derivation of $(\neg c, i)$ from $(\top, c)$ and (c, i) from Chisholm's set in the previous section indicates that we have to extend the language or introduce other machinery to combine cumulative transitivity with strengthening of the input. In [13] we therefore propose to formalize reusable output by the following adapted cumulative transitivity rule: if we have (a, x) and $(a \wedge x, y)$ then we can derive $(a, x \wedge y)$ instead of (a, y). This rule can be combined with strengthening of the input (conflict tolerant) and extended with the conjunction rule (conflict intolerant) by adding a local consistency constraint in the following way.[2]

Definition 4 (Logic of imperatives) *The consistent conditional base G implies (a, x) in the logic of imperatives, written as $G \vdash_{\mathrm{I}} (a, x)$, if and only if (a, x) can be derived from G with replacements by logical equivalents (for input and output) and the following two rules, extended with the following condition R_{IO} with a consistency constraint on the conjunction of the input and output.*

R_{IO}: *(a, x) may only be derived if $a \wedge x$ is consistent: it must always be possible to fulfill a derived conditional.*

$$\mathrm{SI}_{R_{IO}} : \frac{(a, x), R_{IO}}{(a \wedge b, x)} \qquad \mathrm{CTA}_{R_{IO}} : \frac{(a, x), (a \wedge x, y), R_{IO}}{(a, x \wedge y)}$$

Moreover, this minimal logic of imperatives can be extended with the unrestricted conjunction rule.

$$\mathrm{AND} : \frac{(a, x), (a, y)}{(a, x \wedge y)}$$

Obviously the logic of imperatives is a closure operator and thus monotone, compact and idempotent. Moreover, reconsider our variant of Chisholm's paradox with the robot's goals 'to get me some coffee' $(\top, c)$, 'to inform me that coffee will arrive if it gets me some coffee' (c, i), and 'not to inform me

[2] The deontic logic proposed in [13] differs from the logic of imperatives in this paper in several aspects. For example, it satisfies the axiom (a, a) to facilitate the semantic (preference-based) presentation, and it satisfies the so-called 'deontic' axiom $\neg(a, \bot)$, where $\bot$ stands for any contradiction like $p \wedge \neg p$, to express that 'ought implies can.' We do not consider the former because in this paper we are not interested in semantics, and we do not consider the latter because we cannot express this axiom in our simple formal language. Moreover, this deontic logic has the conjunction rule for the output, but it only has a restricted strengthening of the input rule to deal with a particular problem of reasoning about normative conflicts: the consistency of 'the window should be open if the sun shines' and 'it should be closed if it rains' [16].

that coffee will arrive if it does not get me coffee' $(\neg c, \neg i)$. In the logic of imperatives the goal 'to get me some coffee and inform me about it' $(\top, c \wedge i)$ can be derived, but $(\top, i)$ and $(\neg c, i)$ cannot be derived.

$$\frac{(\top, c) \quad (c, i)}{(\top, c \wedge i)} \text{ CTA}$$

The logic of imperatives cannot be extended with WO, because the counter-intuitive derivation of Chisholm's paradox would directly be reinstated, and the extension with OR also leads to counterintuitive consequences. However, these two derivations are in most cases intuitive. For example, in a decision-theoretic context WO can be used to infer from a goal for $x \wedge y$ that there are two subgoals, one for x and one for y. Moreover, the OR rule can be used to reason by cases. The formal language has therefore been extended with labels [11, 4, 8, 9, 10] in versions of a labelled deductive system as it was introduced by Gabbay in [2]. This extension can be based on materializations, outputs or fulfilments, just like the consistency constraints in the logic of reusable output. Again we only consider the latter most cautious option.

In the following definition we introduce the first labelled deductive system of this paper. Roughly speaking, the label L of a conditional $(a, x)_L$ consists of a record of the fulfilments (F) of the premises that are used in the derivation of $(a, x)_L$. Where there is no application of reasoning by cases, F can be taken to be a boolean formula, that grows by conjunction as premises are combined. But in general, to cover the parallel tracks created through reasoning by cases, we need to consider sets of boolean formulas [4]. Each formula $(a, x)_L$ occurring as a premise has a label that consists of its own (propositionally consistent) fulfilment. The label of a derived formula is the union (OR) or the product (SI, CTA) of the fulfilments of the premises used in this inference rule, where the product is defined by

$$\{a_1, \ldots, a_n\} \times \{b_1, \ldots, b_m\} = \{a_1 \wedge b_1, \ldots, a_1 \wedge b_m, \ldots, a_n \wedge b_m\}$$

The labels are used to check that fulfilments are consistent.

Definition 5 (Labelled deductive system with fulfilment constraint) *Let $\mathcal{L}$ be a propositional base logic. The language of the labelled system consists of the labelled conditionals $(a, x)_L$, with a and x sentences of $\mathcal{L}$, and L a set of sentences of $\mathcal{L}$ (fulfilments). A premise of the labelled system is a formula $(a, x)_{\{a \wedge x\}}$, where $a \wedge x$ is consistent in $\mathcal{L}$. The labelled deductive system with the fulfilment constraint consists of the inference rules below, extended with the following condition R_F.*

R_F: *$(a, x)_F$ may only be derived if each $f \in F$ is consistent: it must always be possible to fulfill a derived conditional and each of the conditionals* *it is derived from, though – for reasoning by cases – not necessarily all of them at the same time.*

The inference rules of the labelled system are replacements by logical equivalents (for input and output) and the following four rules.

$$\mathrm{SI}_{R_F} : \frac{(a,x)_F, R_F}{(a \wedge b, x)_{F \times \{b\}}} \qquad \mathrm{CTA}_{R_F} : \frac{(a,x)_{F_1}, (a \wedge x, y)_{F_2}, R_F}{(a, x \wedge y)_{F_1 \times F_2}}$$

$$\mathrm{WO}_{R_F} : \frac{(a,x)_F, R_F}{(a, x \vee y)_F} \qquad \mathrm{OR}_{R_F} : \frac{(a,x)_{F_1}, (b,x)_{F_2}, R_F}{(a \vee b, x)_{F_1 \cup F_2}}$$

We say $(a_1, x_1), \ldots, (a_n, x_n) \vdash_{\mathrm{LDSF}} (a, x)$ *if there is a conditional* $(a, x)_L$ *that can be derived from* $\{(a_1, x_1)_{\{a_1 \wedge x_1\}}, \ldots, (a_n, x_n)_{\{a_n \wedge x_n\}}\}$.

To show the equivalence between the logic of reusable propositional output with the fulfilment constraint and the labelled deductive system, we have to show that they validate the same proof rules and that it is equivalent to check at each node of the proof tree (as in the labelled deductive system) or only at the root (as in the logic). We therefore prove several lemmas. The first one shows that the logic implies the counterparts of the proof rules of the labelled system.

Lemma 6 *The following rule is implied by the logic of reusable output.*

$$\mathrm{CTA} : \frac{(a,x), (a \wedge x, y)}{(a, x \wedge y)}$$

Proof We have the following derivation.

$$\dfrac{(a,x) \qquad \dfrac{(a,x) \quad (a \wedge x, y)}{(a,y)}\,\mathrm{CT}}{(a, x \wedge y)}\,\mathrm{AND}$$

The fulfilment constraint holds this derivation if it holds for CTA. ■

Before we show that the labelled deductive system implies the proof rules of the logic, we prove the following useful lemma.

Lemma 7 *For each conditional* $(a, x)_F$ *derived in the labelled deductive system we have that each element of* F *classically implies* $a \wedge x$.

Proof By induction on the structure of the proof tree. The property trivially holds for the premises, and it is easily seen that the proof rules retain the property. ■

The third lemma shows that the labelled deductive system, and in particular the unusual rule CTA, implies counterparts of the proof rules of the logic, such as the cumulative transitivity rule and the conjunction rule.

Lemma 8 *The following cumulative transitivity rule and conjunction rule are implied by the labelled deductive system with the fulfilment constraint.*

$$\mathrm{CT}_{R_F}: \frac{(a,x)_{F_1}, (a \wedge x, y)_{F_2}, R_F}{(a,y)_{F_1 \times F_2}} \qquad \mathrm{AND}_{R_F}: \frac{(a,x)_{F_1}, (a,y)_{F_2}, R_F}{(a, x \wedge y)_{F_1 \times F_2}}$$

Proof We have the following derivations.

$$\dfrac{\dfrac{(a,x)_{F_1} \quad (a \wedge x, y)_{F_2}}{(a, x \wedge y)_{F_1 \times F_2}}\ \mathrm{CTA}}{(a,y)_{F_1 \times F_2}}\ \mathrm{WO} \qquad \dfrac{(a,x)_{F_1} \quad \dfrac{(a,y)_{F_2}}{(a \wedge x, y)_{F_2 \times \{x\}}}\ \mathrm{SI}}{(a, x \wedge y)_{F_1 \times F_2 \times \{x\}}}\ \mathrm{CTA}$$

Moreover, from Lemma 7 follows that in the second derivation each element of F_1 implies x, and therefore $F_1 \times F_2 \times \{x\}$ is equivalent with $F_1 \times F_2$. From the consistency of each element of $F_1 \times F_2$ (root of the left side derivation) follows the consistency of each element of $F_1 \times F_2 \times \{x\}$ (root of the right side derivation) and the consistency of each element of $F_2 \times \{x\}$ (other node of the right side derivation). ∎

The fourth lemma below shows – as the last step of the previous proof already suggested – that it is equivalent to check the consistency of the labels at each node or only at the root.

Lemma 9 *Consider any potential derivation of the labelled system, not necessarily satisfying R_F. Then the following two conditions are equivalent:*

1. *The final derivation satisfies condition R_F,*
2. *The derivation satisfies R_F everywhere.*

Proof The labels are, in a suitable sense, cumulative. Every element of every label in the derivation is classically implied by some element of the label of the final conclusion. ∎

The first theorem shows the equivalence between the global constraint on proofs and the local constraint on labels of formulas.

Theorem 10 (Equivalence ROF and LDSF) *Let G be a consistent conditional base. We have $G \vdash_{\mathrm{ROF}} (a,x)$ if and only if $G \vdash_{\mathrm{LDSF}} (a,x)$.*

Proof For each proof of (a, x) from G in one system, we can give a proof in the other system. This proof contains two parts. First we show how a proof tree of the second system can be constructed from the proof tree of the first system, and then we show that if the fulfilment constraint holds for the first proof, then it must also hold for the second proof.

$\Rightarrow$ Given a proof of (a, x) in the logic of reusable output with the fulfilment constraint. To construct a proof in the labelled deductive system, first replace occurrences of CT and AND by equivalent proofs using SI, CTA and WO, as indicated in Lemma 8. Then label the premises with the conjunction of their input and output, and label the other nodes as indicated by the rules of the labelled system. We now show that if the consistency constraint holds for the first proof, then it also holds for the second proof. First, the consistency of all cases implies the consistency for all elements of the label of the root. Second, the consistency of all elements of the label of the root implies the consistency of all elements of the label of each node, see Lemma 9.

$\Leftarrow$ Given a proof of (a, x) in the labelled deductive system. To construct a proof in the logic of reusable output, first replace occurrences of CTA by equivalent proofs using CT and AND, see Lemma 6, and then remove all labels. The consistency of each element of the label of the root in the first proof implies the consistency of a case, because each element of F in the label implies the fulfilments of all nodes of a case, and we are done. ∎

The first theorem and its proof has been instructive but relatively straightforward. We now proceed with the third way of expressing the fulfilment constraint.

5 PHASING OF PROOF RULES

There is another way in which the logic of imperatives in the previous section has been extended to incorporate the proof rules WO and OR: by so-called phasing of proof rules [6, 7, 9].[3] The idea of phasing is that the proof rules can only be applied in a restricted order. The counterintuitive derivations cannot be derived when SI and CTA cannot be applied once either WO or OR has been applied. In this section the strict order SI, CTA, WO and OR is implemented in the second version of a labelled deductive system.

The allowed order of application is represented by a phasing function ρ, defined by $\rho(\text{SI}) = 1$, $\rho(\text{CTA}) = 2$, $\rho(\text{WO}) = 3$ and $\rho(\text{OR}) = 4$. The label L of a premise $(a, x)_L$ is 0, and the label of a derived formula is the phase of the

---

[3] Phasing has been introduced in a two-phased preference-based deontic logic [6], an extension of Prohairetic Deontic Logic [13] (see the discussion on the logic of imperatives in the previous section) with Hansson's logic DSDL3 [3]. Again this logic differs in several aspects from the logic presented here, for example it contains the axiom (a, a) to facilitate the semantic preference-based presentation.

last proof rule used to derive (a, x). Thus, the label represents a record of the proof rules that are used in the derivation of $(a, x)_L$. The labels are used to check that the phase of reasoning is non-decreasing.

Definition 11 (Labelled deductive system with phasing) *Let $\mathcal{L}$ be a propositional base logic and let ρ be the phasing function from proof rules to integers defined by $\rho(\text{SI}) = 1$, $\rho(\text{CTA}) = 2$, $\rho(\text{WO}) = 3$ and $\rho(\text{OR}) = 4$. The language of the labelled system consists of the labelled conditionals $(a, x)_L$, with a and x sentences of $\mathcal{L}$, and L an integer (the phase). A premise is a formula $(a, x)_0$, where $a \wedge x$ is consistent in $\mathcal{L}$. The labelled deductive system with phasing consists of the inference rules below, extended with the following condition $R_{IOp} = R_{IO} + R_p$.*

R_{IO}: *$(a, x)_p$ may only be derived if $a \wedge x$ is consistent: it must always be possible to fulfill a derived conditional.*

R_p: *$(a, x)_p$ may only be derived if $p \geq p_i$ for all $(a_i, x_i)_{p_i}$ it is derived from.*

The inference rules of the labelled system are replacements by logical equivalents (for input and output) and the following four rules.

$$\text{SI}_{R_{IOp}} : \frac{(a, x)_p, R_{IOp}}{(a \wedge b, x)_{\rho(\text{SI})}} \qquad \text{CTA}_{R_{IOp}} : \frac{(a, x)_{p_1}, (a \wedge x, y)_{p_2}, R_{IOp}}{(a, x \wedge y)_{\rho(\text{CTA})}}$$

$$\text{WO}_{R_{IOp}} : \frac{(a, x)_p, R_{IOp}}{(a, x \vee y)_{\rho(\text{WO})}} \qquad \text{OR}_{R_{IOp}} : \frac{(a, x)_{p_1}, (b, x)_{p_2}, R_{IOp}}{(a \vee b, x)_{\rho(\text{OR})}}$$

We say $(a_1, x_1), \ldots, (a_n, x_n) \vdash_{\text{LDSP}} (a, x)$ if there is a $(a, x)_L$ that can be derived from $\{(a_1, x_1)_0, \ldots, (a_n, x_n)_0\}$.

Before we show the equivalence of the two labelled systems, we again prove lemmas for proof rewriting and consistency checks.

Lemma 12 *For each derivation of $(a, x)_L$ of G in the labelled deductive system with the fulfilment constraint there is an alternative derivation of $(a, x)_L$ of G that respects the order of the labelled deductive system with phasing.*

Proof All possible sequences of two steps of the labelled system that do not respect the order of the labelled deductive system with phasing are given below (left side derivations), together with the alternative derivations that do respect it (right side derivations). Moreover, we show that the label of the root of the left derivation is equivalent to the label of the root of the right derivation, and thus that if the fulfilment constraint hold for the left derivation, then it also holds for the right derivation. This follows from the fact that each element of the label implies the fulfilment of the conditional (Lemma 7) and some elementary operations on $\cup$ and $\times$. For example, consider the two proofs of e.2. concerning a sequence of an OR and a CTA step. The label of the first derived conditional is

$$F_1 \times (F_2 \cup F_3) = (F_1 \times F_2) \cup (F_1 \times F_3)$$

and the label of the second derived conditional is

$$(F_1 \times \{b\} \times F_2) \cup (F_1 \times \{\neg b \vee c\} \times F_3)$$

The two sets contain equivalent formulas, because each element of F_2 classically implies b and each element of F_3 classically implies $\neg b \vee c$ (Lemma 7). The other proofs are analogous.

a. Reversing the order of CTA$_2$ and SI$_1$

$$\dfrac{\dfrac{(a,x)_{F_1} \quad (a\wedge x,y)_{F_2}}{(a,x\wedge y)_{F_1\times F_2}}\;\text{CTA}}{(a\wedge b,x\wedge y)_{F_1\times F_2\times\{b\}}}\;\text{SI} \qquad \dfrac{\dfrac{(a,x)_{F_1}}{(a\wedge b,x)_{F_1\times\{b\}}}\;\text{SI} \quad \dfrac{(a\wedge x,y)_{F_2}}{(a\wedge b\wedge x,y)_{F_2\times\{b\}}}\;\text{SI}}{(a\wedge b,x\wedge y)_{F_1\times F_2\times\{b\}}}\;\text{CTA}$$

b. Reversing the order of WO$_3$ and SI$_1$

$$\dfrac{\dfrac{(a,x)_F}{(a,x\vee y)_F}\;\text{WO}}{(a\wedge b,x\vee y)_{F\times\{b\}}}\;\text{SI} \qquad \dfrac{\dfrac{(a,x)_F}{(a\wedge b,x)_{F\times\{b\}}}\;\text{SI}}{(a\wedge b,x\vee y)_{F\times\{b\}}}\;\text{WO}$$

c.1. Reversing the order of WO$_3$ and CTA$_2$, case 1

$$\dfrac{\dfrac{(a,x)_{F_1}}{(a,x\vee y)_{F_1}}\;\text{WO} \quad (a\wedge(x\vee y),z)_{F_2}}{(a,(x\vee y)\wedge z)_{F_1\times F_2}}\;\text{CTA} \qquad \dfrac{\dfrac{(a,x)_{F_1} \quad \dfrac{(a\wedge(x\vee y),z)_{F_2}}{(a\wedge x,z)_{F_2\times\{x\}}}\;\text{SI}}{(a,x\wedge z)_{F_1\times F_2\times\{x\}}}\;\text{CTA}}{(a,(x\vee y)\wedge z)_{F_1\times F_2\times\{x\}}}\;\text{WO}$$

c.2. Reversing the order of WO$_3$ and CTA$_2$, case 2

$$\dfrac{(a,x)_{F_1} \quad \dfrac{(a\wedge x,y)_{F_2}}{(a\wedge x,y\vee z)_{F_2}}\;\text{WO}}{(a,x\wedge(y\vee z))_{F_1\times F_2}}\;\text{CTA} \qquad \dfrac{\dfrac{(a,x)_{F_1} \quad (a\wedge x,y)_{F_2}}{(a,x\wedge y)_{F_1\times F_2}}\;\text{CTA}}{(a,x\wedge(y\vee z))_{F_1\times F_2}}\;\text{WO}$$

d. Reversing the order of OR$_4$ and SI$_1$

$$\dfrac{\dfrac{(a,x)_{F_1} \quad (b,x)_{F_2}}{(a\vee b,x)_{F_1\cup F_2}}\;\text{OR}}{((a\vee b)\wedge c,x)_{(F_1\cup F_2)\times\{c\}}}\;\text{SI} \qquad \dfrac{\dfrac{(a,x)_{F_1}}{(a\wedge c,x)_{F_1\times\{c\}}}\;\text{SI} \quad \dfrac{(b,x)_{F_2}}{(b\wedge c,x)_{F_2\times\{c\}}}\;\text{SI}}{((a\wedge c)\vee(b\wedge c),x)_{(F_1\times\{c\})\cup(F_2\times\{c\})}}\;\text{OR}$$

e.1. Reversing the order of OR$_4$ and CTA$_2$, case 1

$$\dfrac{\dfrac{(a,x)_{F_1} \quad (b,x)_{F_2}}{(a\vee b,x)_{F_1\cup F_2}}\,\text{OR} \qquad ((a\vee b)\wedge x,y)_{F_3}}{(a\vee b, x\wedge y)_{(F_1\cup F_2)\times F_3}}\,\text{CTA}$$

$$\dfrac{\dfrac{(a,x)_{F_1} \quad \dfrac{((a\vee b)\wedge x,y)_{F_3}}{(a\wedge x,y)_{F_3\times\{a\}}}\,\text{SI}}{(a,x\wedge y)_{F_1\times F_3\times\{a\}}}\,\text{CTA} \qquad \dfrac{(b,x)_{F_2} \quad \dfrac{((a\vee b)\wedge x,y)_{F_3}}{(b\wedge x,y)_{F_3\times\{b\}}}\,\text{SI}}{(b,x\wedge y)_{F_2\times F_3\times\{b\}}}\,\text{CTA}}{(a\vee b,x\wedge y)_{(F_1\times F_3\times\{a\})\cup(F_2\times F_3\times\{b\})}}\,\text{OR}$$

e.2. Reversing the order of OR$_4$ and CTA$_2$, case 2

$$\dfrac{(a,x)_{F_1} \quad \dfrac{(a\wedge x\wedge b,y)_{F_2} \quad (a\wedge x\wedge(\neg b\vee c),y)_{F_3}}{(a\wedge x,y)_{F_2\cup F_3}}\,\text{OR}}{(a,x\wedge y)_{F_1\times(F_2\cup F_3)}}\,\text{CTA}$$

$$\dfrac{\dfrac{\dfrac{(a,x)_{F_1}}{(a\wedge b,x)_{F_1\times\{b\}}}\,\text{SI} \quad (a\wedge x\wedge b,y)_{F_2}}{(a\wedge b,x\wedge y)_{F_1\times\{b\}\times F_2}}\,\text{CTA} \qquad \dfrac{\dfrac{(a,x)_{F_1}}{(a\wedge(\neg b\vee c),x)_{F_1\times\{\neg b\vee c\}}}\,\text{SI} \quad (a\wedge x\wedge(\neg b\vee c),y)_{F_3}}{(a\wedge(\neg b\vee c),x\wedge y)_{F_1\times\{\neg b\vee c\}\times F_3}}}{(a,x\wedge y)_{(F_1\times\{b\}\times F_2)\cup(F_1\times\{\neg b\vee c\}\times F_3)}}\,\text{OR}$$

f. Reversing the order of OR$_4$ and WO$_3$

$$\dfrac{\dfrac{(a,x)_{F_1}\quad(b,x)_{F_2}}{(a\vee b,x)_{F_1\cup F_2}}\,\text{OR}}{(a\vee b,x\vee y)_{F_1\cup F_2}}\,\text{WO} \qquad\qquad \dfrac{\dfrac{(a,x)_{F_1}}{(a,x\vee y)_{F_1}}\,\text{WO}\quad\dfrac{(b,x)_{F_2}}{(b,x\vee y)_{F_2}}\,\text{WO}}{(a\vee b,x\vee y)_{F_1\cup F_2}}\,\text{OR}$$

■

The second lemma shows that for phased proofs a consistency check on the fulfilments is equivalent with a consistency check on the conjunction of the input and output of the phase 1 conditionals.

Lemma 13 *Consider any potential derivation of the labelled system with phasing, satisfying R_p but not necessarily R_{IO}, as well as its counterpart in the labelled system with the fulfilment constraint, not necessarily satisfying R_F. Then the following four conditions are equivalent:*

1. *The first derivation satisfies condition R_{IO} throughout phase 1.*
2. *The first derivation satisfies condition R_{IO} everywhere.*

3. *The second derivation satisfies condition R_F throughout phase 1.*

4. *The second derivation satisfies R_F everywhere.*

Proof *Clearly* $(2) \Rightarrow (1)$ *and* $(4) \Rightarrow (3)$. *Through phase 1, for each formula in the derivation with fulfilments the conjunction of input and output is equivalent to the unique element of its label. Hence* $(1) \Leftrightarrow (3)$. *In phase 2 the conjunction of input and output is also equivalent to the unique element of its label, which is equivalent to the label of the second premise of each derivation step. In phase 3 and 4 the rules preserve the consistency of the conjunction of input and output, and they also preserve the property that each element of the label is consistent. From this we have* $(1) \Rightarrow (2)$ *and* $(3) \Rightarrow (4)$. *Putting this together gives us* $(1) \Leftrightarrow (2) \Leftrightarrow (3) \Leftrightarrow (4)$ *and we are done.*

In Theorem 14 below we show that for each derivation in the labelled deductive system with the fulfilment constraint there is an equivalent derivation in the labelled deductive system with phasing.

Theorem 14 (Equivalence LDSF **and** LDSP**)** *Let G be a consistent conditional base. We have* $G \vdash_{\text{LDSF}} (a, x)$ *if and only if* $G \vdash_{\text{LDSP}} (a, x)$.

Proof For each proof of (a, x) from G in one system, we can give a proof in the other system. First we show how a proof tree of the second system can be constructed from the proof tree of the first system, and then we show that if the fulfilment constraint holds for the first proof, then it must also hold for the second proof.

$\Rightarrow$ Given a proof of (a, x) in the labelled deductive system with the fulfilment constraint. To construct a proof in the labelled deductive system with the phasing constraint, first iteratively replace two subsequent steps in the wrong order by several steps in the right order, see Lemma 12. For a finite tree, after a finite number of steps all derivation steps are ordered, because no set of replacements cycles (and can be used to construct infinite proof trees). Then label the premises with 0, and label the other nodes as indicated by the rules of the labelled system. It follows from Lemma 13 that if the consistency constraint holds for the first proof, then it also holds for the second proof.

$\Leftarrow$ Given a proof of (a, x) in the labelled deductive system with the phasing constraint. To construct a proof in the labelled deductive system with the fulfilment constraint, label the premises with the conjunction of input and output, and label the other nodes as indicated by the rules of the labelled system. It follows from Lemma 13 that if the consistency constraint holds for the first proof, then it also holds for the second proof. ■

Corollary 15 *Let G be a consistent conditional base. We have* $G \vdash_{\text{ROF}} (a, x)$ *if and only if* $G \vdash_{\text{LDSP}} (a, x)$.

Proof Follows directly from Theorem 10 and 14. ∎

6 PHASING: SOME EXAMPLES

In this final section we illustrate phasing by several examples. To emphasize the intuitions behind phasing, we introduce some new terminology. We call phase-1, phase-2 and phase-4 conditionals respectively atomic, compound and combined conditionals. Note that the logic of imperatives covers compounding but not combining; the other logics discussed in this paper cover both.

Definition 16 *Let G be a consistent conditional base.*

- *The conditional (a, x) is an* atomic conditional *of G if $a \wedge x$ is consistent and there is a conditional $(b, x) \in G$ such that $a \vdash b$.*
- *The conditional $(a, x_1 \wedge \ldots \wedge x_n)$ is a* compound conditional *of G if there is a set $\{(a, x_1), (a \wedge x_1, x_2), \ldots, (a \wedge x_1 \wedge \ldots \wedge x_{n-1}, x_n)\}$ of atomic conditionals of G.*
- *The conditional $(a_1 \vee \ldots \vee a_n, x)$ is a* combined conditional *of G if there is a set $\{(a_1, x_1), \ldots, (a_n, x_n)\}$ of compound conditionals of G such that $x_i \vdash x$ for $i = 1 \ldots n$.*

Finally, all sets of conditionals are closed under logical equivalence for input and output.

The new terminology is illustrated by the following example, inspired from preference-based reasoning.[4] Consider the set of atomic propositions a_i for $i = 1, \ldots, n$ and $x_{i,j}$ for $i = 1, \ldots, n$ and $j = 1, \ldots, m_n$. We write b_i when a_i is true and the other a_j are false, and we write $y_{i,j}$ when b_i is true as well as the $x_{i,1}, \ldots, x_{i,j}$.

$$\begin{array}{lll} b_i & =_{def} & a_i \wedge \bigwedge_{j \neq i} \neg a_j \\ y_{i,j} & =_{def} & b_i \wedge x_{i,1} \wedge \ldots \wedge x_{i,j}. \end{array}$$

Now consider the following set of atomic conditionals.

$$G = \left\{ \begin{array}{cccc} (b_1, y_{1,1}), & (b_2, y_{2,1}), & \ldots, & (b_n, y_{n,1}) \\ (y_{1,1}, y_{1,2}), & (y_{2,1}, y_{2,2}), & \ldots, & (y_{n,1}, y_{n,2}) \\ \ldots, & \ldots, & & \ldots \\ (y_{1,m_1-1}, y_{1,m_1}), & (y_{2,m_2-1}, y_{2,m_2}), & \ldots, & (y_{n,m_n-1}, y_{n,m_n}) \end{array} \right\}$$

[4] In preference-based reasoning compounding and combining correspond to the construction of a preference ordering and minimization in this constructed ordering [6, 7].

Each column of atomic conditionals can be used to derive the compound conditional (b_i, y_{i,m_i}), and together they can be used to derive the combined conditional $(b_1 \vee \ldots \vee b_n, y_{1,m_1} \vee \ldots \vee y_{1,m_n})$. This derivation is illustrated in Figure 10.1. Each column is a case: compounding is reasoning within a case and combining is reasoning by cases.

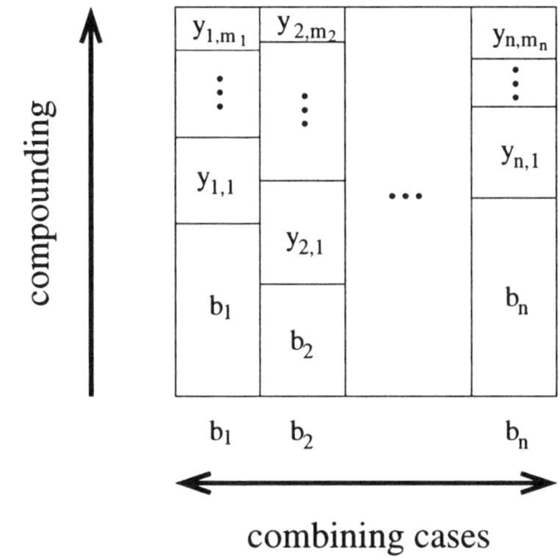

Figure 10.1 Compounding and combining

Before we discuss more examples, we show the straightforward equivalence of the new phasing definition with the labelled deductive system with phasing. The following lemma shows that in the labelled system with phasing the successive applications of the same proof rule can be replaced by a single application of a generalized rule.

Lemma 17 *A set of successive applications of* SI *can be replaced by single* SI *step, and a set of successive applications of* WO *can be replaced by single* WO *step. A set of successive applications of* CTA *can be replaced by the following generalized cut rule, and a set of successive applications of* OR *can be replaced by the following generalized disjunction rule.*

$$\text{GCTA}_{R_{IOp}} : \frac{(a, x_1)_{p_1}, \ldots, (a \wedge x_1 \wedge \ldots \wedge x_{n-1}, x_n)_{p_n}, R_{IOp}}{(a, x_1 \wedge \ldots \wedge x_n)_{\rho(\text{CTA})}}$$

$$\text{GOR}_{R_{IOp}} : \frac{(a_1, x)_{p_1}, \ldots, (a_n, x)_{p_n}, R_{IOp}}{(a_1 \vee \ldots \vee a_n, x)_{\rho(\text{OR})}}$$

Proof For SI and WO the proofs are trivial. For CTA, it follows from the fact that the order of CTA steps is irrelevant, as an inductive generalization of the case for three premises below.

$$\dfrac{(a,x)_{p1} \qquad \dfrac{(a\wedge x,y)_{p2} \quad (a\wedge x\wedge y,z)_{p3}}{(a\wedge x,y\wedge z)_{\rho(\mathrm{CTA})}}\,\mathrm{CTA}}{(a,x\wedge y\wedge z)_{\rho(\mathrm{CTA})}}\,\mathrm{CTA}$$

$$\dfrac{\dfrac{(a,x)_{p1} \quad (a\wedge x,y)_{p2}}{(a,x\wedge y)_{\rho(\mathrm{CTA})}}\,\mathrm{CTA} \qquad (a\wedge x\wedge y,z)_{p3}}{(a,x\wedge y\wedge z)_{\rho(\mathrm{CTA})}}\,\mathrm{CTA}$$

Similarly, for OR the proof follows from the fact that the order of OR steps is irrelevant.

$$\dfrac{(a,x)_{p1} \qquad \dfrac{(b,x)_{p2} \quad (c,x)_{p3}}{(b\vee c,x)_{\rho(\mathrm{OR})}}\,\mathrm{OR}}{(a\vee b\vee c,x)_{\rho(\mathrm{OR})}}\,\mathrm{OR}$$

$$\dfrac{\dfrac{(a,x)_{p1} \quad (b,x)_{p2}}{(a\vee b,x)_{\rho(\mathrm{OR})}}\,\mathrm{OR} \qquad (c,x)_{p3}}{(a\vee b\vee c,x)_{\rho(\mathrm{OR})}}\,\mathrm{OR}$$

■

This gives us the final theorem of this paper.

Theorem 18 (Equivalence LDSP and phasing) *Let G be a consistent conditional base. We have $G \vdash_{\mathrm{LDSP}} (a,x)$ if and only if (a,x) is a combined conditional of G.*

Proof Lemma 17 shows how a proof in the deductive system with phasing can be rewritten into a phased proof tree of depth of at most four. From Lemma 13 follows that a consistency check on all steps is equivalent to a consistency check on atomic conditionals only, and we are done. ■

As a further illustration of phasing we consider the conditional bases discussed in Section 2 and 3 We assume – without mentioning it explicitly – that the sets also contain conditionals where inputs and outputs are replaced by logical equivalents, and we write $cons(a)$ for the consistency check in $\mathcal{L}$ on a. The first base $G = \{(a,b),(b,c),(c,d)\}$ illustrates that derivations go as far as possible by compounding atomic conditionals.

- Atomic conditionals are (a,b), $(a\wedge b,c)$ and $(a\wedge b\wedge c,d)$.

- A compound conditional is $(a, b \wedge c \wedge d)$.
- A combined conditional is (a, d).

The second conditional base $G = \{(a, x \wedge y), (\neg a, x \wedge \neg y)\}$ illustrates that reasoning by cases is done by combining compound conditionals.

- The atomic and compound conditionals are

$$\{(a \wedge b, x \wedge y) \mid cons(a \wedge b \wedge x \wedge y)\}$$

$$\{(\neg a \wedge b, x \wedge \neg y) \mid cons(\neg a \wedge b \wedge x \wedge \neg y)\}$$

- The combined conditionals are

$$\{(a \wedge b, (x \wedge y) \vee z) \mid cons(a \wedge b \wedge x \wedge y)\}$$

$$\{(\neg a \wedge b, (x \wedge \neg y) \vee z) \mid cons(\neg a \wedge b \wedge x \wedge \neg y)\}$$

$$\{(\top, x \vee z) \mid cons(a \wedge b \wedge x \wedge y) \wedge cons(\neg a \wedge b \wedge x \wedge \neg y)\}$$

The third base $G = \{(\top, c), (c, i), (\neg c, \neg i)\}$ illustrates context-sensitivity, because $(\neg c, i)$ is not a conditional of G. The conditional $(\top, i)$ is a compound conditional but not an atomic conditional; therefore it cannot be used to derive this counterintuitive conditional. Note that the combined conditionals $\{(a \wedge c, i \vee x) \mid cons(a \wedge c \wedge i)\})$ are implied by the combined conditionals $\{(a, (c \wedge i) \vee x) \mid cons(a \wedge c \wedge i)\}$.

- The atomic conditionals are

$$\{(a, c) \mid cons(a \wedge c)\}$$

$$\{(a \wedge c, i) \mid cons(a \wedge c \wedge i)\}$$

$$\{(a \wedge \neg c, \neg i) \mid cons(a \wedge \neg c \wedge \neg i)\}$$

- The compound conditionals are

$$\{(a, c) \mid cons(a \wedge c)\}$$

$$\{(a \wedge c, i) \mid cons(a \wedge c \wedge i)\}$$

$$\{(a, c \wedge i) \mid cons(a \wedge c \wedge i)\}$$

$$\{(a \wedge \neg c, \neg i) \mid cons(a \wedge \neg c \wedge \neg i)\}$$

- The combined conditionals are

$$\{(a, c \vee x) \mid cons(a \wedge c)\}$$

$$\{(a, (c \wedge i) \vee x) \mid cons(a \wedge c \wedge i)\}$$

$$\{(a \wedge \neg c, \neg i \vee x) \mid cons(a \wedge \neg c \wedge \neg i)\}$$

The fourth base $G = \{(\top, x), (\top, \neg x)\}$ illustrates conflict tolerance, because $(\top, y)$ is not a conditional of G if y is not classically implied by x or $\neg x$. The conditional $(\top, x \vee y)$ is a combined conditional but not an atomic conditional; therefore it cannot be used to derive this counterintuitive conditional.

- The atomic and compound conditionals are

$$\{(a, x) \mid cons(a \wedge x)\}$$

$$\{(a, \neg x) \mid cons(a \wedge \neg x)\}$$

- The combined conditionals are

$$\{(a, x \vee y) \mid cons(a \wedge x)\}$$

$$\{(a, \neg x \vee y) \mid cons(a \wedge \neg x)\}$$

The examples illustrate that for small bases (such as the lists of benchmark examples in [4] and [13]) it is much easier to write the set of all combined conditionals than the set of all implied conditionals of the logic of reusable propositional output with the fulfilment constraint. Surprisingly, we have shown in this paper that (a, x) can be derived from the base G if and only if there is a combined conditional for (a, x) of G.

Acknowledgment

Thanks to Patrick van der Laag, David Makinson and Emil Weydert for discussions on the issues raised in this paper, and comments on earlier versions of it. Many of the issues and examples discussed in this paper were raised during discussions with David Makinson.

Leendert van der Torre
Department of Artificial Intelligence
Vrije Universiteit
De Boelelaan 1081a
1081 HV, Amsterdam
The Netherlands
torre@cs.vu.nl

References

[1] R.M. Chisholm. Contrary-to-duty imperatives and deontic logic. *Analysis*, 24:33–36, 1963.

[2] D. Gabbay. *Labelled Deductive Systems*, volume 1. Oxford University Press, 1996.

[3] B. Hansson. An analysis of some deontic logics. In R. Hilpinen, editor, *Deontic Logic: Introductory and Systematic Readings*, pages 121–147. D. Reidel Publishing Company, Dordrecht, Holland, 1971. A reprint from *Noûs*, 1969.

[4] D. Makinson. On a fundamental problem of deontic reasoning. In P. McNamara and H. Prakken, editors, *Norms, Logics and Information Systems. New Studies on Deontic Logic and Computer Science*, pages 29–53. IOS Press, 1999.

[5] D. Makinson and L. van der Torre. Input-output logics. 1999. Submitted.

[6] Y. Tan and L. van der Torre. How to combine ordering and minimizing in a deontic logic based on preferences. In *Deontic Logic, Agency and Normative Systems. Proceedings of the ΔEON'96*, Workshops in Computing, pages 216–232. Springer, 1996.

[7] L. van der Torre. *Reasoning about Obligations: Defeasibility in Preference-based Deontic Logic*. PhD thesis, Erasmus University Rotterdam, 1997.

[8] L. van der Torre. Labeled logics of conditional goals. In *Proceedings of the Thirteenth European Conference on Artificial Intelligence (ECAI'98)*, pages 368–369. John Wiley & Sons, 1998.

[9] L. van der Torre. Phased labeled logics of conditional goals. In *Logics in Artificial Intelligence*, LNAI 1489, pages 92–106. Springer, 1998.

[10] L. van der Torre. Defeasible goals. In *Symbolic and Quantitative Approaches to Reasoning and Uncertainty*, LNAI 1638. Springer, 1999.

[11] L. van der Torre and Y. Tan. The many faces of defeasibility in defeasible deontic logic. In D. Nute, editor, *Defeasible Deontic Logic*, volume 263 of *Synthese Library*, pages 79–121. Kluwer, 1997. Extended and revised version of L. van der Torre and Y. Tan, Cancelling and overshadowing: two types of defeasibility in defeasible deontic logic. In *Proceedings of the Fourteenth International Joint Conference on Artificial Intelligence (IJCAI'95)*, pages 1525–1532. Morgan Kaufman, 1995.

[12] L. van der Torre and Y. Tan. Diagnosis and decision making in normative reasoning. *Artificial Intelligence and Law*, 7:51–67, 1999. Revised and extended version of L. van der Torre and Y. Tan. Distinguishing different

roles in normative reasoning. In *Proceedings of the Sixth International Conference on AI and Law (ICAIL'97)*, pages 225–232. ACM Press, 1997.

[13] L. van der Torre and Y. Tan. Contrary-to-duty reasoning with preference-based dyadic obligations. *Annals of Mathematics and Artificial Intelligence*, 2000. To appear. Revised and extended version of L. van der Torre and Y. Tan. Prohairetic Deontic Logic (PDL). In *Logics in Artificial Intelligence*, LNAI 1489, pages 77-91. Springer, 1998.

[14] L. van der Torre and E. Weydert. Goals, desires, utilities and preferences. In *Proceedings of the ECAI'98 Workshop on Decision Theory meets Artificial Intelligence*, 1998.

[15] L. van der Torre and E. Weydert. Parameters for utilitarian desires in a qualitative decision theory. *Applied Intelligence*, 2000. To appear. Revised and extended version of L. van der Torre and E. Weydert. Parameters for utilitarian desires (extended abstract). In *Proceedings of the IJCAI'99 Workshop on Practical Reasoning and Rationality (PRR'99)*, 1999.

[16] G.H. von Wright. A new system of deontic logic. In R. Hilpinen, editor, *Deontic Logic: Introductory and Systematic Readings*, pages 105–120. D. Reidel Publishing Company, Dordrecht, Holland, 1971. A reprint of 'A New System of Deontic Logic,' *Danish Yearbook of Philosophy* 1:173–182, 1964, and 'A Correction to a New System of Deontic Logic', *Danish Yearbook of Philosophy* 2:103–107, 1965.

APPLIED LOGIC SERIES

1. D. Walton: *Fallacies Arising from Ambiguity*. 1996 ISBN 0-7923-4100-7
2. H. Wansing (ed.): *Proof Theory of Modal Logic*. 1996 ISBN 0-7923-4120-1
3. F. Baader and K.U. Schulz (eds.): *Frontiers of Combining Systems*. First International Workshop, Munich, March 1996. 1996 ISBN 0-7923-4271-2
4. M. Marx and Y. Venema: *Multi-Dimensional Modal Logic*. 1996 ISBN 0-7923-4345-X
5. S. Akama (ed.): *Logic, Language and Computation*. 1997 ISBN 0-7923-4376-X
6. J. Goubault-Larrecq and I. Mackie: *Proof Theory and Automated Deduction*. 1997 ISBN 0-7923-4593-2
7. M. de Rijke (ed.): *Advances in Intensional Logic*. 1997 ISBN 0-7923-4711-0
8. W. Bibel and P.H. Schmitt (eds.): *Automated Deduction - A Basis for Applications*. Volume I. Foundations - Calculi and Methods. 1998 ISBN 0-7923-5129-0
9. W. Bibel and P.H. Schmitt (eds.): *Automated Deduction - A Basis for Applications*. Volume II. Systems and Implementation Techniques. 1998 ISBN 0-7923-5130-4
10. W. Bibel and P.H. Schmitt (eds.): *Automated Deduction - A Basis for Applications*. Volume III. Applications. 1998 ISBN 0-7923-5131-2
(Set vols. I-III: ISBN 0-7923-5132-0)
11. S.O. Hansson: *A Textbook of Belief Dynamics*. Theory Change and Database Updating. 1999 Hb: ISBN 0-7923-5324-2; Pb: ISBN 0-7923-5327-7
Solutions to exercises. 1999. Pb: ISBN 0-7923-5328-5
Set: (Hb): ISBN 0-7923-5326-9
Set: (Pb): ISBN 0-7923-5329-3
12. R. Pareschi and B. Fronhöfer (eds.): *Dynamic Worlds from the Frame Problem to Knowledge Management*. 1999 ISBN 0-7923-5535-0
13. D.M. Gabbay and H. Wansing (eds.): *What is Negation?* 1999 ISBN 0-7923-5569-5
14. M. Wooldridge and A. Rao (eds.): *Foundations of Rational Agency*. 1999 ISBN 0-7923-5601-2
15. D. Dubois, H. Prade and E.P. Klement (eds.): *Fuzzy Sets, Logics and Reasoning about Knowledge*. 1999 ISBN 0-7923-5911-1
16. H. Barringer, M. Fisher, D. Gabbay and G. Gough (eds.): *Advances in Temporal Logic*. 2000 ISBN 0-7923-6149-0
17. D. Basin, M.D. Agostino, D.M. Gabbay, S. Matthews and L. Viganò (eds.): *Labelled Deduction*. 2000 ISBN 0-7923-6237-3
18. P.A. Flach and A.C. Kakas (eds.): *Abduction and Induction*. Essays on their Relation and Integration. 2000 ISBN 0-7923-6250-0
19. S. Hölldobler (ed.): *Intellectics and Computational Logic*. Papers in Honor of Wolfgang Bibel. 2000 ISBN 0-7923-6261-6

KLUWER ACADEMIC PUBLISHERS – DORDRECHT / BOSTON / LONDON